Progress in Precision Agriculture

Series Editor
Margaret A. Oliver, Soil Research Centre, University of Reading, Berkshire, Berkshire, UK

This book series aims to provide a coherent framework to cover the multidisciplinary subject of Precision Agriculture (PA), including technological, agronomic, economic and sustainability issues of this subject. The target audience is varied and will be aimed at many groups working within PA including agricultural design engineers, agricultural economists, sensor specialists and agricultural statisticians. All volumes will be peer reviewed by an international advisory board.

Davide Cammarano • Frits K. van Evert
Corné Kempenaar

Editors

Precision Agriculture: Modelling

 Springer

Editors
Davide Cammarano (iD)
Department of Agroecology, iClimate,
Centre for Circular Bioeconomy (CBIO)
Aarhus University
Tjele, Denmark

Frits K. van Evert
Agrosystems Research
Wageningen University & Research
Wageningen, The Netherlands

Corné Kempenaar
Agrosystems Research
Wageningen University & Research
Wageningen, The Netherlands

ISSN 2511-2260   ISSN 2511-2279   (electronic)
Progress in Precision Agriculture
ISBN 978-3-031-15260-3   ISBN 978-3-031-15258-0   (eBook)
https://doi.org/10.1007/978-3-031-15258-0

This Springer imprint is published by the registered company Springer Nature Switzerland AG
The registered company address is: Gewerbestrasse 11, 6330 Cham, Switzerland

Preface

This book is part of a series on Precision Agriculture established by Springer with Margaret Oliver as the series editor. We believe this is the first book that addresses specifically the question of how crop growth modeling can be used in precision agriculture. The separate domains of crop growth modeling and precision agriculture share an interest in understanding and managing arable crops to achieve profit and productivity with minimal negative impacts on the environment. During the past two to three decades, crop modelers have focused almost exclusively on evaluating the impact of climate change. Precision agriculture scientists and engineers have focused on identifying management zones using data from remote sensing, soil sensors, and yield monitors. But only a few scientists and engineers have been working on using crop growth models for precision agriculture. This book aims to highlight these efforts and encourage collaboration between crop growth modelers and precision agriculture scientists.

The seeds for this book were sown at the European Conference on Precision Agriculture which took place 8–11 July 2019 in Montpellier (France) (with many presentations about sensors but very few about crop growth modeling) when Margaret Oliver invited Davide Cammarano and Frits van Evert to work on this book. Corné Kempenaar joined soon after. This book describes how models can be used to monitor crops and soils in precision agriculture and how simulation outcomes can be used to support farmers' decisions. The book aims to provide useful information to graduate-level professionals that want to broaden their knowledge of precision agriculture; to scientists who want to learn about using academic knowledge in practical farming; and to farmers, farm consultants, and extension workers who want to increase their understanding of the science behind some of the commercial software available to the farming community. Therefore, the book has three parts. Part I provides an introduction to modeling and precision agriculture. Part II explores the state-of-the-art in modeling for precision agriculture, with chapters focusing on the major processes in precision agriculture: water use, nitrogen, and other amendments, together with weeds, pests, and diseases. Part III contains a number of short chapters that each describe a commercial, model-based service that is currently available to farmers.

We thank the authors and the series editor for working with us in preparing this book during a period in which we all faced difficulties due to the coronavirus pandemic.

West Lafayette, Denmark   Davide Cammarano

Wageningen, The Netherlands   Frits K. van Evert

Wageningen, The Netherlands   Corné Kempenaar

Contents

Part III Case Studies

Part I
Modelling for Precision Agriculture

Introduction

Frits K. van Evert, Davide Cammarano ⓘD, and Corné Kempenaar

1 Rationale

Modern agriculture is in many ways a success story. The number of people in the world has increased markedly during the last 200 years, and it is projected to reach about 9.8 billion by 2050 (Fig. 1a), while the productivity of many crops from the beginning of the Green Revolution has kept up until now (Fig. 1b). As a result of these two opposing trends, the prevalence of undernourishment is at a historic low and still decreasing (Fig. 1). Of course it remains to be seen if these trends can continue in the face of more frequent exceptional climate events (IPCC, 2021), the coronavirus pandemic and breakout of war.

The increases in productivity have been possible because of a number of contributing factors ranging from technical to socioeconomic. From a technical point of view, plant breeding, better crop rotation and fertilization, mechanization of most field operations and crop protection against pests and diseases have helped to increase yields (Aggarwal et al., 2019). From a socioeconomic point of view, many governments also started to support agricultural research, as well as education for farmers and extension services, to ensure that the knowledge produced by research can be used in practice. There were also many systemic changes, such as

F. K. van Evert (✉) · C. Kempenaar
Agrosystems Research, Wageningen University & Research, Wageningen, The Netherlands
e-mail: frits.vanevert@wur.nl

D. Cammarano
Department of Agroecology, iClimate, Centre for Circular Bioeconomy (CBIO),
Aarhus University, Tjele, Denmark

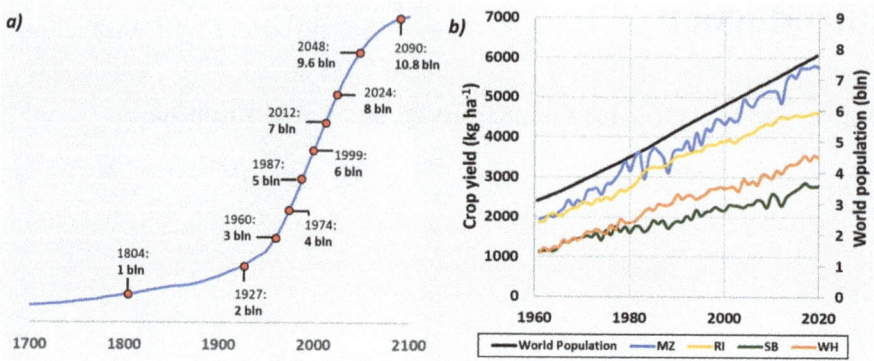

Fig. 1 Patterns of (**a**) population growth from 1700 to a projected growth in 2100. The blue line shows the continuous patterns, while the red dots report the world's population at given points in time. The data from 1700 to 2100 were from (https://ourworldindata.org/grapher/population? time=1500), while the world's population from 2010 to 2100 were downloaded from https:// ourworldindata.org/future-population-growth; (**b**) crop unit yield of maize (MZ, blue line), rice (RI, yellow line), soybean (SB, dark green line), wheat (WH, orange line) and the world's population growth (black line) from 1960 to 2020. The data on crop yields were obtained from https://www. fao.org/faostat/en/#data, while the world's population data were from https://ourworldindata.org/ grapher/population?time=1500..latest

subsidies on fertilizers, the establishment of cooperative banks, the development of road networks and the opening of borders as in the European Union.

Even when several technological innovations helped to increase yield, it may be possible to attribute yield increases to specific factors (Fig. 2). However, in most cases, socioeconomic and technical factors interact to determine yield, together with decision-making (e.g. farmers' decisions on when to plant and how much fertilizer to apply). However, as the wheat yield trend in Australia shows (Fig. 2), recent climate extremes can still cause deviation from a long-term trend no matter what the level of technology. This is something to consider in future planning because it is likely that climate extremes will be more frequent and affect yield (IPCC, 2021).

At the same time, modern agricultural practices are having a large negative impact on the environment. The use of large amounts of fertilizers has led to problems with nitrate leaching to groundwater, phosphates causing eutrophication of surface water and a rise in greenhouse gas emission (during production of fertilizers as well as during crop production) from the whole farming sector. For some specific issues (e.g. nitrate in groundwater), the European Union (EU) has regulated the amount of organic and inorganic nitrogen farmers can apply in specific areas deemed nitrate vulnerable zones (NVZs) to limit the environmental impacts of fertilization (EU, 1991).

In addition, crop protection agents affect more than just the pests and diseases that are targeted. Chemical crop protection agents can also affect nontarget insects, birds and mammals, as can herbicides used to control weeds (Voltz et al., 2022).

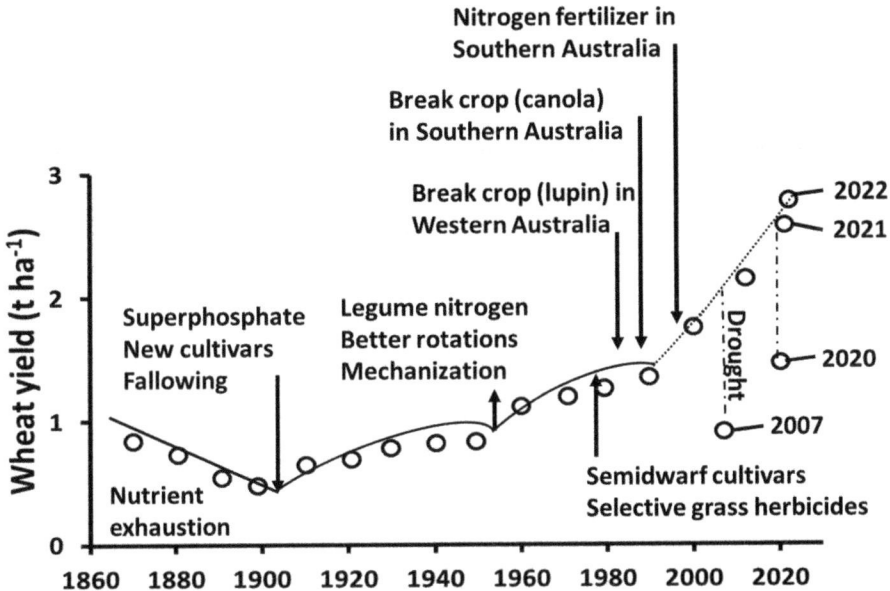

Fig. 2 A modified Australian wheat yield patterns from 1860 to 2022 from Angus (2001). The data from 1860 to 2000 were extracted from Angus (2001), while those from 2000 to 2022 were obtained from ABARES (2022)

The common practice of having rotations with only a few different crops grown in large fields is an important contributing factor to a reduction in biodiversity.

Finally, the use of heavy machinery leads to loss of soil structure and increased soil compaction (Hamza & Anderson, 2005). Among other things, this means that the soil has difficulty absorbing rainwater which in turn may lead to surface runoff and erosion.

All the above factors could be exacerbated by the predicted changes in climate patterns and extremes. Therefore, for agriculture to be economically profitable and minimize environmental effects, it must become more sustainable. Digital agriculture is one of the novel scientific fields that may help to make agriculture more sustainable. Precision agriculture (PA) is a branch of this which could benefit from the insight gained from such new discipline. Precision agriculture is one of the technologies (among breeding, agronomy and so on) that can help to make agriculture more sustainable from an economic and environmental point of view. And accounting for dynamic interactions between soil-plant-atmosphere-crop management with crop growth models can support decision-making in precision agriculture. Figure 3 shows how the links between deterministic crop growth models, machine learning and other technologies in PA can help sustain the upward trend of technology.

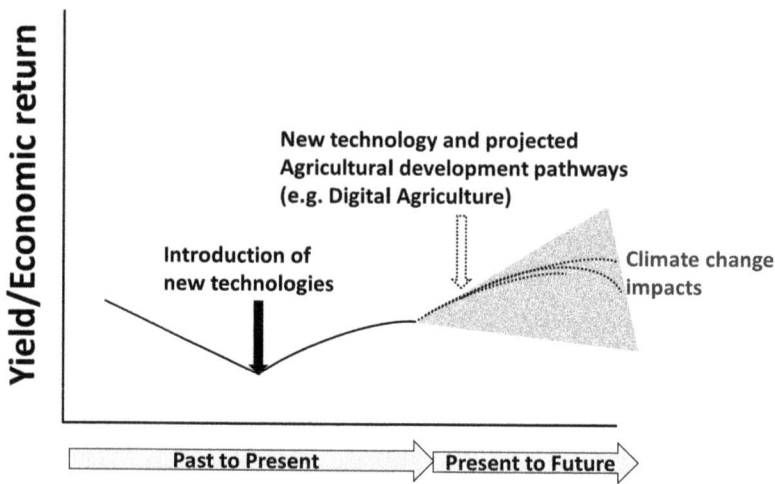

Fig. 3 A theoretical diagram that draws from the work of Angus (2001) in which the value of introducing new technologies can be reduced by projected changes in the climate

2 Crop Modelling

A model is a simplified representation of a part of reality. Models are used to understand and manage our surroundings and can take many different forms. For example, it is a model if one expresses in poetic language the fact that including a legume crop in the rotation leads to higher yields.[1] Another example of a model is a linear (or curvilinear) regression relationship between the amount of manure applied and crop yield. A model can also take into account time, for example, when a curve is used to relate crop nitrogen uptake to the number of days since emergence (Steltenpool & van Erp, 1995).

In this book we concern ourselves primarily with complex dynamic models. These models describe discrete event systems, continuous state systems (differential equations) and hybrid continuous state and discrete event systems. A foundational theory of dynamic systems and models is available (Zeigler, 1976; Zeigler et al., 2000).

Crop simulation models, or crop growth models, are a formal representation of mathematical algorithms that describe the interaction of crops with the environment (Jones et al., 2003; Van Ittersum et al., 2003). Crop models are dynamic models (time is a factor) and usually simulate daily (or hourly, depending on the models) interactions between the soil, plant, weather and crop management. They are constructed with many subroutines that simulate specific processes (e.g. crop phenology, soil organic carbon and so on) that are then interrelated. Therefore, crop growth models comprise a mix of simple empirical and mechanistic subroutines.

---

[1] The Georgics, by Virgil. For example, https://gutenberg.org/ebooks/232

Table 1 The main category of input data needed for running a crop simulation model

Input	Common input among crop models
Weather	Solar radiation, maximum and minimum temperature, rainfall
Soil	Clay, silt, sand; organic carbon, nitrogen
Management	Planting date, density and depth; fertilizer (or irrigation) application date and amount; fertilizer type (usually for nitrogen only)
Boundary conditions (or initial conditions)	Value of soil water and nitrogen at the start of the simulation (usually prior to sowing)

They require a certain number of inputs to be able to run as highlighted in Table 1; although the number of those inputs depends on the complexity of the crop model. Furthermore, when inputs are not available, they are usually estimated. Kersebaum et al. (2015) classified the amount of detail needed in terms of input data and for model calibration.

Nevertheless, crop models have been used in different branches of research to extrapolate experimental agronomic information beyond field experimentation and to test and assess climate impacts and changes (Asseng et al., 2015, 2019; Rötter et al., 2016). In recent years crop models have been included in frameworks and support systems that also include economics and human behaviour (Cammarano et al., 2020).

Crop modelling is not new; it started in the 1960s in the Netherlands under De Wit (Van Ittersum et al., 2003) as a mathematical relationship was developed between biomass growth and solar radiation. Since then, the number of modelling schools and modelling approaches has grown until we have now reached a point of having multiple models for simulating the same crops. For example, there are about 30 wheat crop models, 19 for maize and 13 for rice (www.agmip.org).

The key processes simulated in those models are development, growth, yield, nutrient and water uptake. For each of these processes, external environmental factors control the main processes. Phenological development, which simulates the passage through different stage of the plant's life, like flowering or maturity, is generally controlled by air temperature and day length, and crop growth by solar radiation and temperature. The external factors for nutrient and water supply are rainfall, irrigation, temperature and nutrient supply.

The majority of crop models utilize De Wit's (Van Ittersum et al., 2003) concept to simulate crop growth. First, in a potential production situation, crop growth is limited only by light, temperature and the crop's genetics. Water-limited production takes place when in addition to the above, growth is limited at least part of the time by the availability of water; nitrogen-limited production takes place when growth is limited by the availability of nitrogen. In order to simulate water- and nitrogen-limited production, rainfall, irrigation, nutrient supply and soil conditions need to be known.

Another advantage of the use of crop models is that they can extrapolate the interactions between plant-soil-agronomy through time. In this instance, crop growth simulations with historical long-term weather data should not be interpreted as trying to simulate a particular year; rather, this kind of simulation can help to increase our understanding of the crop's sensitivity to different weather patterns in a given area (e.g. what is the probability of crop failure because of heat stress).

Despite being considered 'point-based' models, crop models have been applied at different spatial scales. For instance, at the global level, they have been used to quantify the effects of climate change on crop production (Elliott et al., 2015; Jägermeyr et al., 2021) and the values of adaptation (Asseng et al., 2019). At the regional level, they have been coupled with economic modelling to predict how climate change affects rates of poverty (Cammarano et al., 2020). Other applications are their use in gene-based modelling and to design crop ideotypes for better crop adaptation to the future climate (Bustos-Korts et al., 2019; Zheng et al., 2012).

3   Precision Agriculture

It could be argued that the earliest farmers already engaged in precision agriculture. If you are farming for subsistence on a small piece of land, and the work is done by hand, then by definition, you are working with a great level of precision, for example, removing weeds close to the crop plants and applying fertilizer where it counts the most. This precision was lost with the advent of mechanization, when manufactured fertilizer and crop protection agents were applied uniformly over large areas.

Modern precision agriculture was initially developed, from a theoretical point of view, in the early 1980s with the aim of understanding factors that lead to spatial variation in crop growth within a field and how to manage it. However, PA was not very popular until technologies such as global navigation satellite systems (GNSS), remote sensing and better agricultural equipment (e.g. variable-rate application, yield monitors) became widely available after 2000. Despite an increase in popularity, the adoption among farmers is still not widespread (Lowenberg-DeBoer & Erickson, 2019).

There have been many definitions of precision agriculture, but the International Society of Precision Agriculture (ISPAG, www.ispag.org) recently defined it as:

> Precision Agriculture is a management strategy that gathers, processes and analyzes temporal, spatial and individual data and combines it with other information to support management decisions according to estimated variability for improved resource use efficiency, productivity, quality, profitability and sustainability of agricultural production.

Nowadays precision agriculture, pushed by new data analytical tools and better high-resolution images, is generating a lot of data which are often used in models in a 'black box' situation. It is not that straightforward to assume that a given drone/satellite image represents nutritional stress without understanding what is happing at the soil-plant interface and how weather affects this relationship (Colaço & Bramley, 2018; Colaço et al., 2021).

4   Use of Models in Precision Agriculture

In precision agriculture the goal of models has been to understand and explain the spatial variation of crop growth or yield (as mapped by drones, satellites, yield monitoring) and to help address the management decisions related to the site-specific input application (Basso et al., 2001). A variety of modelling approaches has been used in precision agriculture. Of particular interest are regression models, simple dynamic models, crop growth models, digital twins and machine learning.

4.1   Regression Models

The models that were first used in precision agriculture were not crop growth models, but much more simple decision rules that related a measurement on the crop or the soil to a specific action. They are still used successfully for developing functional relationships between agronomic input and yield. Several manufacturers market a sensor that measures crop reflectance in two or more wavebands and use this approach to derive an immediate rate for a side dressing of nitrogen. Such a system can be delivered in a single package, it bypasses the need for a data infrastructure, and it addresses an important problem, namely, the over-use of nitrogen fertilizers.

Nitrogen uptake and plant reflectance values are affected by factors other than soil nitrogen supply (Colaço & Bramley, 2018); therefore normalization procedures like well-fertilized reference strips, non-N-fertilized strips, and relative yield are usually used to gain understanding. In addition, over the years, researchers have developed relationships between sensor readings and nitrogen rates to determine the optimal amount of nitrogen. This was done by developing a variety of indicators like the NSI (nitrogen sufficiency index) and INSEY (in-season estimated yield) or the relationship between chlorophyll meter readings, grain yield and chlorophyll index as reviewed by Samborski et al. (2009).

A major limitation of the use of such algorithms, however, is that they can be highly site- and year-specific and it could be difficult to extrapolate them to other genotypes, soils and environments. Ransom et al. (2019) evaluated statistical and machine learning (ML) algorithms for including soil and weather information to improve N recommendation tools for maize.

Puntel et al. (2019) identified static and dynamic variables that affect the algorithms for the economic optimal rate of nitrogen and found that some static variables (e.g. soil depth) and some dynamic ones (e.g. number of days with precipitation >20 mm, residue amount, soil nitrate at planting and heat stress around silking) helped in improving such algorithms.

Over the years, however, successful commercial products were developed using this type of approach (Samborski et al., 2009; Tremblay et al., 2011). An example of a successful application of such models using a combination of remote sensing and nitrogen response is with the passive or active remote sensors and algorithms developed for linking crop N response to the normalized difference vegetation index (NDVI) (Butchee et al., 2011; Holland et al., 2004, 2012; Scharf et al., 2011).

The Yara N-Sensor provides a real-time variable rate of N for different crops (e.g. wheat (*Triticum aestivum* L.), oilseed (*Brassica napus* L.), barley (*Hordeum vulgare* L.) and maize (*Zea mays* L.)) given the crop-specific algorithms developed. The advantage of the system is that the prediction of N uptake in real time enables a correction for N on the go. However, this system does not provide absolute recommendations but focused on applying more fertilizer where the crop does not grow well (Jasper et al., 2009; Reusch, 2003).

Another agronomic practice that has been optimized through the use of models is herbicide application. Pre-emergence herbicides are applied to the soil between sowing and emergence of the crop to kill weeds in the topsoil layer. The amount of herbicide needed to control the weeds depends on the clay and organic matter contents of the soil. These soil properties can be estimated with measurements of apparent electric conductivity of the soil. Thus, the rate of soil herbicide application can be derived from an instantaneous measurement of the soil (Kempenaar et al., 2014) in the same way that a measurement of crop reflectance can be used to determine a rate of nitrogen application. The above technology when used for variable-rate application of herbicide for potato haulm killing (PHK) results in a reduction of herbicide use of up to 50% compared to a uniform rate application (Kempenaar et al., 2004, 2008, 2017).

Several approaches have been proposed for irrigation, and those that fit the scope of this chapter are the thermal-based algorithms for scheduling irrigation based on canopy temperature, such as the crop water stress index (CWSI) (Jackson et al., 1981, 1988) and the time-temperature threshold (TTT) (Wanjura et al., 1992, 1995). The latter index was used to set up automatic irrigation on cotton and to automate the amount of irrigation for soybean (O'Shaughnessy & Evett, 2010; Peters & Evett, 2008).

A good overview of the differences between statistical and process-based crop models is discussed in Lobell and Asseng (2017).

4.2   Simple Dynamic Models

A dynamic model takes time into account, and in its simplest form, with thermal time as one of the independent variables, has been successful in recommending the rate of N for side dressing in potatoes (Booij et al., 2017; Van Evert et al., 2012). Soil nitrogen supply in the Netherlands varies widely from year to year and from field to field. Therefore, the standard practice of applying the recommended amount of N around planting frequently results in either an over- or undersupply of N. A recommendation system was developed where some N is applied at planting, and then around 1 July, a simple dynamic model is used to determine how much N the crop would have taken up in the absence of N limitation. The ideal amount of N is then compared with the actual amount of N measured by canopy reflectance; the difference between the two numbers is applied as side dress N. This system maintains yield and reduces N on average by 15% (van Evert et al., 2012; Kempenaar et al., 2017).

4.3   Crop Growth Models

An important limitation of the N side dress system for potatoes mentioned above is that it does not consider nitrate leaching, mineralization of organic matter and other processes that influence the availability of N in the soil. For example, after several weeks with limited rain, the side dress system will recommend a large rate of N for side dressing even though crop growth is limited by lack of water and not nitrogen. This is where crop growth models become useful.

One of the advantages of running a crop model in precision agriculture is that they take into account the effect of climate variation at each point in space as highlighted in the scientific literature (Basso et al., 2021; Batchelor et al., 2002; Cammarano et al., 2021; Maestrini & Basso, 2018; Martinez-Feria & Basso, 2020; Paz et al., 1999; Puntel et al., 2018; Thorp et al., 2008). Basso et al. (2001) demonstrated how to minimize the technical costs for running a crop model in the precision agriculture context.

Early efforts to use complex dynamic models in PA include EPIPRE (Zadoks, 1981), GOSSYM/COMAX (McKinion et al., 1989) and Tipstar (Jansen, 2008; Jansen et al., 2003). Those early efforts were difficult to implement because of the limited processing power of early computers which meant that it required hours or even days of runtime to obtain a result. The practical application of these models was also limited by obstacles related to acquisition, storage and distribution of the relevant crop, soil and farm management data needed to run the models.

Batchelor and Paz (1997) and Paz et al. (1998, 1999) then used two crop models to evaluate how the spatial variation of soil moisture affects crop yield. Thorp et al. (2008) developed a prototype of a decision-support tool based on a crop growth model to use in precision agriculture. Their aim was to let a crop growth model simulate a homogeneous unit of land to improve understanding of the effect of spatiotemporal variation on crop growth and development and to adopt better site-specific management. The research was based on the concept of defining homogeneous management zones, a common theme in PA which has been subjected to extensive scientific work (Nawar et al., 2017). Basso et al. (2001) integrated the information from remote sensing with a crop growth model to identify management zones and have an improved understanding of the causes leading to the variation in yield. An improved approach was adopted by Cammarano et al. (2021) using a farmer's field and demonstrated that the use of crop models can help to identify the cause of spatial variation in yield but also to quantify the effects of soil and weather on crop quality.

McNunn et al. (2019) used a crop growth model to predict site-specific subfield optimum seeding density and N rates based on publicly available data sources. They also used the modelled outputs to estimate the environmental footprint in terms of N leaching and greenhouse gas emissions. This example shows how a crop growth model can aid the estimation of economically optimal rates of N by also taking into account the dynamic interactions of the N cycling and how it affects the environment. Nowadays the advances in computing technologies and digital data of soil and weather have simplified the use of crop models (from simpler crop models to more mechanistic ones). For example, a global simulation study on the effect of climate change on wheat and maize production made in the early 1990s (Rosenzweig & Parry, 1994) produced an overall message that is comparable to current research (Elliott et al., 2015; Jägermeyr et al., 2021; Müller et al., 2019). Given that the same crop models were run, the main technical differences between the two studies were the speed and computational capability and the availability of input with a finer spatial resolution of the crop models (see the global figures of the two studies). Therefore, simple 'apps' and technological progress (computers and data) have led to an interest in full-fledged crop models as showcased in Part III of this book.

4.4   Digital Twins

Recently there has been interest in using crop growth models as a digital twin. A digital twin (DT) is a model of a physical object with emphasis on (1) the connection between the real-world object and its virtual counterpart and (2) the use of real-time data from sensors to keep the model synchronized (Grieves, 2014). DTs are widely used in engineering and the concept has recently started to receive attention in agriculture (Pylianidis et al., 2021; Van Evert et al., 2021).

While DT is not a new research field, the integration of crop growth models in DT is still in development. There are only a few examples on how crop growth models can benefit from DT. The integration of sensor information to help to improve model simulations is being developed for field crops as well as for livestock (Van Evert et al., 2021).

4.5   Machine Learning

Another expanding research field is machine learning (ML) applied to PA. Machine learning is a field of computer science comprising algorithms that give the computer the ability to learn without explicit programming (Samuel, 1959). To simplify the ML approach, the steps can be divided into three parts: (I) input data, (II) building the model and (III) generalization (which is like a crop model evaluation, e.g. testing the model with dataset not used for training the model).

Ransom et al. (2019) used an ML algorithm to improve a simple regression model for optimizing N fertilization. But ML algorithms are also suited to analysing the output from real-time sensors, such as the data from soil moisture and weather sensors. Machine learning has potential for estimating soil properties as reported in an extensive review by Nasirahmadi and Hensel (2022) and other research where maps of the apparent electrical conductivity of bulk soil were used together with some targeted soil samples to estimate soil properties spatially. However, if the interest is, for example, the conservation of soil organic carbon, ML might fail to predict the dynamic changes of organic carbon and how soil type, weather and human management affect it. In this case, a dynamic soil carbon model would be more suited than ML because it takes into account explicitly the effect of microbial activities and the role of different fractions of soil organic carbon.

Crop yield estimation is another field that has been extensively studied in which ML has been applied (Nasirahmadi & Hensel, 2022). Although there is potential in this area for ML, yield prediction using remote sensing and simple regression has also produced satisfactory results (Bean et al., 2018; Puntel et al., 2019; Raun et al., 2005). An issue with yield prediction is that ML models, once tested and trained in a given environment/crop/soil/year, cannot be applied easily in other contexts (or other years) because they are highly site-specific. The main limitation, in this case, is not ML per se but the agronomic understanding as to why yield prediction is needed. It has often been reported that the integration of ML and remote sensing improves yield, quality and agronomic management. However, in most cases yield prediction is done too late when farmers cannot do anything to correct eventual problems. It is important that scientists using ML for yield prediction should be aware of the agronomic implications of their applications.

On pest-disease and weed recognition, ML has shown considerable potential in applicability with images from digital cameras and the application of convolutional neural networks (CNNs) (Alibabaei et al., 2022). A high accuracy (about 99%) has been reported with respect to manual procedures that might make future applications of ML in weed-disease detection profitable. But it has also been reported that the accuracy depends strongly on the quality of training datasets and that overtraining the ML models could also hamper the approach (Alibabaei et al., 2022; Nasirahmadi & Hensel, 2022).

Machine learning applied to irrigation has been shown to improve the optimization of irrigation water while minimizing the use of resources (e.g. water and electricity). In particular, the increased number of sensors deployed in the field has made ML algorithms robust, but this is also their main limitation in applicability because of the associated costs for the farmers.

5   Chapters of the Book

The book contains three parts. Part I gives a broad overview of precision agriculture, modelling and issues related to the use of models in agricultural practice. Part II explores the state-of-the-art modelling for precision agriculture. Part III contains a series of short chapters that each describe a commercial, model-based service that is currently available to farmers.

In Part I, Kersebaum and Wallor (chapter "Process-Based Modelling of Soil–Crop Interactions for Site-Specific Decision Support in Crop Management") discuss the use of process-based models for soil-crop interactions and how those modelling approaches can help to describe the spatiotemporal dynamics of soil-crop-atmosphere relationships. Knowledge about within-field spatial variation of temporally stable soil properties can be used to make timely and spatially adapted management decisions. In chapter "Models in Crop Protection" Fedele, Bove and Rossi describe different modelling approaches for crop protection, focusing on process-based approaches. It is interesting to note that the crop growth modelling community is working closely with the crop protection community to include these approaches in the crop growth models (Bregaglio et al., 2021; Donatelli et al., 2017). The chapter also discusses how the plant disease models can be used to improve decision-making in integrated pest management because they can be used to analyse complex relationships among factors and improve tactical decision-making and strategic management. Chapter "Development and Adoption of Model-Based Practices in Precision Agriculture" (Akaka, Gallego, Georgantzis, Rhan and Tisserand) discusses adoptions and limitations of model-based practices in PA. In particular, the authors explore those issues from a scientific point of view and a sociocultural point of view. They introduce the concept of 'co-creation' and conclude that this is important: users need to be involved from the start. Scientifically we are able to create very sophisticated models, but their adoption is somewhat lagging behind. A model for application in PA should also take into account stakeholders' needs and address their relevant questions.

Part II starts with the chapter "Process-Based Models and Simulation of Nitrogen Dynamics" (Cammarano, Miguez and Puntel), which gives an overview of the modelling approaches used to simulate soil and plant nitrogen. The chapter is a non-exhaustive description of nitrogen modelling (which are often subroutines to more complex crop models), but it gives the reader an idea of the complexity of process-based crop growth models. The authors describe two case studies in which crop growth models are used in a spatial context to optimize nitrogen management through an improved understanding of the soil-plant-weather-agronomy interactions. The chapter by Heinen (chapter "Modelling Soil Water Dynamics") describes the theory and application of different water modelling approaches. The author describes the main theory of water movement and how those models can be used to predict current and projected changes in soil water content. The chapter also consider how to model the spatial variation of soil processes and properties. In chapter "Data Fusion in a Data-Rich Era", Castrignanò and Belmonte introduced the topic of data fusion in precision agriculture. They highlight how geostatistical models can be useful in data fusion for PA applications. The authors discuss the implications of spatial and/or temporal autocorrelation and the effect of different supports on agricultural operations. Keiji, Kozan and De Wit (chapter "Data Assimilation of Remote Sensing Data into a Crop Growth Model") follow with a discussion on the assimilation of remote sensing data into a crop growth model. The chapter gives an overview and history of data assimilation between remote sensing and crop growth models. Then it points out the potential and limitations of such an approach and how current technologies might facilitate such integration.

Part III comprises a series of chapters describing model-based solutions available for farmers from commercial companies. Each chapter is organized in a structured way in which the contributors introduce their solution and give a brief generic overview of the technology, application, benefits and any additional information. In this section, we have contributions from Adapt-N ® (Yara International) which demonstrates how their process-based solution is developed into a commercial tool for optimizing N fertilization. Granular Agronomy's software also deals with the spatial optimization of N fertilization in which a crop growth model is used to estimate N requirements. The Xarvio ® Digital Farming Solutions illustrates how their crop model is integrated in their digital farming platform and also describes a case study on how to integrate model output into agronomic decision-making. The Kubota Corporation discussed their Kubota Smart Agri System used for precision farming and autonomous unmanned agricultural machinery with a working example on a rice transplanter with automatic steering, a combine harvester with an automatic operating assistance function and a fully autonomous tractor. The TSC Research and Innovation described the AgSkyNet platform used to provide growers with information during crop growing and post-harvest. They discuss case studies of solutions for integrated pest assessment, crop growth simulation to estimate yield and to assess the severity of effects of stubble burning and pesticide residue detection. The Dacom Farm Intelligence highlights their information system and how it is used to connect data and models and how that information is processed to make informed decisions on crop protection, crop growth, irrigation and precision agriculture decisions. Finally, **farm**maps, a platform made for precision agriculture, comprises multiple models (for potato, soil water dynamics, late blight infection and nematode management) and can be used to identify field and within-field agronomic management solutions. The user interface of **farm**maps is explicitly designed with farmers in mind in order to facilitate uptake of the platform.

6   Current State of Modelling for Precision Agriculture and Work Needed

Please recall that PA is *...a management strategy that gathers...and analyzes...data...to support management decisions.* This simple statement can break down in several places. In many cases there is a lack of sensors for some applications of precision agriculture. For example, it is currently not possible to measure potato tuber size and weight in the field in the same way that aboveground biomass can be measured. It may also be the case that a suitable actuator (such as a precision planter, spreader or sprayer) is not available. For example, few farmers today own a sprayer with section control. In this book we focus on the transformation of data into agronomic advice that can be relayed to an actuator. In the market this step is often left unspecified. You can see this in the many soil and crop sensors that are marketed today for more sustainable nutrient use or crop protection, but that lack models for making agronomic advice and an application rate prescription map. They leave this decision to the expertise of the user or their advisor. This hampers the adoption of precision agriculture. Successful applications of precision agriculture consist of three parts: data, decisions and actuators. A chain is only as strong as its weakest link.

It is starting to be recognized that successful application of precision agriculture requires that knowledge on PA technology and decision-support models are combined in order to develop and align the three components of a precision agriculture application. In the Netherlands, this is done, e.g. on the National Field Lab Precision Agriculture (www.proeftuinprecisielandbouw.nl) and Farm of the Future (www.farmofthefuture.nl). These are partnerships that are supported by the national government and which involve farmers, companies and scientists. All relevant parties are brought together to integrate technologies and know-how into fully functional applications. This approach is also applied in international projects in Brazil (Soybean Brazil NL[2]) and Japan (TTADDA[3]). Precision agriculture is all about cooperation and integration.

6.1   Data

Access to data is a major obstacle for the application of model-based technologies in precision agriculture. Both technical and socioeconomic obstacles exist.

From a technical point of view, it can be noted that a large number of farm management information systems (FMIS) is on the market, including systems offered by the major tractor companies. However, there is no easy way to transfer data from one FMIS to another or combine data from two FMIS. Some of the major manufacturers offer application programmer's interfaces (APIs) built around the data model of their system, but the difficult task of translating from the source data model to the target data model is left to the user. A generic approach is taken by AgGateway, a non-profit consortium with more than 100 members, whose Agricultural Data Application Programming Toolkit (ADAPT)[4] proposes a unified data model which can exchange data between any two FMIS. The Open Ag Data

---

[2] https://www.wur.nl/nl/Onderzoek-Resultaten/Onderzoeksprojecten-LNV/Expertisegebieden/kennisonline/Smart-technology-for-soybean-production.htm

[3] https://www.wur.nl/nl/onderzoek-resultaten/onderzoeksprojecten-lnv/expertisegebieden/kennisonline/transition-to-a-data-driven-agriculturen-ttadda-for-a-new-dutch-japanese-potato-circular-value-chain-.htm

[4] https://adaptframework.org/

Alliance[5] and DKE agrirouter[6] are similar efforts. Several vocabularies and data standards have been available for many years (Agro-XML,[7] EDI-Teelt++[8]), but they have not gained widespread acceptance.

Interoperability of equipment is another technical barrier frequently encountered by farmers (Van Evert et al., 2018). Connections between machines typically use the ISO 11783 (ISOBUS)[9] standard which defines the communication between agricultural machinery and also the data transfer between these machines and farm software applications. Unfortunately, the standard leaves room for interpretation so that in practice many incompatibilities arise when farmers buy equipment from different manufacturers. Competition between commercial providers precludes in many cases the formation of successful networks.

Social and legal issues related to agricultural data include data ownership, exchange, control and security (Kritikos, 2017). Farmers are often hesitant to share data because of social and legal challenges stemming from the wide range of actors involved in the farm data chain and the fragmented and uneven character of the data ecosystem. While it is clear that farmers' personal data is protected by current personal data regulations,[10] the ownership of equipment-generated data raises concerns among farmers and other agricultural stakeholders. Most companies state that farmers own the data they produce, but when they are aggregated with other farmers' data companies will consider the aggregated data their property. For example, measuring yield with a drone or a combine harvester yields data that are not personal (and therefore not protected), but may nevertheless give an indication of farm income (which is protected, personal data). Farmers may also waive data ownership rights by signing service agreements that they have not read (Rasmussen, 2016) or which are imposed on them unfairly because they are the weaker party.[11]

Data security in agriculture and privacy implications resulting from a security breach are a major concern for the digitization process of agriculture. The large number of devices and connections used in precision agriculture renders the complete system vulnerable, and the users of such systems should be aware of these and obtain protection.

Data-based decision tools can bring benefits to farming, but ethical questions must be asked about the consequences of changes in power relations between farmers and the other stakeholders that may follow the introduction of modelling in precision agriculture. Mepham's ethical matrix (Mepham, 2005) has been

---

[5] https://openag.io/

[6] https://my-agrirouter.com/en/

[7] www.agroxml.de

[8] https://www.agroconnect.nl/

[9] https://www.isobus.net/isobus/

[10] The Data Protection Directive (95/46/EC). From 25/05/18: The General Data Protection Regulation (GDPR) (Regulation (EU) 2016/679).

[11] Bertolo, S. Building a European Data economy. EIP-AGRI Workshop. Bratislava, 4–5 April 2017.

proposed as a tool to examine ethical questions (well-being, autonomy and fairness) in agriculture and biomedical practice. An ethical issue is that in some cases data are provided by citizens (farmers), yet the information primarily benefits the commercial actors. Another ethical issue is that individual farmers may not be able to make the investments that are necessary to reap the benefits of data-based tools. The organisation of farmers around data cooperatives that collect and sell data can be a solution for ensuring smallholder farmers also benefit from the data economy.

Finally, farmers may be willing to exchange their data if they see the benefit and understand the risks (Van Evert et al., 2018). To this end, a consortium of agricultural associations has developed the EU Code of Conduct on agricultural data sharing.[12]

6.2 Models

Combining models and data provide useful information for farmers. The benefits of digital twin technology will only be realized if a sensible combination of model calibration, model initialization and data assimilation is used. If that is done, it will enable high-quality fertilizer recommendations and in a broader sense better agronomic management. The main limitation is not the data analysis or the computational capability, but rather a critical and improved understanding of what we need and can obtain from the model. Farmers have specific agronomic needs and any technology (remote sensing, modelling, etc.) needs to be adapted to answer those questions. The third section of this book shows clearly that it is possible to use crop growth models in precision agriculture, but in the basic and applied research field, it seems that this issue is less developed.

There is often a diffidence towards crop models (e.g. arguments used against them are that they are too hard to use or they require too many data) which is sometimes due to users' limited knowledge of the agricultural system (soil-plant-atmosphere or agronomy per se) rather than the limitation of the models.

Another practical issue to solve in the near future is the frequency of data that are needed to drive/parameterize/calibrate a crop model during data assimilation. Nowadays it is possible to obtain very detailed data at high spatial and temporal resolutions from drones and satellites. However, does a crop model need all those data? Can the same accuracy be obtained with less data? This is not a trivial question because it might affect the willingness of farmers to adopt new technologies.

Finally, despite the models used (crop growth models or ML), it is important to translate the data into agronomic decisions. It is possible that models and modelling approaches used to estimate yield or grain quality are applied too late during the growing season, which makes the utility of those simulations less relevant because it might be too late to correct for any deficiency. Identifying the right time is therefore

---

[12]https://www.cema-agri.org/images/publications/brochures/EU_Code_of_conduct_on_agricul
tural_data_sharing_by_contractual_agreement_2020_ENGLISH.pdf

important. In addition to that, applied research should focus on involving farmers and agronomists in the design of the experimental approach because they are the ones working on those issues on a daily basis.

6.3 Actuation

Actuation is not the biggest bottleneck, even though many farmers do not own precision application equipment. But, the next generation of machines, the new generations of farmers cultivating the land and better data analytics linking sensors and machinery will enable further actuation of digitalization.

In some contexts, for example, in developing countries, actuation is occurring at a fast pace (MaMo, 2019). Rather than the modelling per se, it is the actuation of digital and precision agriculture that is happening because in most cases the farmers are young and open to new technologies and therefore their main entry point is digital agriculture (e.g. smart phone usage in agricultural management).

7 Conclusion

Crop growth models have great potential to be of benefit in precision agriculture. A combination of models and data from sensors (i.e. DT) can generate information to support farmers' management decisions. When a suitable crop growth model is not available or does not offer the required functionality, statistical techniques like ML can be used instead.

The greatest bottleneck is represented by data availability, usage and quality. Farmers are weary to invest in collecting the real-time (or near real-time) data needed to run simulations as long as the transformation of these data into agronomic decision is uncertain. Acceptance of models by farmers and other practitioners is also a bottleneck in many cases. Lack of trust in models and feeling overwhelmed by complexity play a role. This points to the need for better connections between researchers, extension workers, farmers and advisors to leverage the benefits of data and modelling.

Further Reading *Models*: the Agricultural Model Intercomparison and Improvement Project (https://agmip.org/) contains information and papers about crop modeling activities at different scales and for different purposes. Some specific websites such as https://www.apsim.info/, https://dssat.net/, and https://www.wur.nl/en/Research-Results/Research-Institutes/Environmental-Research/Facilities-Tools/Software-models-and-databases/WOFOST.htm contain information about specific crop growth models.

Precision agriculture: The International Society of Precision Agriculture's webpage (https://ispag.org/) contains links to scientific publications from the scientific Journal Precision Agriculture and additional material.

Conflict of Interest The authors declare that there are no conflicts of interest.

References

ABARES, A.B.o.A.a.R.E.a.S. (2022). *Australian crop report*.

Aggarwal, P., Vyas, S., Thornton, P., Campbell, B. M., & Kropff, M. (2019). Importance of considering technology growth in impact assessments of climate change on agriculture. *Global Food Security, 23*, 41–48. https://doi.org/10.1016/j.gfs.2019.04.002

Alibabaei, K., Gaspar, P. D., Lima, T. M., Campos, R. M., Girão, I., Monteiro, J., et al. (2022). A review of the challenges of using deep learning algorithms to support decision-making in agricultural activities. *Remote Sensing, 14*, 638.

Angus, J. F. (2001). Nitrogen supply and demand in Australian agriculture. *Australian Journal of Experimental Agriculture, 41*, 277–288. https://doi.org/10.1071/EA00141

Asseng, S., Ewert, F., Martre, P., Rötter, R. P., Lobell, D. B., Cammarano, D., et al. (2015). Rising temperatures reduce global wheat production. *Nature Climate Change, 5*, 5. https://doi.org/10.1038/nclimate2470

Asseng, S., Martre, P., Maiorano, A., Rötter, R. P., O'Leary, G. J., Fitzgerald, G. J., et al. (2019). Climate change impact and adaptation for wheat protein. *Global Change Biology, 25*, 155–173. https://doi.org/10.1111/gcb.14481

Basso, B., Ritchie, J. T., Pierce, F. J., Braga, R. P., & Jones, J. W. (2001). Spatial validation of crop models for precision agriculture. *Agricultural Systems, 68*, 97–112. https://doi.org/10.1016/S0308-521X(00)00063-9

Basso, B., Martinez-Feria, R. A., Rill, L., & Ritchie, J. T. (2021). Contrasting long-term temperature trends reveal minor changes in projected potential evapotranspiration in the US Midwest. *Nature Communications, 12*, 1476. https://doi.org/10.1038/s41467-021-21763-7

Batchelor, & Paz. (1997). The role of water stress in creating spatial yield variability in soybeans. In *Proceedings of the 9th integrated crop management conference*.

Batchelor, W. D., Basso, B., & Paz, J. O. (2002). Examples of strategies to analyze spatial and temporal yield variability using crop models. *European Journal of Agronomy, 18*, 141–158. https://doi.org/10.1016/S1161-0301(02)00101-6

Bean, G. M., Kitchen, N. R., Camberato, J. J., Ferguson, R. B., Fernandez, F. G., Franzen, D. W., et al. (2018). Improving an active-optical reflectance sensor algorithm using soil and weather information. *Agronomy Journal, 110*, 2541–2551. https://doi.org/10.2134/agronj2017.12.0733

Booij, J. A., Van Evert, F. K., van Geel, W. C. A., Kroonen-Backbier, B. M. A., & Kempenaar, C. (2017). *Roll-out of online application for N sidedress recommendations in potato*. EFITA. Available online at http://library.wur.nl/WebQuery/wurpubs/fulltext/445495

Bregaglio, S., Willocquet, L., Kersebaum, K. C., Ferrise, R., Stella, T., Ferreira, T. B., et al. (2021). Comparing process-based wheat growth models in their simulation of yield losses caused by plant diseases. *Field Crops Research, 265*, 108108. https://doi.org/10.1016/j.fcr.2021.108108

Bustos-Korts, D., Boer, M. P., Malosetti, M., Chapman, S., Chenu, K., Zheng, B., et al. (2019). Combining crop growth modeling and statistical genetic modeling to evaluate phenotyping strategies. *Frontiers in Plant Science, 10*. https://doi.org/10.3389/fpls.2019.01491

Butchee, K. S., May, J., & Arnall, B. (2011). Sensor based nitrogen management reduced nitrogen and maintained yield. *Crop Management, 10*, 1–5. https://doi.org/10.1094/CM-2011-0725-01-RS

Cammarano, D., Valdivia, R. O., Beletse, Y. G., Durand, W., Crespo, O., Tesfuhuney, W. A., et al. (2020). Integrated assessment of climate change impacts on crop productivity and income of commercial maize farms in Northeast South Africa. *Food Security, 12*, 659–678. https://doi.org/10.1007/s12571-020-01023-0

Cammarano, D., Basso, B., Holland, J., Gianinetti, A., Baronchelli, M., & Ronga, D. (2021). Modeling spatial and temporal optimal N fertilizer rates to reduce nitrate leaching while improving grain yield and quality in malting barley. *Computers and Electronics in Agriculture, 182*, 105997. https://doi.org/10.1016/j.compag.2021.105997

Colaço, A. F., & Bramley, R. G. V. (2018). Do crop sensors promote improved nitrogen manage-ment in grain crops? *Field Crops Research, 218*, 126–140. https://doi.org/10.1016/j.fcr.2018. 01.007

Colaço, A. F., Richetti, J., Bramley, R. G. V., & Lawes, R. A. (2021). How will the next-generation of sensor-based decision systems look in the context of intelligent agriculture? A case-study. *Field Crops Research, 270*, 108205. https://doi.org/10.1016/j.fcr.2021.108205

Donatelli, M., Magarey, R. D., Bregaglio, S., Willocquet, L., Whish, J. P. M., & Savary, S. (2017). Modelling the impacts of pests and diseases on agricultural systems. *Agricultural Systems, 155*, 213–224. https://doi.org/10.1016/j.agsy.2017.01.019

Elliott, J., Müller, C., Deryng, D., Chryssanthacopoulos, J., Boote, K. J., Büchner, M., et al. (2015). The global gridded crop model intercomparison: Data and modeling protocols for Phase 1 (v1.0). *Geoscientific Model Development, 8*, 261–277.

EU, E.U. (1991). *Council Directive 91/676/EEC of 12 December 1991 concerning the protection of waters against pollution caused by nitrates from agricultural sources*. In Union, E. (Ed.), p. 8.

Grieves, M. (2014). Digital twin: Manufacturing excellence through virtual factory replication. *White paper*, 1–7.

Hamza, M. A., & Anderson, W. K. (2005). Soil compaction in cropping systems: A review of the nature, causes and possible solutions. *Soil and Tillage Research, 82*, 121–145. https://doi.org/ 10.1016/j.still.2004.08.009

Holland, K. H., Schepers, J. S., Shanahan, J. F., & Horst, G. L. (2004, July 25–28). Plant canopy sensor with modulated polychromatic light source. In *Proceedings of the 7th international conference on precision agriculture and other precision resources management*. Hyatt Regency.

Holland, K. H., Lamb, D. W., & Schepers, J. S. (2012). Radiometry of proximal Active Optical Sensors (AOS) for agricultural sensing. *IEEE Journal of Selected Topics in Applied Earth Observations and Remote Sensing, 5*, 1793–1802. https://doi.org/10.1109/jstars.2012.2198049

IPCC, I.P.o.C.C. (2021). *Climate change 2021: The physical science basis. Contribution of Working Group I to the Sixth Assessment Report of the Intergovernmental Panel on Climate Change*. In V. Masson-Delmotte, P. Zhai, A. Pirani, S. L. Connors, C. Péan, S. Berger, N. Caud, Y. Chen, L. Goldfarb, M. I. Gomis, M. Huang, K. Leitzell, E. Lonnoy, J. B. R. Matthews, T. K. Maycock, T. Waterfield, O. Yelekçi, R. Yu, & B. Zhou (Eds.). Cambridge University Press.

Jackson, R. D., Idso, S. B., Reginato, R. J., & Pinter, P. J., Jr. (1981). Canopy temperature as a crop water stress indicator. *Water Resources Research, 17*, 1133–1138. https://doi.org/10.1029/ WR017i004p01133

Jackson, R. D., Kustas, W. P., & Choudhury, B. J. (1988). A reexamination of the crop water stress index. *Irrigation Science, 9*, 309–317. https://doi.org/10.1007/BF00296705

Jägermeyr, J., Müller, C., Ruane, A. C., Elliott, J., Balkovic, J., Castillo, O., et al. (2021). Climate impacts on global agriculture emerge earlier in new generation of climate and crop models. *Nature Food, 2*, 873–885. https://doi.org/10.1038/s43016-021-00400-y

Jansen, D. M. (2008). *Beschrijving van TIPSTAR: hét simulatiemodel voor groei en productie van zetmeelaardappelen* (Nota 547). Research International. Available online at http://edepot.wur. nl/27135.Plant

Jansen, D. M., Davies, J. A., & Steenhuizen, J. W. (2003). *Testen van Tipstar in de praktijk* (Report 244). Plant Research International.

Jasper, J., Reusch, S., & Link, A. (2009). *Active sensing of the N status of wheat using optimized wavelength combination: Impact of seed rate, variety and growth stage* (pp. 23–30). JIAC.

Jones, J. W., Hoogenbom, G., Porter, C. H., Boote, K. J., Batchelor, W. D., Hunt, L. A., et al. (2003). The DSSAT cropping system model. *European Journal of Agronomy, 18*, 31.

Kempenaar, C., Groeneveld, R. M. W., & Uenk, D. (2004, August 31–September 2). An innovative dosing system for potato haulm killing herbicides. In *Proceedings of the XII international conference on weed biology* (pp. 511–518).

Kempenaar, C., Achten, V. T. J., Van Evert, F. K., Van der Lans, A. M., Olijve, A. J., Van der Schans, D. A., et al. (2008). Biomassa-afhankelijk doseren van gewasbeschermingsmiddelen. *Gewasbescherming, 39*, 177–182.

Kempenaar, C., Hejting, S., & Michielsen, J. M. (2014). Perspectives for site specific application of soil herbicides in arable farming. In *12th International Conference on Precision Agriculture (ICPA)*.

Kempenaar, C., Been, T., Booij, J., van Evert, F., Michielsen, J.-M., & Kocks, C. (2017). Advances in variable rate technology application in potato in the Netherlands. *Potato Research, 60*, 295–305. https://doi.org/10.1007/s11540-018-9357-4

Kersebaum, K. C., Boote, K. J., Jorgenson, J. S., Nendel, C., Bindi, M., Frühauf, C., et al. (2015). Analysis and classification of data sets for calibration and validation of agro-ecosystem models. *Environmental Modelling and Software, 72*, 402–417.

Kritikos, M. (2017). *Precision agriculture in Europe: Legal, social and ethical considerations*. European Parliamentary Research Service. https://doi.org/10.2861/278

Lobell, D. B., & Asseng, S. (2017). Comparing estimates of climate change impacts from process-based and statistical crop models. *Environmental Research Letters, 12*, 015001. https://doi.org/10.1088/1748-9326/aa518a

Lowenberg-DeBoer, J., & Erickson, B. (2019). Setting the record straight on precision agriculture adoption. *Agronomy Journal, 111*, 1552–1569. https://doi.org/10.2134/agronj2018.12.0779

Maestrini, B., & Basso, B. (2018). Drivers of within-field spatial and temporal variability of crop yield across the US Midwest. *Scientific Reports, 8*, 14833. https://doi.org/10.1038/s41598-018-32779-3

MaMo, M. M. P. (2019). *Byte by byte: Policy innovation for transforming Africa's food system with digital technologies*.

Martinez-Feria, R. A., & Basso, B. (2020). Unstable crop yields reveal opportunities for site-specific adaptations to climate variability. *Scientific Reports, 10*, 2885. https://doi.org/10.1038/s41598-020-59494-2

McKinion, J. M., Baker, D. N., Whisler, F. D., & Lambert, J. R. (1989). Application of the GOSSYM/COMAX system to cotton crop management. *Agricultural Systems, 31*, 55–65. https://doi.org/10.1016/0308-521X(89)90012-7

McNunn, G., Heaton, E., Archontoulis, S., Licht, M., & VanLoocke, A. (2019). Using a crop modeling framework for precision cost-benefit analysis of variable seeding and nitrogen application rates. *Frontiers in Sustainable Food Systems, 3*. https://doi.org/10.3389/fsufs.2019.00108

Mepham, B. (2005). *Bioethics: An introduction for the biosciences*. Oxford University Press.

Müller, C., Elliott, J., Kelly, D., Arneth, A., Balkovic, J., Ciais, P., et al. (2019). The global gridded crop model intercomparison phase 1 simulation dataset. *Scientific Data, 6*, 50. https://doi.org/10.1038/s41597-019-0023-8

Nasirahmadi, A., & Hensel, O. (2022). Toward the next generation of digitalization in agriculture based on digital twin paradigm. *Sensors, 22*, 498.

Nawar, S., Corstanje, R., Halcro, G., Mulla, D., & Mouazen, A. M. (2017). Chapter four – Delineation of soil management zones for variable-rate fertilization: A review. In D. L. Sparks (Ed.), *Advances in agronomy* (pp. 175–245). Academic Press.

O'Shaughnessy, S. A., & Evett, S. R. (2010). Canopy temperature based system effectively schedules and controls center pivot irrigation of cotton. *Agricultural Water Management, 97*, 1310–1316. https://doi.org/10.1016/j.agwat.2010.03.012

Paz, O. J., Batchelor, D. W., Colvin, T. S., Logsdon, S. D., Kaspar, T. C., & Karlen, D. L. (1998). Analysis of water stress effects causing spatial yield variability in soybeans. *Transactions of the ASAE, 41*, 1527–1534. https://doi.org/10.13031/2013.17284

Paz, J. O., Batchelor, W. D., Babcock, B. A., Colvin, T. S., Logsdon, S. D., Kaspar, T. C., et al. (1999). Model-based technique to determine variable rate nitrogen for corn. *Agricultural Systems, 61*, 69–75. https://doi.org/10.1016/S0308-521X(99)00035-9

Peters, R. T., & Evett, S. R. (2008). Automation of a center pivot using the temperature-time-threshold method of irrigation scheduling. *Journal of Irrigation and Drainage Engineering, 134*, 286–291. https://doi.org/10.1061/(ASCE)0733-9437(2008)134:3(286)

Puntel, L. A., Sawyer, J. E., Barker, D. W., Thorburn, P. J., Castellano, M. J., Moore, K. J., et al. (2018). A systems modeling approach to forecast corn economic optimum nitrogen rate. *Frontiers in Plant Science, 9.* https://doi.org/10.3389/fpls.2018.00436

Puntel, L. A., Pagani, A., & Archontoulis, S. V. (2019). Development of a nitrogen recommendation tool for corn considering static and dynamic variables. *European Journal of Agronomy, 105*, 189–199. https://doi.org/10.1016/j.eja.2019.01.003

Pylianidis, C., Osinga, S., & Athanasiadis, I. N. (2021). Introducing digital twins to agriculture. *Computers and Electronics in Agriculture, 184*, 105942. https://doi.org/10.1016/j.compag.2020.105942

Ransom, C. J., Kitchen, N. R., Camberato, J. J., Carter, P. R., Ferguson, R. B., Fernández, F. G., et al. (2019). Statistical and machine learning methods evaluated for incorporating soil and weather into corn nitrogen recommendations. *Computers and Electronics in Agriculture, 164*, 104872. https://doi.org/10.1016/j.compag.2019.104872

Rasmussen, N. (2016). From precision agriculture to market manipulation: A new frontier in the legal community. *Minnesota Journal of Law, Science & Technology, 17*, 489–516.

Raun, W. R., Solie, J. B., Stone, M. L., Martin, K. L., Freeman, K. W., Mullen, R. W., et al. (2005). Optical sensor-based algorithm for crop nitrogen fertilization. *Communications in Soil Science and Plant Analysis, 36*, 2759–2781. https://doi.org/10.1080/00103620500303988

Reusch, S. (2003). *Optimisation of oblique-view remote measurement of crop N-uptake under changing irradiance conditions.*

Rosenzweig, C., & Parry, M. L. (1994). Potential impact of climate change on world food supply. *Nature, 367*, 133–138. https://doi.org/10.1038/367133a0

Rötter, et al. (2016). *Analysis of crop yield variability and yield gaps for maize and wheat in diverse climatic zones.*

Samborski, S. M., Tremblay, N., & Fallon, E. (2009). Strategies to make use of plant sensors-based diagnostic information for nitrogen recommendations. *Agronomy Journal, 101*, 800–816. https://doi.org/10.2134/agronj2008.0162Rx

Samuel, A. L. (1959). Some studies in machine learning using the game of checkers. *IBM Journal of Research and Development, 3*, 210–229. https://doi.org/10.1147/rd.33.0210

Scharf, P. C., Shannon, D. K., Palm, H. L., Sudduth, K. A., Drummond, S. T., Kitchen, N. R., et al. (2011). Sensor-based nitrogen applications out-performed producer-chosen rates for corn in on-farm demonstrations. *Agronomy Journal, 103*, 1683–1691. https://doi.org/10.2134/agronj2011.0164

Steltenpool, J. A. N., & van Erp, P. J. (1995). Schatting van de actuele N-opname door aardappelen. *Meststoffen, 1995*, 45–50.

Thorp, K. R., DeJonge, K. C., Kaleita, A. L., Batchelor, W. D., & Paz, J. O. (2008). Methodology for the use of DSSAT models for precision agriculture decision support. *Computers and Electronics in Agriculture, 64*, 276–285. https://doi.org/10.1016/j.compag.2008.05.022

Tremblay, N., Bouroubi, M. Y., Vigneault, P., & Belec, C. (2011). Guidelines for in-season nitrogen application for maize (Zea mays L.) based on soil and terrain properties. *Field Crops Research, 122*, 273–283. https://doi.org/10.1016/j.fcr.2011.04.008

Van Evert, F. K., Booij, R., Jukema, J. N., Ten Berge, H. F. M., Uenk, D., Meurs, E. J. J., et al. (2012). Using crop reflectance to determine sidedress N rate in potato saves N and maintains yield. *European Journal of Agronomy, 43*, 58–67. https://doi.org/10.1016/j.eja.2012.05.005

Van Evert, F. K., Wolters, S., Van Boheemen, K., & Van Dijk, C. (2018). *Final report on research results, research projects and industry solutions. Deliverable 1.6 of the Horizon 2020 project "Smart-AKIS".* Available online at https://www.smart-akis.com/wp-content/uploads/2019/01/Report-on-Research-Results-Research-Projects-and-Industry-Solutions.pdf. Accessed 22 Apr 2022.

Van Evert, F. K., Berghuijs, H. N. C., Hoving, I. E., De Wit, A. J. W., & Been, T. H. (2021). A digital twin for arable and dairy farming. In J. Stafford (Ed.), *Precision agriculture '21* (pp. 919–925). Wageningen Academic Publishers.

Van Ittersum, M. K., Leffelaar, P. A., Van Keulen, H., Kropff, M. J., Bastiaans, L., & Goudriaan, J. (2003). On approaches and applications of the Wageningen crop models. *European Journal of Agronomy, 18*, 201–234.

Voltz, M., Guibaud, G., Dagès, C., Douzals, J.-P., Guibal, R., Grimbuhler, S., et al. (2022). Pesticide and agro-ecological transition: Assessing the environmental and human impacts of pesticides and limiting their use. *Environmental Science and Pollution Research, 29*, 1–5. https://doi.org/10.1007/s11356-021-17416-3

Wanjura, R. D., Upchurch, R. D., & Mahan, R. J. (1992). Automated irrigation based on threshold canopy temperature. *Transactions of the ASAE, 35*, 1411–1417. https://doi.org/10.13031/2013.28748

Wanjura, F. D., Upchurch, R. D., & Mahan, R. J. (1995). Control of irrigation scheduling using temperature-time thresholds. *Transactions of the ASAE, 38*, 403–409. https://doi.org/10.13031/2013.27846

Zadoks, J. C. (1981). EPIPRE: A disease and pest management system for winter wheat developed in the Netherlands. *EPPO Bulletin, 11*, 365–369. https://doi.org/10.1111/j.1365-2338.1981.tb01945.x

Zeigler, B. P. (1976). *Theory of modelling and simulation.* Wiley.

Zeigler, B. P., Praehofer, H., & Kim, T. G. (2000). *Theory of modeling and simulation* (2nd ed.). Academic Press.

Zheng, B., Chenu, K., Dreccer, M. F., & Chapman, S. C. (2012). Breeding for the future: What are the potential impacts of future frost and heat events on sowing and flowering time requirements for Australian bread wheat (Triticum aestivum) varieties? *Global Change Biology, 18*, 2899–2914. https://doi.org/10.1111/j.1365-2486.2012.02724.x

Process-Based Modelling of Soil–Crop Interactions for Site-Specific Decision Support in Crop Management

K. C. Kersebaum and E. Wallor

Abstract Spatial variation within fields influences soil water, nutrient and crop growth dynamics. Depending on climatic conditions, spatial patterns of yield are often not stable over time because weather conditions favour different processes from year to year. The same holds for rapidly fluctuating soil processes like water and nitrogen dynamics, while other soil properties remain relatively stable over a few years. Estimation of an appropriate irrigation or fertilizer amount in relation to site-specific yield expectations is essential to optimize water or nutrient use efficiency. On the other hand, farmers' aims are often in conflict with the interest of water suppliers to keep nutrient loads below the drinking water standard. Process-based agro-ecosystem models can help to reflect the spatio-temporal dynamics in soil–crop–atmosphere relations to transcribe spatial within-field variation of temporally stable soil properties into timely and spatially adapted management decisions.

Keywords Crop growth · Nitrogen · Yield variation · Nitrate leaching

1 Introduction

Soil properties often vary considerably across landscapes and within fields influencing soil water, nutrient and crop growth dynamics. Spatial variation of soil properties are determined by soil genesis, surface topography, land use history and previous and current management (Moore et al., 1993; Trangmar et al., 1985; Pennock & Frick, 2001; Manning et al., 2001; Kersebaum, 1995). Furthermore, topography and hydrology of agricultural fields influence crop growth and biomass production (e.g. Timlin et al., 1998; Kravchenko & Bullock, 2000, 2002; Reuter, 2004; Si &

K. C. Kersebaum (✉)
Leibniz Centre for Agricultural Landscape and Land Use Research (ZALF), Müncheberg, Germany
e-mail: ckersebaum@zalf.de

E. Wallor
University of Applied Science Eberswalde, Eberswalde, Germany

© The Author(s), under exclusive license to Springer Nature Switzerland AG 2023
D. Cammarano et al. (eds.), *Precision Agriculture: Modelling*, Progress in Precision Agriculture, https://doi.org/10.1007/978-3-031-15258-0_2

25

Farrell, 2004; Basso et al., 2009; Martinez-Feria & Basso, 2020) through topographic shading (Reuter et al., 2005), capillary rise (Kersebaum & Nendel, 2014; Bormann, 2012) or oxygen deficiency caused by topographic wetness (Kersebaum et al., 2002; Martinez-Feria & Basso, 2020). Figure 1 shows examples of relevant elements which cause spatial variation in crop production and their effects on the main processes in the soil–crop interaction.

During the last two decades, technical solutions have been provided within the framework of precision agriculture (PA) to consider these spatial patterns for site-specific management. The aim is to improve productivity and nutrient use efficiency; reduce pollution to the environment, enabling high yields at suitable locations and avoid over-fertilization in zones with unfavourable conditions for plant growth.

From the practical aspect, farmers look for management zones, which have a more or less stable behaviour relative to other parts of the field and a similar response to management. Methods have been described to identify such management zones, which are based on the analysis of multi-year yield maps (e.g. Stafford et al., 1999; Heermann et al., 2003; Nawar et al., 2017). Spatial patterns that are stable in time can often be observed under conditions where one factor (e.g. water availability) is dominantly limiting. In temperate regions, water can be the limiting factor on field sites with low water-holding capacity, but may also favour these sites in wet years, where oxygen deficiency occurs in heavier clay soils due to excessive rainfall. Spatial yield patterns are often not stable over time because weather conditions favour different processes from year to year (e.g. Stafford et al., 1999; Kersebaum et al., 2005b; Maestrini & Basso, 2018). Interaction between the temporal dynamics of soil water and nutrients and their spatial variation results in a pronounced temporal change in the range of their spatial representativeness (Greminger et al., 1985; Wendroth et al., 1999). Hence, at different times of the year with diverging soil water status, variable characteristic spatial patterns of soil water determine crop growth patterns across the field. Depending on whether it is a rather dry or wet year, different underlying characteristics affect the spatial variation of crop production and grain yield (Timlin et al., 1998; Kaspar et al., 2003; Delin & Berglund, 2005; Li et al., 2019). Kumhálová et al. (2011) found a significant relation between yield and crop nutrient concentration with topography. However, the correlation coefficients between flow accumulation and yield were weak for the wetter years and strong for the drier years.

Figure 2 shows a subjective assessment of some key effects on variables and processes caused by different spatial properties for two different annual weather situations. The diversity of responses between the two contrasting weather conditions indicates why spatial pattern in crop yield may be stable or not across years. However, the uncertainty of these subjective assessments is high, and seasonal weather will vary gradually among the assumed extremes. This makes an assessment of the spatio-temporal behaviour difficult to predict just from expert knowledge. To integrate and understand the various interactions between processes and to quantify how crops grow in response to weather, soil and management early crop models were developed in the 1960s pioneered by the work of de Wit (1965) and Monteith (1965). Presently, a range of models of differing degrees of model complexity with

Property	Origin of variability	Processes affected
Topography 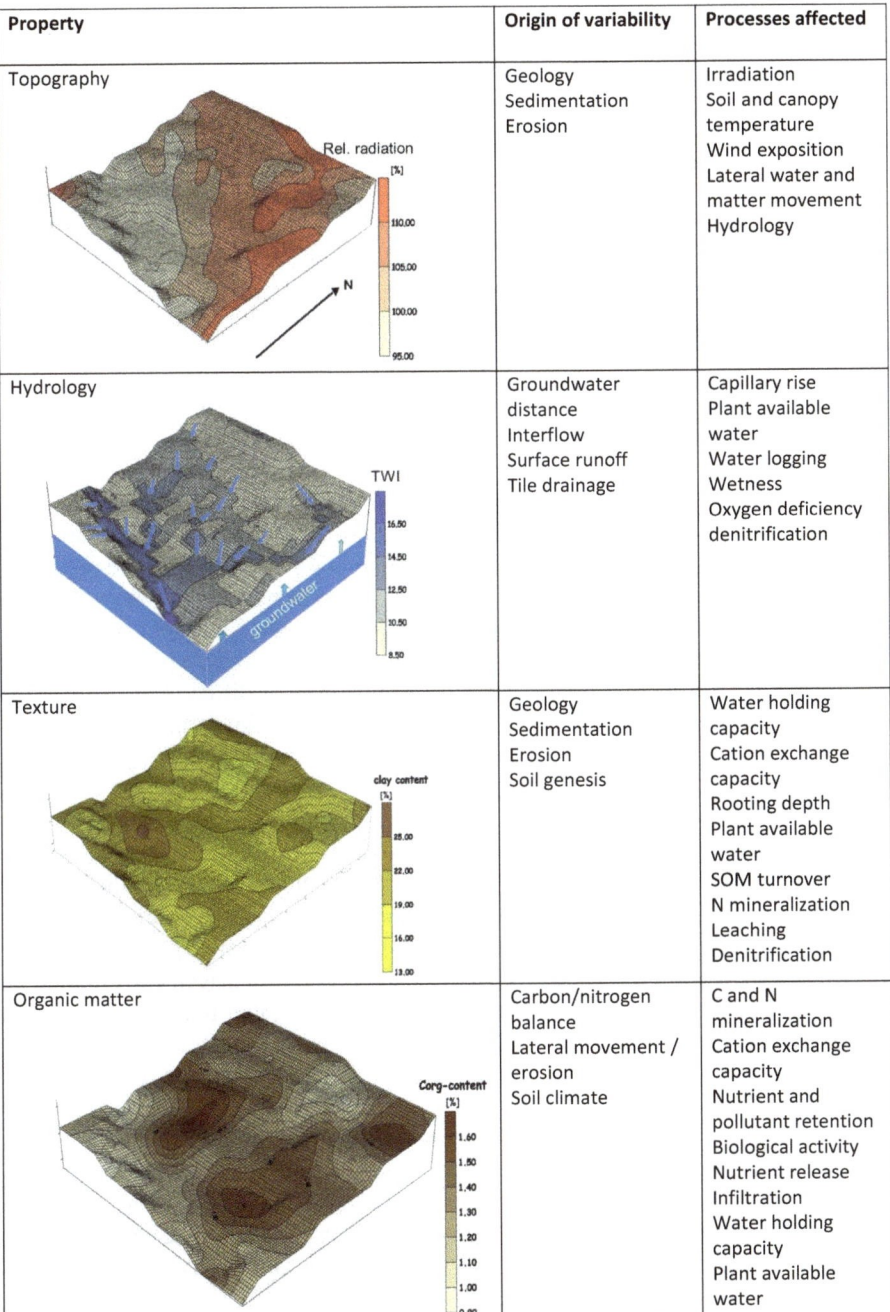	Geology Sedimentation Erosion	Irradiation Soil and canopy temperature Wind exposition Lateral water and matter movement Hydrology
Hydrology	Groundwater distance Interflow Surface runoff Tile drainage	Capillary rise Plant available water Water logging Wetness Oxygen deficiency denitrification
Texture	Geology Sedimentation Erosion Soil genesis	Water holding capacity Cation exchange capacity Rooting depth Plant available water SOM turnover N mineralization Leaching Denitrification
Organic matter	Carbon/nitrogen balance Lateral movement / erosion Soil climate	C and N mineralization Cation exchange capacity Nutrient and pollutant retention Biological activity Nutrient release Infiltration Water holding capacity Plant available water

Fig. 1 Spatial properties on a field or landscape scale, their origin and effect on different processes

Site property / Impact	Relief position				Groundwater		Texture			SOM	
Impact	Hill	Foot	N slope	S slope	Shallow	Deep	Sand	Loam	Clay	High	Low
Irradiation	↑↑	↓↓	↓↓	↑↑	○○	○○	○○	○○	○○	↗↗	↙↙
Soil temperature	↑↕	↓↓	↓↓	↑↑	↓○	↑○	↑○	↙↗	↓↗	↑↑	↓↓
ETa	↗↑	↑↙	↙○	↗○	↑○	↓○	↓↙	↑↗	↗○	↗○	↙○
Soil moisture	↓↙	↗↑	↗○	↘○	↑↗	↓○	↓↙	↑↗	↑↑	○↗	○↙
Rooting depth	↙↙	↗↗	○○	○○	○↙	↗○	↙↙	↑↗	↗↙	↗○	↙○
Plant avail. water	↓↙	↑↗	○○	↙○	↑↗	↓↙	↓↓	↑↑	↓↙	↗○	↙○
Runoff/erosion	↗↑	↓↓	↗↗	↗↗	○↗	○○	↓↓	↙↗	↑↑	↙↙	↗↗
Deep percolation	↓↙	↗↑	↙↗	↙○	↗↑	↙↓	↗↑	↓↓	○↗	○○	○○
N leaching	↙○	○↑	↙○	↙↗	↙↗	↑↗	↑↑	○↓	↙↓	↗↗	↙↙
N mineralization	↙↙	↑↗	↙↓	↙↓	↑↗	↓○	↓↙	↗○	○↙	↑↗	↓↓
Denitrification	↓↙	↑↗	↙↓	○↗	↑↗	○↙	↓↓	○↗	↑↑	○↗	↙○
Crop phenology	↗○	↙↓	↙↓	↑↗	↗○	○○	↗○	○○	↗↙	↗○	○○
Plant growth	↓↗	↑↙	↗↙	↑↕	↑↕	○↙	↓↙	↗↑	↗↙	↑↑	↙↗

Red = hot/dry year; blue= cold/wet year; ↑ = higher/faster; ↓ = lower/slower; ↗= slightly higher/faster; ↙ = slightly lower/slower;
↕ = direction depend on other factors; ○ = no significant effect

Fig. 2 Subjectively estimated impacts of different site properties on processes for cold/wet and hot/dry seasonal conditions

emphasis on different research questions, crops or regions are available (Ewert et al., 2015). While statistical models aim to use empirical relations among different input variables without considering the temporal behaviour of the system's state variables, process-based agro-ecosystem models simulate the most relevant processes in soil–crop–atmosphere system simultaneously and consider their multiple interactions. These models are usually developed as one-dimensional models, which operate at a plot scale with well-defined site properties. The overlay of different property maps can be used to create multiple pedons or ecotopes. These can form units of homogeneous properties, which can be simulated to reflect the spatial variation of crop responses within a field or regions.

Zones of similar crop responses delineated as uniform management zones (UMZs) are an integral part of the decision process at the farm level, especially for nitrogen fertilization. For the farmer, it is important to apply an appropriate amount of fertilizer corresponding to the site-specific yield expectations to increase nutrient use efficiency. From the aspect of water protection, of paramount interest are the nutrient concentration of seepage water and appropriate management limitations, which are required to maintain groundwater quality. The latter requires a long-term assessment of fertilization effects on the average nitrogen concentration to determine site-specific fertilizer limitations to meet the drinking water standard (Kersebaum et al., 2005b).

This chapter describes the general concept of process-based agro-ecosystem models and their potential to be applied in PA to support spatio-temporal management decisions. Examples are provided to demonstrate that integrated modelling of crop growth interactions with soil water and nutrient dynamics can be a valuable tool for such assessments using spatially variable site conditions and long-term weather data.

2   General Character of Process-Based Agro-ecosystem Models

Process-based agro-ecosystem models consider the most relevant processes in soil–crop–atmosphere systems and their interactions. Usually, the soil–plant–atmosphere continuum is subdivided into compartments and sub-compartments. For instance, crops are separated into organ compartments and soil profiles into layers of defined thickness. Changes within a compartment are the result of energy or matter inputs and outputs across the compartment boundaries following the principle of energy or matter balance. A central characteristic of a process-based dynamic model is that they simulate changes within a compartment in time increments using the status at the beginning of the time step. Changes during the time step are simulated as the result of multiple processes in parallel considering various fluxes, transformations and energy, water and matter balances. Consequently, dynamic systems depend on their history because a state variable results from its previous status, which affects the processes involved during the time step. For example, the water content and nitrate concentration in a soil compartment at the start of a time step determines how much water and nitrogen can be taken up by plants from this compartment or determines if additional water, e.g. from rainfall, entering the compartment just fills up the water content in the compartment or if a certain amount of water and nitrogen flow to deeper compartments. Therefore, a dynamic model requires a definition of the initial status of a state variable, which is usually measured at one or several points across a field. For instance, without any knowledge of the initial water contents in the soil compartments, the input of precipitation to the surface becomes meaningless since it is not possible to calculate a balance for the next time step or a wrong assumption of the initial conditions might have serious consequences for other processes, e.g. germination of crop seeds. Figure 3 provides a simplified scheme of a process-based model for plant–nutrient dynamics in a soil–crop system.

Several models exist to simulate soil–crop interactions, which differ in their approaches to handle different processes. For water dynamics, two main approaches are commonly used: a more mechanistic approach based on the Richards equation and a more functional approach, which defines the capacity of a layer to store water against gravity and how much water is available to plants. The mechanistic approach simulates water flow along a potential gradient using soil water pressure heads and functions to describe the dependence of water conductivity on water content or pressure head. Estimation of soil water characteristics (water content vs. pressure head and water conductivity vs. pressure head) is laborious, and they are spatially very variable in the field; therefore, the so-called pedotransfer functions are sometimes used to derive these functions from easily available soil information, e.g. soil texture. The functional capacity approach, also called tipping bucket approach, is often used and seemed appropriate to be used at the field as well as the regional scale because model parameters are more robust and can be derived easily from basic soil information (Addiscott & Wagenet, 1985). The required capacity parameters for water-holding capacity (field capacity or upper drain limit) and lower limit of plant

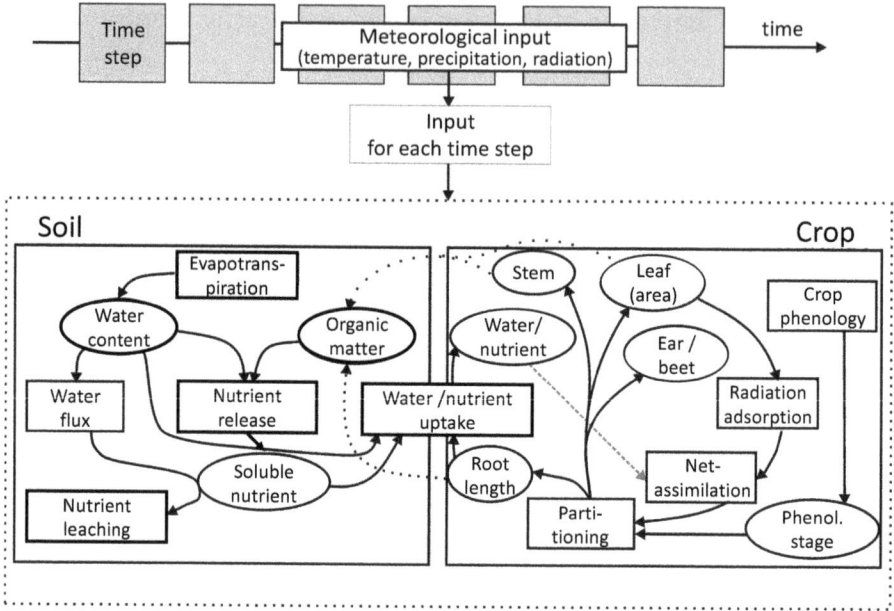

Fig. 3 Simplified scheme of interactive processes in the soil–crop system represented in process-based agro-ecosystem models

extractable water (wilting point) can be derived from soil texture information or simply by measuring soil water content in soil a few days after rainfall or after longer dry periods.

For crop models, two main approaches are applied. The radiation use efficiency (RUE) approach transforms incoming and adsorbed radiation into biomass production, while the photosynthesis minus respiration approach determines gross carbon (C) assimilation from photosynthesis and calculates separately the amount of C loss by crop respiration. Crop models are either designed for a specific crop or follow a generic approach. The latter uses the same model structure and defines different crops by using crop-specific parameter files. Crop parameters can be divided into crop-specific and genotype-specific parameters. Parameters have to be calibrated using detailed data on crop phenology, biomass partitioning to crop organs, leaf area and other crop state variables. Once calibrated, the model can be applied to the same crop or cultivar to evaluate the validity of the parameters on independent data sets. Data requirements for calibration are higher than for evaluation. A guide of which data are required for both is provided by Kersebaum et al. (2015).

Regarding nitrogen dynamics, some models simulate only nitrogen processes, while others couple nitrogen to carbon mineralization. Models also differ in the number of pools with different decay behaviour and in the complexity to simulate the various processes of N-dynamics, e.g. denitrification and NH_4 volatilization.

Table 1 gives an overview of selected integrated soil–crop models and their main characteristics for the processes implemented, which were developed for different

Table 1 Characteristics of selected crop models and their requirements for meteorological data input

Model reference	Light to biomass[a]	Yield formation[b]	Crop phenology[c]	Crop stress[d]	Soil water[e]	ET[f]	Soil organic matter[g]	Input weather variables[h]
APSIM Keating et al. (2003)	RUE	Prt, B, (Gn)	T, DL, V	W, N, A, H	C	PT	CN, P(3), B	R, Tx, Tn, Rd
CROPSYST Stöckle et al. (2003)	RUE	HI, B	T, DL, V	W, N, H	C/R	PM	N, P(4)	R, Tx, Tn, Rd, RH, W
COUPMODEL Jansson and Moon (2001)	RUE	Prt, B	T	W, N	R	PM	CN, P(2)	R, Ta, Rd, RH, W
DAISY Abrahamsen and Hansen (2000)	P-R	Prt	T, DL, V	W, N, H	R	PM	CN, P(6), B	R, Tx, Tn, Rd, e, RH, W
DSSAT/CERES Jones et al. (2003)	RUE	HI, B, Gn	T, DL, V	W, N	C	PT	CN, P(4), B	R, Tx, Tn, Rd
DSSAT/IXIM Lizaso et al. (2011)	P-R	Gn	T, DL	W, N, H	C	PT	CN, P(2), B	R, Tx, Tn, Rd
DSSAT/Cropsim Hunt and Pararajasingham (1995)	RUE	Prt	T, DL, V	W, N	C	PT	CN, P(4), B	R, Tx, Tn, Rd
EPIC Williams et al. (1989)	RUE	HI	T, V	W N, H	C	PM, PT, O	N, P(5), B	R, Tx, Tn, Rd, RH, W
EXPERT-N Biernath et al. (2011)	RUE/P- R	Prt, B, Gn	T, DL, V	W, N	R	PM	CN, P(3), B	R, Tx, Tn, Rd, RH, W
HERMES Kersebaum (2011)	P-R	Prt	T, DL, V, O	W, N, O, (H)	C, CR	PM, PT, O	N, (P2)	R, Tx, Tn, e, Rd, RH, W
MAIZESIM Kim et al. (2012)	P-R	HI, Prt	T, DL	W, N, H	R	P, O	N, P(1), B	R, Tx, Tn, Rd, RH, W
MONICA Nendel et al. (2011)	P-R	Prt	T, DL, V	W, N, O, H	C, CR	PM	CN, P(6), B	R, Tx, Tn, Rd, RH, W

(continued)

Table 1 (continued)

Model reference	Light to biomass[a]	Yield formation[b]	Crop phenology[c]	Crop stress[d]	Soil water[e]	ET[f]	Soil organic matter[g]	Input weather variables[h]
RZWQM2 Sadhukhan et al. (2019)	RUE	HI, B, Gn	T, DL, V	W, N, H	R	SW	CN, P(1), B	R, Tx, Tn, Rd, RH, W
SALUS Basso et al. (2010)	RUE	Prt, HI	T, DL, V	W, N, H	C	PT	CN, P(3), B	R, Tx, Tn, Rd
SIMPLACE-Lintul Gaiser et al. (2013)	RUE	Prt	T, DL	W, N, H	C	PM, HAR	CN,P(7), B	R, Tx, Tn, Rd, RH, W
Sirius Jamieson et al. (1998)	RUE	Prt, B	T, DL, V	W, N	C	P, PT	N, P(2)	R, Tx, Tn, Rd, e, W
SPACSYS (3D) Wu et al. (2007)	P-R	Prt	T, DL, V	W, N	R	PM	CN,P(4), B	R, Tx, Tn, Rd, RH, W
STICS Brisson et al. (2003)	RUE	B, Gn	T, DL, V, O	W, N, H	C	P, PT, SW	N, P(3)	R, Tx, Tn, Rd, e, W

[a]*RUE* Radiation use efficiency, *P-R* Gross respiration minus respiration
[b]*HI* fixed harvest index, *B* total above ground biomass, *Gn* number of grains, *Prt* partitioning during reproductive phase
[c]*T* temperature, *DL* photoperiod (daylength), *V* vernalization, *O* water/nutrient stress effects
[d]*W* water limited, *N* nitrogen limited, *H* heat stress, *O* oxygen limited (aeration deficit)
[e]*C* capacity approach, *R* Richards approach, *CR* capacity approach with capillary rise
[f]*PM* Penman-Monteith, *PT* Priestley-Taylor, *P* Penman, *SW* Shuttleworth-Wallace, *HAR* Hargreaves, *O* other ET formula options
[g]*CN* coupled carbon and nitrogen turnover, *N* nitrogen net mineralisation, *P(x)* number of soil organic matter (SOM) pools, *B* microbial biomass pool
[h]*R* precipitation, *Tx* daily maximum temperature, *Tn* daily minimum temperature, *Rd* global radiation, *RH* relative humidity, *W* wind speed, *e* vapor pressure deficit

purposes. Additionally, the required daily meteorological input data are listed. Some models can be configured by the user, e.g. different formulae for potential evapotranspiration or the user can select if a mechanistic or a simpler model approach for water should be used.

At the field and regional scales, the majority of applied models are vertical one-dimensional models which do not consider lateral fluxes. In principle, they are point models and can be applied to spatially variable conditions simulating multiple soil columns in parallel. However, a few models exist that simulate, at least partly, processes in two or three spatial dimensions (e.g. Timlin et al., 2001; Wu et al., 2007; Ward et al., 2018; Roy et al., 2019), which are mostly limited to the single crop or crop row scale.

Depending on their complexity and purpose, models have different input requirements. In general, all models require data inputs for meteorological variables, soil properties and management data.

3 Modelling Spatial Variation

In the simple case of a flat field, spatial processes are dominantly in the vertical direction. To consider spatial variation within a field, spatial information on soil properties is required as an input for agro-ecosystem models. Spatial soil information is usually provided by soil maps of different scales. However, conventional soil maps are primarily not dedicated to provide data for modelling. Conventional soil classification and mapping are often based on principles of soil genesis and include some generalization to group combinations of stable soil properties to a specific soil type (Kempen et al., 2012). Delineated soil map units are considered representative for that specific soil type; however, it is well known that there is considerable spatial variation of properties within the units that is mostly disregarded. Furthermore, the input required by models is more related to functional properties of soils (Stoorvogel et al., 2015). Therefore, delineated soil map units and functional diversity, e.g. derived from multiyear yield maps, often do not coincide. This limits the use of conventional soil maps in PA (Stoorvogel et al., 2015). Precise knowledge of soil properties and their spatial distribution at the field scale is still limited (Adamchuk et al., 2004; Schmidhalter et al., 2008; Geesing et al., 2014). Conventional soil sampling by augering or digging profiles at a relatively small spatial resolution for spatial investigation is costly and time-consuming. However, applied at a low resolution capturing the contrasting areas of a field, they are required as basis of modern techniques of digital soil mapping (DSM). This includes identification of spatial relations using variogram analysis and geostatistical interpolation such as kriging and linking auxiliary data to observed soil properties by co-kriging. Proximal sensing enables information about soil properties to be obtained at much higher spatial resolution to generate quasi-continuous property maps. Stoorvogel et al. (2015) provide an excellent overview of the various opportunities to obtain soil spatial data.

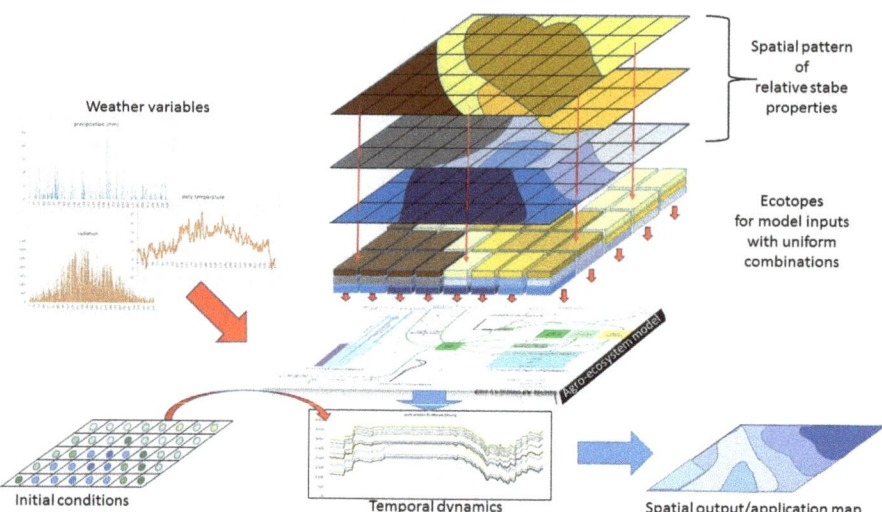

Fig. 4 Scheme of intersecting multiple spatial patterns of input variables, weather data, and initial conditions to translate spatial information into a model based spatio-temporal dynamic for specific spatial outputs and decision support

Modelling requires the overlay of spatial patterns of the various soil properties relevant in determining yield, such as texture, organic C and N and related model parameters such as water-holding capacity, plant available water or C and N pools to form units of their combination for modelling, which show consistent functional responses (Basso et al., 2009). Figure 4 shows schematically how models can be used to transfer spatial information of relatively stable soil information to a spatio-temporal dynamic state variable, e.g. soil water content.

As described previously, dynamic process-based models require an initial status to start simulation and temporally variable inputs of controlling variables such as weather and management. Since some of the state variables show strong spatio-temporal dynamics, initialization should be ideally during a phase when spatial patterns are governed by other more stable soil properties. Soil water content, for example, can be initialized during winter in humid environments when soils are usually at field capacity or in summer in arid or semi-arid regions, when soils are depleted of moisture and are close to wilting point (Kersebaum et al., 2019). Spatial measurements of soil water during these periods can also be used to derive the spatial patterns of the above-mentioned capacity parameters.

Although modern sensor techniques can provide spatial patterns of soil properties at a high resolution, not all differences in soil properties create a significant difference in model outputs depending on the model sensitivity. Therefore, some clustering or classification for each individual state variable used as model input might be advisable based on a sensitivity analysis of the model applied.

4   Site Sensitivity of Agro-ecosystem Models

A large number of agro-ecosystem models have been developed and used for risk assessment studies in agricultural production, e.g. under climate change. These models differ in complexity and functionality to simulate and predict the major physiological processes that determine crop growth and yield and their responses to environmental and management factors. Models come with an inherent prediction error when used to predict the response of crop growth and soil water flux dynamics to climatic conditions such as changing seasonal rainfall patterns, rising temperatures and elevated atmospheric CO_2 concentrations.

Depending on the climate conditions, crop models are mostly responsive to plant available water (Aggarval, 1995; Pachepsky & Acock, 1998; Wassenaar et al., 1999), especially under conditions of temporal dry spells. However, models do not respond equally to variable soil conditions as demonstrated on a regional scale by Hoffmann et al., 2016 or by Wallor et al. (2018) for spatially variable field data sets. Differences in model responses are related to (i) processes considered by the models; (ii) functions to transfer basic soil information into model parameters, e.g. pedotransfer functions; (iii) response functions, which determine how crop growth is affected by limiting factors; and (iv) different assumptions or interpretations on boundary conditions.

Hydrological boundary conditions, e.g. the presence of a shallow groundwater table, can have a considerable influence on water availability (Bormann, 2009, 2012). Depending on soil texture, additional water can be supplied to the root zone when a shallow groundwater level is present. However, not all models are able to consider capillary rise (Kimball et al., 2019). While models using a mechanistic approach for water dynamics based on the Richards equation are principally able to simulate capillary rise, only a few models using a tipping bucket approach can consider this by simplified algorithms and empirical data (Kersebaum & Richter, 1991). However, information of groundwater level is often not very precise and models using the Richards equation might be too sensitive in relation to the uncertainty of the available data. Even more complicated to handle are lateral fluxes due to surface runoff, interflow or groundwater flow, which would require 3D approaches and an enormous effort to obtain the required input data (e.g. Egea et al., 2015). Simplified approaches such as the use of topographic wetness index can be applied to modify parameters of one-dimensional models (Kersebaum et al., 2002).

Response to dry and wet soil conditions vary among models. While not all models include algorithms for oxygen deficiency underwater logging, the response to drought is mostly considered by models. However, the response functions for water stress in crop models may differ depending on the climatic conditions for which the model was originally developed since critical thresholds for limitations on water uptake depend on climatic conditions. Wallor et al. (2018) showed that models developed primarily under semi-arid to arid conditions might overreact to moderately dry conditions in a temperate climate zone. Water balance models commonly include plant water stress functions (e.g. Feddes et al., 1978), where critical limits of soil water supply can be adapted according to the vegetation characteristics. Although it is commonly acknowledged that root water uptake is defined by water potential gradients and hydraulic resistances in the soil–plant system (Steudle & Peterson, 1998), this principle is rarely included in models (Cai et al., 2017). In many crop models, the root water uptake is described as a sink term in the soil compartments without solving water flow mechanistically towards roots. Vertical distribution of water uptake is typically assumed to be proportional to root length densities and local water availability. The parameters and concepts used in these models are to a large extent empirical. Feddes et al. (2001) showed that the onset of water stress in crops varies depending on the evaporative demand of the atmosphere resulting in an onset at larger water contents under high evaporative demand compared to moderate vapour pressure deficit. Empirical relations implemented in models may include inherently the conditions under which the models were primarily developed. Therefore, applying models to spatially variable data under climate conditions different from their origin revealed that consistency between soil water simulations and crop yield estimation are sometimes weak.

Figure 5 shows the relation between deviations of three models on soil water contents at 60 locations in a field in Germany and the related differences in yield prediction (Wallor et al., 2018). While model AS (APSIM) developed in Australia showed a very steep yield relation with a pronounced response of crop yields to moderate water deviations, model CO (COUP) developed in Scandinavia showed relative small yield deviations although soil water was generally overestimated. Model HE (Hermes) showed consistent water and yield simulations with only moderate deviation at most locations since it was developed under similar conditions.

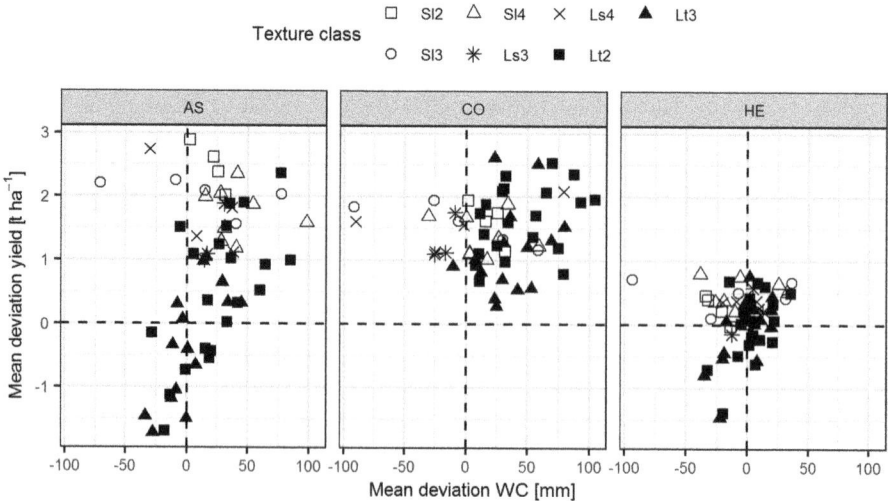

Fig. 5 Deviations between simulated and observed soil water contents in the root zone of a heterogeneous field and their relation to model deviations from observed crop yields for three different models (*AS* APSIM, *CO* COUPMODEL, *HE* HERMES). (Data from Wallor et al. 2018)

Wallor et al. (2018) showed that subsoil information, especially under water-limited conditions, is indispensable to get a realistic simulation of spatial crop growth patterns, which has been emphasized earlier by several authors. McDonald et al. (2013) demonstrated that subsoil properties restricting rooting depth can limit water use and yield in many Australian soils. The strong impact of subsoil water availability at and after heading for wheat yield formation was emphasized by Baier and Robertson (1968). Kirkegaard et al. (2007) quantified the beneficial effect of available subsoil water even under moderate water stress at an efficiency of 59 kg ha^{-1} grain yield per mm subsoil water uptake and concluded that relatively small amounts of subsoil water can be highly valuable to grain yield. However, obtaining subsoil information requires considerable sampling effort, and it is also less easily detectable by proximal sensors.

Another relevant source of differences among models is the different consideration of root distribution and related water and nutrient uptake (Wu & Kersebaum, 2008). While few models use two- or three-dimensional root distribution (Wu et al., 2007; Postma et al., 2008; Bingham & Wu, 2011), the majority use one-dimensional vertical approaches. Vertical root penetration and distribution might be considered using empirical functions of different shape (Wu & Kersebaum, 2008) related to the phenological development of the crops and may consider soil temperature and or bulk density to modify the vertical exploration of the root zone. While soil temperature can be modelled using daily weather input and soil properties, bulk density usually needs to be determined in the field. Additionally, it may vary with time.

5   Examples of Model Use for Precision Agriculture

5.1   Identification of Site-Specific Management Using Long-Term Simulations

Knowledge of the spatial patterns within a field is critical not only to farmers for potential variable–rate applications but also for water managers to control water body pollution. Selection of homogenous zones within the field to run crop models with site-specific input can be useful to understand and predict better the impact of weather, soil and landscape characteristics on spatial and temporal patterns of crop yields to enhance resource use efficiency at field level. Stable zones may be managed using a zone-specific management strategy, selected before the season, often called strategic (Basso et al., 2011). Crop simulation models can be applied in simulating the crop response to irrigation and crop management (Delgoda et al., 2016). They provide the opportunity to evaluate the benefit of several precision management strategies, e.g., irrigation, as they reduce the need for time-consuming field experiments (Jones et al., 2003).

Site-specific response curves, e.g. for fertilizer input, can be derived from virtual fertilization experiments through running crop models with site-specific input and calibrated over long-time weather and yield records (Basso et al., 2011).

Additionally, the long-term simulations provide information on yield potentials to derive realistic yield goals, cost-benefit calculations and related environmental impacts, which can be used for multi-criteria optimization of sustainable management practice. DeJonge et al. (2007) investigated the effect of variable–rate irrigation management on maize production in Iowa using the CERES-maize model. They compared crop yields for a period of 28 years under simulated scenarios of no irrigation, scheduled uniform irrigation and precision irrigation. Thorp et al. (2008) described a methodology for applying the Decision Support System for Agrotechnology Transfer (DSSAT) crop growth model in analysing variable–rate management practices including irrigation on crop growth and yield.

5.2   Inverse Modelling to Derive Unknown Properties

Estimating soil properties at a high resolution might be too expensive if traditional soil sampling and laboratory methods are applied. However, use of crop yield maps from several years can be used to estimate yield-relevant model parameters by inverse modelling. The diversity of individual responses of zones to the variation in inter-annual weather creates a footprint based on different limiting factors in each year. This can be used to optimize soil properties by matching simulated yield with yield map data. Timlin et al. (2001) and Morgan et al. (2003) used a simple water balance model that uses the relative transpiration ratio to calculate relative yield from available water in the soil profile. Soil water-holding capacity was optimized to match the spatial pattern of relative yields to the observed yield map. Florin et al. (2011) derived an inverse meta-model from the APSIM model using neural networks to estimate soil available water capacity (AWC) from available yield data. Alternatively, remote sensing data of NDVI may be used (Araya et al., 2016). He et al. (2021) derived plant available water (PAWC) using the APSIM model to simulate wheat yield on synthetic soils with contrasting PAWC and climates. The simulated results were used to develop an empirical model to relate simulated yield to PAWC (He & Wang, 2019). The empirical model was used inversely to predict PAWC from observed crop yield. However, the number of seasons required to achieve a certain identifiability of relevant parameters and to reduce equifinality may depend on the strength of limiting factors under a given environment.

5.3   Operational Use for Site-Specific Management Operations

There are several options to use process-based simulation models for operational decisions. Irrigation is a typical application for models in PA since crops are quite sensitive to water supply and deficits require quick action to avoid water stress. The advantage of using a well-calibrated model is that it may use weather forecast data to predict the risk over a range of a few days of probable water deficiency and to assess the amount required in advance. However, extensive calibration and validation are required to establish model accuracy (Adeyemi et al., 2017).

Simulation models can be linked to real-time soil or plant sensors and weather data to determine daily irrigation requirements of crops. For example, the simulation framework VARIwise is capable of real-time decision support in precision irrigation (McCarthy et al., 2010). It can incorporate real-time data input from field sensors in arriving at irrigation decisions. The combination of different sensor inputs into the simulation framework enables adaptive decision support with the system being able to readjust irrigation decisions based on plant feedback and also explore optimal control strategies.

The application of simulation models for variable-rate fertilization of nitrogen is much more challenging and complex since many processes and interactions have to be considered. Moreover, inter-annual variation in N demand and supply hamper the use of average response functions. Additionally, fertilizer recommendation at the time of N application has to cover the future demand of crops within a field for the next few weeks or even months. Since weather forecasts presently provide reliable predictions for 3–5 days, model predictions have an inherent uncertainty. Basso et al., 2011 derived tactical recommendations based on N response curves derived from long-term simulations of management zones with different yield expectations by combining them with current seasonal observations, e.g. the available soil water in the root zone at the time of N dressing. Assuming recent production costs and grain prices, the simulations helped identify an optimal N rate for each of the zones based on agronomic, economic and environmental sustainability of N management.

Real-time applications of process-oriented simulation models for tactical N fertilization in PA are rare so far. Kersebaum and Beblik (2001) tested real-time application of the HERMES model for entire field recommendations for winter cereals in Germany. They used subsequent applications of the model during the growing season by combining actual current weather with typical site-specific weather from long-term weather data for prediction. Nitrogen fertilizer application was split into four doses per season. The general scheme of the procedure is shown in Fig. 6.

The scheme was tested for several years across Germany (Kersebaum & Beblik, 2001) and also in China (Michalczyk et al., 2014), and in comparison with measurement-based N fertilization recommendations, they showed better nitrogen use efficiency than the other methods. Real-time model application for PA considering within-field variability showed on average a reduction in variable N rates of 40 kg N ha^{-1} compared to measured soil mineral N combined with in-season recommendations of a proximal crop N sensor (Kersebaum et al., 2003, 2005a). However, an inherent uncertainty of the prediction remains because of inter-annual weather variation. The site-specific uncertainty for N recommendations on winter wheat was estimated using long-term simulations for a heterogeneous field (Kersebaum et al., 2007). A maximum difference of 48 kg N ha^{-1} between contrasting (wet and dry) weather scenarios was estimated for the prediction. Among different sites, the uncertainty increased with reduced buffer capacity of

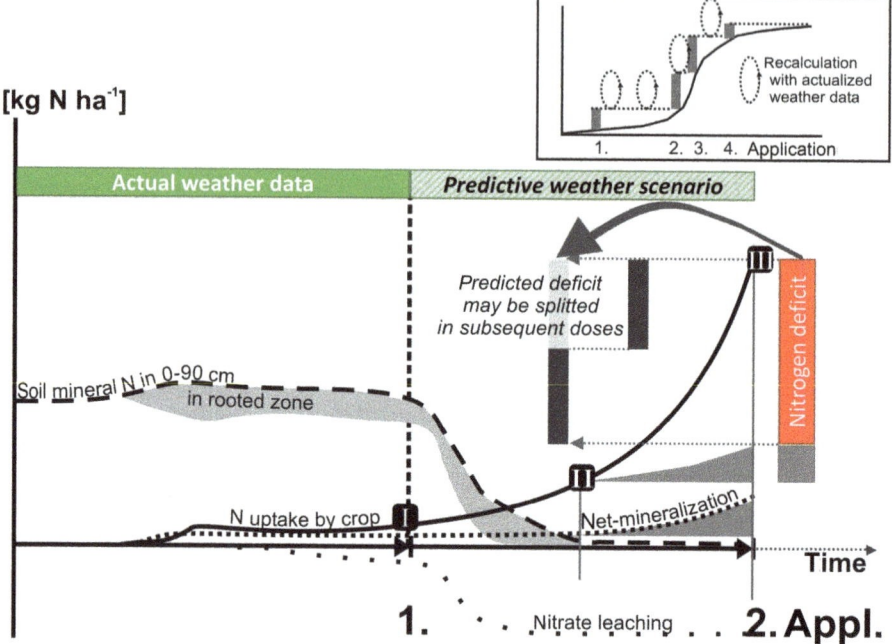

Fig. 6 Scheme of model-based fertilizer recommendations with HERMES. *I* day of recommendation, *II* calculated date, when N deficit occurs, *III* predicted date of predefined development stage for next fertilization. (Kersebaum et al., 2007, modified)

the soil, expressed as water-holding capacity in the root zone. Of the four applications from the whole growing season, the first showed the smallest absolute variation, while the last application (at flowering) had the greatest variation. The smaller difference between the scenarios at the first fertilizer application showed that these recommendations were more dependent on the previous weather than on the prediction. In contrast, water availability during the grain filling phase had a significant influence on the model recommendation resulting in differences of 20 to 30 kg N ha⁻¹ between the scenarios for the third and fourth applications. While N recommendations are mostly oriented to consider the N demand of the crop, modelling also offers the opportunity to estimate limitations of fertilizer application to consider water protection goals (e.g. EU nitrate directive; EEC, 1991 or EU Water Framework Directive; EC, 2000) in nitrate vulnerable zones (NVZs) to meet drinking water standards of groundwater resources (Kersebaum et al., 2005b; Basso et al., 2016). Nitrate vulnerable zones are areas of land that drain into water at risk of pollution, e.g. tile-drained areas, which represent shortcuts for surface water pollution or land over groundwater aquifers, which are not protected by almost impermeable layers (Kersebaum et al., 2006).

While process-oriented modelling has some advantages to estimate variable–rate fertilizer applications, a combination of available sensor techniques together with remote sensing and weather forecasts provide opportunities to reduce uncertainties of each tool and opens new chances for better site-specific management (Morari et al., 2021).

6   Conclusions

Process-oriented simulation models provide various opportunities to analyse and explain spatial patterns of crop growth. However, they require detailed site information such as soil properties, hydrological boundary conditions and weather data at sufficient spatial and temporal resolution. While traditional soil maps often do not match model requirements, new techniques like proximal soil and crop sensing can be used in combination with modelling to derive spatially variable management recommendations for precision agriculture. This requires their implementation in a decision support framework that combines the different tools and implements different targets to achieve greater sustainability through variable-rate applications in Precision Agriculture.

Conflict of Interest   The authors declare that there are no conflicts of interest.

References

Abrahamsen, P., & Hansen, S. (2000). Daisy: An open soil–crop–atmosphere system model. *Environmental Modelling and Software, 15*, 313–330.

Adamchuk, V. I., Hummel, J. W., Morgan, M. T., & Upadhyaya, S. K. (2004). On-the-go soil sensors for precision agriculture. *Computers and Electronics in Agriculture, 44*(1), 71–91.

Addiscott, T. M., & Wagenet, R. J. (1985). Concepts of solute leaching in soils: A review of modelling approaches. *Journal of Soil Science, 36*, 411–424.

Adeyemi, O., Grove, I., Peets, S., & Norton, T. (2017). Advanced monitoring and management systems for improving sustainability in precision irrigation. *Sustainability, 8*(3), 353.

Aggarval, P. K. (1995). Uncertainties in crop, soil and weather inputs used in growth models: Implications for simulated outputs and their applications. *Agricultural Systems, 48*, 361–384.

Araya, S., Lyle, G., Lewis, M., & Ostendorf, B. (2016). Phenologic metrics derived from MODIS NDVI as indicators for plant available water-holding capacity. *Ecological Indicators, 60*, 1263–1272.

Baier, W., & Robertson, G. W. (1968). The performance of soil moisture estimates as compared with the direct use of climatological data for estimating crop yields. *Agricultural Meteorology, 5*(1), 17–31.

Basso, B., Cammarano, D., Chen, D., Cafiero, G., Amato, M., Bitella, G., Rossi, R., & Basso, F. (2009). Landscape position and precipitation effects on spatial variability of wheat yield and grain protein in southern Italy. *Journal of Agronomy and Crop Science, 195*, 301–312.

Basso, B., Cammarano, D., Troccoli, A., Chen, D., & Ritchie, J. T. (2010). Long-term wheat response to nitrogen in a rainfed Mediterranean environment: Field data and simulation analysis. *European Journal of Agronomy, 33*, 132–138.

Basso, B., Ritchie, J. T., Cammarano, D., & Sartori, L. (2011). A strategic and tactical management approach to select optimal N fertilizer rates for wheat in a spatially variable field. *European Journal of Agronomy, 35*, 215–222.

Basso, B., Dumont, B., Cammarano, D., Pezzuolo, A., Marinello, F., & Sartori, L. (2016). Environmental and economic benefits of variable rate nitrogen fertilization in a nitrate vulnerable zone. *Science of the Total Environment, 545–546*, 227–235.

Biernath, C., Gayler, S., Bittner, S., Klein, C., Högy, P., Fangmeier, A., & Priesack, E. (2011). Evaluating the ability of four crop models to predict different environmental impacts on spring wheat grown in open top chambers. *European Journal of Agronomy, 35*, 71–82.

Bingham, I. J., & Wu, L. (2011). Simulation of wheat growth using the 3D root architecture model SPACSYS: Validation and sensitivity analysis. *European Journal of Agronomy, 34*, 181–189.

Bormann, H. (2009). Analysis of possible climate impacts of climate change on the hydrological regimes of different regions in Germany. *Advances in Geosciences, 21*, 3–11.

Bormann, H. (2012). Assessing the soil texture-specific sensitivity of simulated soil moisture to projected climate change by SVAT modelling. *Geoderma, 195*, 73–83.

Brisson, N., Gary, C., Justes, E., Roche, R., Mary, B., Ripoche, D., Zimmer, D., Sierra, J., Bertuzzi, P., Burger, P., Bussière, F., Cabidoche, Y. M., Cellier, P., Debaeke, P., Gaudillère, J. P., Hénault, C., Maraux, F., Seguin, B., & Sinoquet, H. (2003). An overview of the crop model STICS. *European Journal of Agronomy, 18*, 309–332.

Cai, G., Vanderborght, J., Couvreur, V., Mboh, C. M., & Vereecken, H. (2017). Parameterization of root water uptake models considering dynamic root distributions and water uptake compensation. *Vadose Zone Journal, 17*, 160125.

de Wit, C. (1965). *Photosynthesis of leaf canopies*. Wageningen Inst Biol Chem Res Field Crops Herb.

DeJonge, K. C., Kaleita, A. L., & Thorp, K. R. (2007). Simulating the effects of spatially variable irrigation on corn yields, costs, and revenue in Iowa. *Agricultural Water Management, 92*, 99–109.

Delgoda, D., Malano, H., Saleem, S. K., & Halgamuge, M. N. (2016). Irrigation control based on model predictive control (MPC): Formulation of theory and validation using weather forecast data and AQUACROP model. *Environmental Modelling and Software, 78*, 40–53.

Delin, S., & Berglund, K. (2005). Management zones classified with respect to drought and waterlogging. *Precision Agriculture, 6*, 321–340.

EC (European Community). (2000). Directive 2000/60/EC of the European Parliament and of the Council establishing a framework for the Community action in the field of water policy. *Official Journal, L327*, 1–73.

EEC (European Economic Community). (1991). Council Regulation (EEC) No. 2092/91 on organic production of agricultural products and indications referring thereto on agricultural products and foodstuffs. *Official Journal, L198*, 1–15.

Egea, G., Díaz-Espejo, A., & Fernández, J. E. (2015). Numerical simulation of soil water dynamics as a decision support system for irrigation management in drip-irrigated hedgerow olive orchards. In J. Stafford (Ed.), *Precision agriculture '15* (pp. 503–510). Wageningen Academic Publishers.

Ewert, F., Rötter, R. P., Bindi, M., Webber, H., Trnka, M., Kersebaum, K. C., Olesen, J. E., van Ittersum, M. K., Janssen, S., Rivington, M., Semenov, M. A., Wallach, D., Porter, J. R., Stewart, D., Verhagen, J., Gaiser, T., Palosuo, T., Tao, F., Nendel, C., et al. (2015). Crop modelling for integrated assessment of risk to food production from climate change. *Environmental Modelling and Software, 72*, 287–303.

Feddes, R. A., Kowalik, P. J., & Zaradny, H. (1978). *Simulation of field water use and crop yield* (Simulation monograph). Pudoc.

Feddes, R. A., Hoff, H., Bruen, M., Dawson, T., de Rosnay, P., Dirmeyer, P., Jackson, R. B., Kabat, P., Kleidon, A., Lilly, A., & Pitman, A. J. (2001). Modeling root water uptake in hydrological and climate models. *Bulletin of the American Meteorological, 82*, 2797–2809.

Florin, M. J., McBratney, A. B., Whelan, B. M., & Minasny, B. (2011). Inverse meta-modelling to estimate soil available water capacity at high spatial resolution across a farm. *Precision Agriculture, 12*, 421–438.

Gaiser, T., Perkons, U., Küpper, P. M., Kautz, T., Uteau-Puschmann, D., Ewert, F., Enders, A., & Krauss, G. (2013). Modeling biopore effects on root growth and biomass production on soils with pronounced subsoil clay accumulation. *Ecological Modelling, 256*, 6–15.

Geesing, D., Diacono, M., & Schmidhalter, U. (2014). Site-specific effects of variable water supply and nitrogen fertilization on winter wheat. *Journal of Plant Nutrition and Soil Science, 177*, 509–523.

Greminger, P. J., Sud, K., & Nielsen, D. R. (1985). Spatial variability of field-measured soil-water characteristics. *Soil Science Society of America Journal, 49*, 1075–1082.

He, D., & Wang, E. (2019). On the relation between soil water holding capacity and dryland crop productivity. *Geoderma, 353*, 11–24.

He, D., Oliver, Y., & Wang, E. (2021). Predicting plant available water holding capacity of soils from crop yield. *Plant and Soil, 459*, 315–328.

Heermann, D. F., Diker, K., Buchleiter, G. W., & Brodahl, M. K. (2003). The value of additional data to locate potential management zones in commercial corn fields under center pivot irrigation. In J. Stafford & A. Werner (Eds.), *Precision agriculture '03. Proceedings 4th European conference precision agriculture* (pp. 279–284). Wageningen Academic Publishers.

Hoffmann, H., Zhao, G., Asseng, S., Bindi, M., Biernath, C., Constantin, J., Coucheney, E., Dechow, R., Doro, L., Eckersten, H., Gaiser, T., Grosz, B., Heinlein, F., Kassie, B. T., Kersebaum, K. C., Klein, C., Kuhnert, M., Lewan, E., Moriondo, M., et al. (2016). Impact of spatial soil and climate input data aggregation on regional yield simulations. *PLoS One, 11*(4), e0151782.

Hunt, L. A., & Pararajasingham, S. (1995). CROPSIM-wheat – A model describing the growth and development of wheat. *Canadian Journal of Plant Science, 75*, 619–632.

Jamieson, P., Semenov, M., Brooking, I., & Francis, G. (1998). Sirius: A mechanistic model of wheat response to environmental variation. *European Journal of Agronomy, 8*, 161–179.

Jansson, P.-E., & Moon, D. S. (2001). A coupled model of water, heat and mass transfer using object orientation to improve flexibility and functionality. *Environmental Modelling and Software, 16*, 37–46.

Jones, J. W., Hoogenboom, G., Porter, C. H., Boote, K. J., Batchelor, W. D., Hunt, L. A., Wilkens, P. W., Singh, U., Gijsman, A. J., & Ritchie, J. T. (2003). The DSSAT cropping system model. *European Journal of Agronomy, 18*, 235–265.

Kaspar, T. C., Colvin, T. S., Jaynes, D. B., Karlen, D. L., James, D. E., & Meek, D. W. (2003). Relationship between six years of corn yields and terrain attributes. *Precision Agriculture, 4*, 87–101.

Keating, B. A., Carberry, P. S., Hammer, G. L., Probert, M. E., Robertson, M. J., Holzworth, D., Huth, N. I., Hargreaves, J. N. G., Meinke, H., Hochman, Z., McLean, G., Verburg, K., Snow, V., Dimes, J. P., Silburn, M., Wang, E., Brown, S., Bristow, K. L., Asseng, S., et al. (2003). An overview of APSIM, a model designed for farming systems simulation. *European Journal of Agronomy, 18*, 267–288.

Kempen, B., Brus, D. J., Stoorvogel, J. J., Heuvelink, G. B. M., & de Vries, F. (2012). Efficiency comparison of conventional and digital soil mapping for updating soil maps. *Soil Science Society of America Journal, 76*, 2097–2115.

Kersebaum, C. (1995). Application of a simple management model to simulate water and nitrogen dynamics. *Ecological Modelling, 81*, 145–156.

Kersebaum, K. C. (2011). Special features of the HERMES model and additional procedures for parameterization, calibration, validation, and applications In L. R. Ahuja & L. Ma (Eds.), *Methods of introducing system models into agricultural research. Advances in agricultural systems modeling series 2* (pp. 65–94). ASA CSSA SSSA.

Kersebaum, K. C., & Beblik, A. J. (2001). Performance of a nitrogen dynamics model applied to evaluate agricultural management practices. In M. Shaffer, L. Ma, & S. Hansen (Eds.), *Modeling carbon and nitrogen dynamics for soil management* (pp. 551–571). CRC Press.

Kersebaum, K. C., & Nendel, C. (2014). Site-specific impacts of climate change on wheat production across regions of Germany using different CO_2 response functions. *European Journal of Agronomy, 52*, 22–32.

Kersebaum, K. C., & Richter, J. (1991). Modelling nitrogen dynamics in a soil-plant system with a simple model for advisory purposes. *Fertilizer Research, 27*, 273–281.

Kersebaum, K. C., Reuter, H. I., Lorenz, K., & Wendroth, O. (2002). Modelling crop growth and nitrogen dynamics for advisory purposes regarding spatial variability. In L. J. Ahuja, L. Ma, & T. A. Howell (Eds.), *Agricultural system models in field research and technology transfer* (pp. 229–252). Lewis Publishers.

Kersebaum, K. C., Lorenz, K., Reuter, H. I., Wendroth, O., Giebel, A., & Schwarz, J. (2003). Site specific nitrogen fertilisation recommendations based on simulation. In J. Stafford & A. Werner (Eds.), *Precision agriculture. Proceedings 4th European conference on precision agriculture, Berlin* (pp. 309–314). Wageningen Academic Publishers.

Kersebaum, K. C., Lorenz, K., Reuter, H. I., Schwarz, J., Wegehenkel, M., & Wendroth, O. (2005a). Operational use of agrometeorological data and GIS to derive site specific nitrogen fertilizer recommendations based on the simulation of soil and crop growth processes. *Physics and Chemistry of the Earth, 30*, 59–67.

Kersebaum, K. C., Reuter, H. I., Lorenz, K., & Wendroth, O. (2005b). Long term simulation of soil/ crop interactions to estimate management zones and consequences for site specific nitrogen management considering water protection. In J. Stafford (Ed.), *Precision agriculture '05. Proceedings 5th European conference on precision agriculture, Uppsala, Schweden* (pp. 795–802). Wageningen Academic Publishers.

Kersebaum, K. C., Matzdorf, B., Kiesel, J., Piorr, A., & Steidl, J. (2006). Model-based evaluation of agro-environmental measures in the federal state of Brandenburg (Germany) concerning N pollution of groundwater and surface water. *Journal of Plant Nutrition and Soil Science, 169*, 352–359.

Kersebaum, K. C., Reuter, H. I., Lorenz, K., & Wendroth, O. (2007). Model-based nitrogen fertilization considering agro-meteorological data. In T. W. Bruulsema (Ed.), *Managing crop nitrogen for weather. Proceedings of the Soil Science Society of America; Symposium on integrating weather variability into nitrogen recommendations, 15 November 2006, Indianapolis, IN* (pp. 1–9). IPNI.

Kersebaum, K. C., Boote, K. J., Jorgenson, J. S., Nendel, C., Bindi, M., Frühauf, C., Gaiser, T., Hoogenboom, G., Kollas, C., Olesen, J. E., Rötter, R. P., Ruget, F., Thorburn, P. J., Trnka, M., & Wegehenkel, M. (2015). Analysis and classification of data sets for calibration and validation of agro-ecosystem models. *Environmental Modelling and Software, 72*, 402–417.

Kersebaum, K. C., Wallor, E., Lorenz, K., Beaudoin, N., Constantin, J., & Wendroth, O. (2019). Modelling cropping systems with HERMES – Model capability, deficits, and data requirements. In O. Wendroth, R. J. Lascano, & L. Ma (Eds.), *Bridging among disciplines by synthesizing soil and plant processes* (Advances in agricultural systems modeling, Vol. 8, pp. 103–126). ASA, CSSA, and SSSA.

Kim, S.-H., Yang, Y., Timlin, D. J., Fleisher, D. H., Dathe, A., Reddy, V. R., & Staver, K. (2012). Modeling temperature responses of leaf growth, development, and biomass in maize with MAIZSIM. *Agronomy Journal, 104*, 1523–1537.

Kimball, B. A., Boote, K. J., Hatfield, J. L., Ahuja, L. R., Stöckle, C., Archontoulis, S., Baron, C., Basso, B., Bertuzzi, P., Constantin, J., Deryng, D., Dumont, B., Durand, J.-L., Ewert, F., Gaiser, T., Gayler, S., Hoffmann, M. P., Jiang, Q., Kim, S.-H., et al. (2019). Simulation of maize evapotranspiration: An inter-comparison among 29 maize models. *Agricultural and Forest Meteorology, 271*, 264–284.

Kirkegaard, J. A., Lilley, J. M., Howe, G. N., & Graham, J. M. (2007). Impact of subsoil water use on wheat yield. *Australian Journal of Agricultural Research, 58*(4), 303–315.

Kravchenko, A. N., & Bullock, D. G. (2000). Correlation of corn and soybean grain yield with topography and soil properties. *Agronomy Journal, 92*, 75–83.

Kravchenko, A. N., & Bullock, D. G. (2002). Spatial variability of soybean quality data as a function of field topography: I. Spatial data analysis. *Crop Science, 42*(3), 804–812.

Kumhálová, J., Kumhála, F., Kroulík, M., & Matějková, S. (2011). The impact of topography on soil properties and yield and the effects of weather conditions. *Precision Agriculture, 12*, 813–830.

Li, Y., Guan, K., Schnitkay, G. D., DeLucia, E., & Peng, B. (2019). Excessive rainfall leads to maize yield loss of a comparable magnitude to extreme drought in the United States. *Global Change Biology, 25*, 2325–2337.

Lizaso, J. I., Boote, K. J., Jones, J. W., Porter, C. H., Echarte, L., Westgate, M. E., & Sonohat, G. (2011). CSM-IXIM: A new maize simulation model for DSSAT version 4.5. *Agronomy Journal, 103*, 766–779.

Maestrini, B., & Basso, B. (2018). Predicting spatial patterns of within-field crop yield variability. *Field Crops Research, 219*, 106–112.

Manning, G., Fuller, L. G., Eilers, R. G., & Florinsky, I. (2001). Soil moisture and nutrient variation within an undulating Manitoba landscape. *Canadian Journal of Soil Science, 81*(3), 449–458.

Martinez-Feria, R. A., & Basso, B. (2020). Unstable crop yields reveal opportunities for site-specific adaptations to climate variability. *Scientific Reports, 10*, 2885.

McCarthy, A. C., Hancock, N. H., & Raine, S. R. (2010). VARIwise: A general-purpose adaptive control simulation framework for spatially and temporally varied irrigation at sub-field scale. *Computers and Electronics in Agriculture, 70*, 117–128.

McDonald, G. K., Taylor, J. D., Verbyla, A., & Kuchel, H. (2013). Assessing the importance of subsoil constraints to yield of wheat and its implications for yield improvement. *Crop & Pasture Science, 63*, 1043–1065.

Michalczyk, A., Kersebaum, K. C., Roelcke, M., Hartmann, T., Yue, S. C., Chen, X. P., & Zhang, F. S. (2014). Model-based optimisation of nitrogen and water management for wheat–maize systems in the North China Plain. *Nutrient Cycling in Agroecosystems, 98*, 203–222.

Monteith, J. (1965). Light distribution and photosynthesis in field crops. *Annals of Botany, 29*, 17–37.

Moore, I. D., Gessler, P. E., Nielsen, G. A., & Peterson, G. A. (1993). Soil attribute prediction using terrain analysis. *Soil Science Society of America Journal, 57*, 443–452.

Morari, F., Zanella, V., Gobbo, S., Bindi, M., Sartori, L., Pasqui, M., Mosca, G., & Ferrise, R. (2021). Coupling proximal sensing, seasonal forecasts and crop modelling to optimize nitrogen variable rate application in durum wheat. *Precision Agriculture, 22*, 75–98.

Morgan, C. L. S., Norman, J. M., & Lowery, B. (2003). Estimating plant-available water across a field with an inverse yield model. *Soil Science Society of America Journal, 67*, 620–629.

Nawar, S., Corstanje, R., Halcro, G., Mulla, D., & Mouazen, A. M. (2017). Chapter four— Delineation of soil management zones for variable–rate fertilization: A review. *Advances in Agronomy, 143*, 175–245.

Nendel, C., Berg, M., Kersebaum, K. C., Mirschel, W., Specka, X., Wegehenkel, M., Wenkel, K.-O., & Wieland, R. (2011). The MONICA model: Testing predictability for crop growth, soil moisture and nitrogen dynamics. *Ecological Modelling, 222*, 1614–1625.

Pachepsky, Y., & Acock, B. (1998). Stochastic imaging of soil parameters to assess variability and uncertainty of crop yield estimates. *Geoderma, 85*(2), 213–229.

Pennock, D. J., & Frick, A. H. (2001). The role of field studies in landscape-scale applications of process models: An example of soil redistribution and soil organic carbon modeling using CENTURY. *Soil and Tillage Research, 58*(3), 183–191.

Postma, J. A., Jaramillo, R. E., & Lynch, J. P. (2008). Towards modeling the function of root traits for enhancing water acquisition by crops. In L. R. Ahuja, V. R. Reddy, S. A. Saseendran, & Q. Yu (Eds.), *Response of crops to limited water: Understanding and modeling water stress effects on plant growth processes* (pp. 251–276). ASA-CSA-SSSA.

Reuter, H. I. (2004). Spatial crop and soil landscape processes in a loess landscape with respect to relief information [Ph.D. thesis]. University Hannover, Horizonte 16, Der Andere Verlag, Tönning, p. 286.

Reuter, H. I., Kersebaum, K. C., & Wendroth, O. (2005). Modelling of solar radiation influenced by topographic shading—Evaluation and application for precision farming. *Physics and Chemistry of the Earth, 30*(1–3), 143–149.

Roy, P. C., Guber, A., Abouali, M., Nejadhashemi, A. P., Deb, K., & Smucker, A. J. M. (2019). Simulation optimization of water usage and crop yield using precision irrigation. In K. Deb, E. Goodman, C. A. Coello Coello, K. Klamroth, K. Miettinen, S. Mostaghim, & P. Reed (Eds.), *Evolutionary multi-criterion optimization* (Lecture notes in computer science) (Vol. 11411, pp. 695–706). Springer.

Sadhukhan, D., Qi, Z., Zhang, T., Tan, C. S., Ma, L., & Andales, A. A. (2019). Development and evaluation of a phosphorus (P) module in RZWQM2 for phosphorus management in agricultural fields. *Environmental Modelling and Software, 113*, 48–58.

Schmidhalter, U., Maidel, F. X., Heuwinkel, H., Demmel, M., Auernhammer, H., Noack, P. O., et al. (2008). Precision farming—Adaptation of land use management to small scale heterogeneity. In P. Schröder, J. Pfadenhauer, & J. C. Munch (Eds.), *Perspectives for agroecosystem management* (pp. 121–200). Elsevier.

Si, B. C., & Farrell, R. E. (2004). Scale-dependent relationship between wheat yield and topographic indices: A wavelet approach. *Soil Science Society of America Journal, 68*(2), 577–587.

Stafford, J. V., Lark, R. M., & Bolam, H. C. (1999). Using yield maps to regionalize fields into potential management units. In P. C. Robert, R. H. Rust, & W. E. Larson (Eds.), *Proceedings of the 4th conference on precision agriculture* (pp. 225–237). ASA-CSSA-SSSA.

Steudle, E., & Peterson, C. A. (1998). How does water get through roots? *Journal of Experimental Botany, 49*, 775–788.

Stöckle, C., Donatelli, M., & Nelson, R. (2003). CropSyst, a cropping systems simulation model. *European Journal of Agronomy, 18*, 289–307.

Stoorvogel, J. J., Kooistra, L., & Bouma, J. (2015). Managing soil variability at different spatial scales as a basis for precision agriculture. In R. Lal & B. A. Stewart (Eds.), *Soil-specific farming: Precision agriculture* (Advances in soil science, pp. 37–71). CRC Press.

Thorp, K. R., DeJonge, K. C., Kaleita, A. L., Batchelor, W. D., & Paz, J. O. (2008). Methodology for the use of DSSAT models for precision agriculture decision support. *Computers and Electronics in Agriculture, 64*, 276–285.

Timlin, D. J., Pachepsky, Y., Snyder, V. A., & Bryant, R. B. (1998). Spatial and temporal variability of corn grain yield on a hillslope. *Soil Science Society of America Journal, 62*, 764–773.

Timlin, D., Pachepsky, Y., Walthall, C., & Loechel, S. (2001). The use of a water budget model and yield maps to characterize water availability in a landscape. *Soil and Tillage Research, 58*, 219–231.

Trangmar, B. B., Yost, R. J., & Uehara, G. (1985). Application of geostatistic to spatial studies of soil properties. *Advances in Agronomy, 38*, 65–91.

Wallor, E., Kersebaum, K. C., Ventrella, D., Bindi, M., Cammarano, D., Coucheney, E., Gaiser, T., Garofalo, P., Giglio, L., Giola, P., Hoffmann, M. P., Iocola, I., Lana, M., Lewan, E., Maharjan, G. R., Moriondo, M., Mula, L., Nendel, C., Pohankova, E., et al. (2018). The response of process-based agro-ecosystem models to within-field variability in site conditions. *Field Crops Research, 228*, 1–19.

Ward, N. K., Maureir, F., Stöckl, C. O., Brooks, E. S., Painter, K. M., Yourek, M. A., & Gasch, C. K. (2018). Simulating field-scale variability and precision management with a 3D hydrologic cropping systems model. *Precision Agriculture, 19*, 293–313.

Wassenaar, T., Lagacherie, P., Legros, J. P., & Rounsevell, M. D. A. (1999). Modelling wheat yield responses to soil and climate variability at the regional scale. *Climate Research, 11*(3), 209–220.

Wendroth, O., Pohl, W., Koszinski, S., Rogasik, H., Ritsema, C. J., & Nielsen, D. R. (1999). Spatio-temporal patterns and covariance structures of soil water status in two Northeast-German field sites. *Journal of Hydrology, 215,* 38–58.

Williams, J., Jones, C., Kiniry, J., & Spanel, D. (1989). The EPIC crop growth-model. *Transactions of the ASAE. American Society of Agricultural Engineers, 32,* 497–511.

Wu, L., & Kersebaum, K. C. (2008). Modeling water and nitrogen interaction responses and their consequences in crop models. In L. R. Ahuja, V. R. Reddy, S. A. Saseendran, & Q. Yu (Eds.), *Response of crops to limited water: Understanding and modeling water stress effects on plant growth processes* (Advances in agricultural systems modeling 1) (pp. 215–249). ASA-CSA-SSSA.

Wu, L., McGechan, M. B., McRoberts, N., Baddeley, J. A., & Watson, C. A. (2007). SPACSYS: Integration of a 3D root architecture component to carbon, nitrogen and water cycling – Model description. *Ecological Modelling, 200,* 343–359.

Models in Crop Protection

Giorgia Fedele, Federica Bove, and Vittorio Rossi

Abstract A plant disease model is a simplified representation of the relationships between pathogens, crops, and the environment that cause the development of epidemics over time and/or space. This chapter describes the different modelling approaches, focusing on fundamental (process-based) models for application in crop protection. The development of process-based dynamic models consists of four steps: (i) definition of the intended use of the model; (ii) conceptualization of the system; (iii) development of the mathematical framework; and (iv) model evaluation. Plant disease models now have a key role in supporting the decision-making process in integrated pest management (IPM). The advantages of using models in IPM are linked to their ability to process and analyze complex information and to provide outputs supporting decision-making at strategic and tactical levels. In particular, simulation models can support strategic decision-making, and predictive models can assist growers in making tactical decisions. Finally, the chapter briefly considers how different models can be combined with the aim of helping growers to make correct and objective decisions about crop protection.

1 Introduction to Plant Disease Models and Modelling Approaches

1.1 What Is a Model?

A model is a simplified representation of a system (de Wit, 1993; Edminister, 1978; Forrester, 1997; Zadoks, 1971), and a system is a limited part of reality containing

G. Fedele · V. Rossi (✉)
Department of Sustainable Crop Production, Facoltà di Scienze Agrarie, Alimentari e Ambientali, Università Cattolica del Sacro Cuore, Piacenza, Italy
e-mail: vittorio.rossi@unicatt.it

F. Bove
Horta srl, Piacenza, Italy

© The Author(s), under exclusive license to Springer Nature Switzerland AG 2023
D. Cammarano et al. (eds.), *Precision Agriculture: Modelling*, Progress in Precision Agriculture, https://doi.org/10.1007/978-3-031-15258-0_3

interrelated elements (Rabbinge & de Wit, 1989). In the context of crop protection, a plant disease model is a simplified representation of the relationships between pathogens, crops, and environment that cause the development of epidemics; these relationships involve a large number of interactions occurring at different levels of a hierarchy over time and/or space (Rossi et al., 2010).

As a simplification of reality, a model cannot capture all aspects of the system of interest. Consequently, all models have some inaccuracy or incompleteness (Madden et al., 2007), which are accounted for by the definition of model boundaries (i.e., the spatial and temporal conditions, as well as practical constraints, under which the system is observed), and assumptions and/or simplifications made in accordance with the purpose of the modelling exercise (Mulligan & Wainwright, 2013). Simplifications and boundaries require thoughtful consideration, and thus, expertise.

The dualism between model complexity and simplicity is important (Christie et al., 2011). Addition of detailed process descriptions, with increased numbers of variables and parameters, may support a theoretical or biological point of view in that it will more completely describe the complex interactions within the system. The added details, however, may not necessarily improve model performance. A parsimonious model is usually one with the greatest predictive power and the least process complexity (Wainwright & Mulligan, 2004). Although complexity may make the model more accurate, it may also make the model less manageable (Kranz & Hau, 1980). Thus, simplicity should be strived for in developing a model but not at the cost of model performance.

In general, models are used to summarize the essential features of the observations or measurements of interest, and their use enables inferences to be made about large host populations based on more limited direct information (Madden et al., 2007).

1.2   Types of Models and Modelling Approaches

There are different types of models for different modelling objectives. In plant pathology, models have been developed with the following objectives: (i) to understand epidemiological processes and to describe and compare plant disease epidemics as well as their components; (ii) to relate epidemics to crop losses; (iii) to predict plant disease patterns under different crop protection or climatic scenarios; (iv) to support decision-making for management of plant disease; (v) to integrate current knowledge and to identify gaps in knowledge; (vi) to explore hypotheses that cannot be tested in practice to provide a framework for future research; and (vii) to provide an intuitive hands-on analysis of agricultural systems for educational purposes (Campbell & Madden, 1990; Gilligan, 1985; Madden et al., 2007; Norton et al., 1993; Savary & Willocquet, 2014; Shoemaker, 1981; Van Maanen & Xu, 2003). Different types of models require different modelling approaches, whose main features are briefly described in the following sections.

1.2.1 Qualitative vs. Quantitative Models

Models can be either "qualitative" or as "quantitative." A qualitative model represents some essential aspects of reality in a theoretical way, without the use of equations, mathematics, or statistics; descriptions of disease cycles are examples of qualitative models (Madden et al., 2007). A quantitative model uses mathematical notations, symbols, and rules to describe a system and the relationships between variables and the parameters that govern the system (Madden et al., 2007). Therefore, a quantitative model is a mathematical model that can consist of a single function or a complex system of functions (Madden et al., 2007).

1.2.2 Deterministic vs. Stochastic Models

Mathematical models can be "deterministic" or "stochastic." A deterministic model is characterized by variables determined by model parameters and by sets of previous states of these variables (Renard et al., 2013); the model then provides a single solution (or model output) at each run. Because this type of model does not account for the effects of data uncertainty or variability, changes in model outputs are solely due to changes in model components or inputs (Xu & Ridout, 1996). Stochastic models account for random variability in model parameters and/or inputs and thereby provide multiple solutions at each run (Lehoczky, 1990).

1.2.3 Empirical vs. Fundamental Models

Mathematical models can also be "empirical" or "fundamental" (Madden & Ellis, 1988).

Empirical models (also referred to as correlative or data-based models) describe the behaviour of a system on the basis of observations, without reference to cause-and-effect relationships. These models have the potential to provide a good representation of a set of observed data without taking into account the underlying biological mechanisms (molecular, biochemical, population, etc.) (Madden et al., 2007). This lack of relationships makes empirical models easy to develop but weak in accuracy, knowledge, and robustness, so that they require validation and adaptation when used in areas or scenarios different from those in which they were developed (Rossi et al., 2010). It is important to note that new approaches, like artificial neural networks, big data, machine learning, and artificial intelligence, are all empirical. These approaches, which do not require the understanding of complex mechanisms of the pathosystem, use statistics to predict trends by relating patterns with the data supplied. All of these models learn from the experience that is provided in the form of data; the more the experience, the better the model. These empirical models, however, are limited by the range of their input data, i.e., they can only predict results within that range.

Fundamental models (also referred to as explanatory, theoretical, mechanistic, or process-based models) aim to understand how the system works in terms of the underlying processes and the influencing variables (Wainwright & Mulligan, 2004). The development of fundamental models begins with theory, hypothesis, and biological and epidemiological knowledge of the system to be modelled and of the processes involved (Madden et al., 2007). Fundamental models explain the relationships between the elements of a system based on what is known about how the system works in relation to the influencing, external variables (Wainwright & Mulligan, 2004). Because fundamental models analyze components of the epidemic and their changes over time as influenced by external variables, they are dynamic (Teng et al., 1980). These models are based on the assumption that the state of the system can be characterized quantitatively at every moment and that changes in the system can be described with mathematical equations (Rabbinge & de Wit, 1989).

Distinction between empirical and fundamental models, however, can be fuzzy because empirical models can have a fundamental basis, and fundamental models usually have empirical elements (Madden & Ellis, 1988). Moreover, there are models that combine fundamental and empirical approaches (Park et al., 1997).

1.2.4 Simulation vs. Predictive Models

Process-based models can be further defined as "simulation" or "predictive" models. Simulation has been defined as the art of building mathematical models and the study of their properties (Rabbinge & de Wit, 1989); predictive involves the estimation of how the system performs in unknown situations.

Simulation models are often concerned with the theoretical understanding of generic features of epidemic development; they quantify the effects of specific events on epidemic development and allow the identification of problems by distinguishing between cause and effect (Van Maanen & Xu, 2003). Simulation models address specific questions (Zadoks & Rabbinge, 1985). They are often developed to identify the key factors that determine the behaviour of pathosystems to provide disease management options, to quantify the potential efficacy of those options, and to provide a framework for future research (Savary & Willocquet, 2014). Because they allow the numerical integration of data coming from experiments on biological processes, simulation models also enable a numerical visualization of knowledge gaps. Another use of this kind of model is the mobilization of available knowledge to explore the system's behavior. The characteristics of simulation models make them useful for exploring possible future scenarios (e.g., effects of climate change). Simulation models are also effective educational tools, because they provide an intuitive analysis of systems.

Predictive models are mainly developed to predict the behaviour of a pathosystem in the real environment, as affected by a set of external variables, which can be either measured (e.g., weather data) or forecast (e.g., weather forecasts) (Campbell & Madden, 1990; De Vallavieille-Pope et al., 2000; Madden & Ellis, 1988). In modelling, the terms "forecast" and "forecasting" are sometimes reserved for estimates of values at specific future times, while the term "prediction" is used for more general estimates, for instance, to find the value of the target attribute for unknown past or current situations (Camase, 1996).

1.3   A Brief History of Plant Disease Modelling

Models combining environmental data and plant and disease characteristics have been developed since the middle 1900s. The early plant disease models were developed following an empirical approach (Waggoner, 1960) and consisted of simple rules, graphs, or tables showing relationships between particular components of the disease cycle (e.g., infection and sporulation) and the concomitant weather conditions (Baldacci, 1947; Mills, 1944). Dynamic modelling of plant disease epidemics began in the early 1960s with the conceptualization of the temporal progress of diseases by Van der Plank (1960, 1963) and then by Zadoks (1971). After these "milestones," the interest switched to mechanistic models (Fry & Fohner, 1985; Teng, 1985). In 1969, for instance, Waggoner and Horsfall developed a model of early blight epidemics in tomato (*Solanum lycopersicum* L.) and potato (*Solanum tuberosum* L.) caused by *Alternaria solani* and a model of southern corn leaf blight caused by *Helminthosporium maydis* (Waggoner et al., 1969) to investigate the effects of weather and cultural conditions on disease development. Several plant disease models have been developed more recently, with major effort directed at modelling diseases affecting certain cropping systems (e.g., wheat, *Triticum aestivum* L.; potato; and soya bean – *Glycine max* L.) in relation to scientific, economic, and social factors (De Wolf & Isard, 2008; Fedele et al., 2020).

Rather than being used for practical crop protection, plant disease models unfortunately remained primarily as research tools for many years (Butt & Jeger, 1985; De Wolf & Isard, 2008; Krause & Massie, 1975); reasons for the infrequent use of models in practical crop protection were recently considered by Rossi et al. (2019). In the last decade, the imbalance between the number of models developed and deployed has decreased with an increase in research directed toward evaluation and practical application of disease models and especially of predictive models (De Wolf & Isard, 2008; Van Maanen & Xu, 2003). The increase in the practical application of models for disease prediction is closely linked with the advances in weather monitoring and automatic data processing. In addition, there is the ongoing development of information technologies and the Internet, so that models become components of "decision tools" supporting the knowledge-based management of harmful organisms (Rossi et al., 2019). The most advanced decision tools, in which plant disease models are key components, are represented by the decision support systems (DSSs) (Rossi et al., 2019), i.e., computerized information systems that support both daily operational decision-making and long-range strategic decision-making (Rossi et al., 2012; Sonka et al., 1997).

2 Development of Process-Based Models

The development of mechanistic, dynamic models for plant diseases involves the following steps: (i) definition of the intended use of the model; (ii) conceptualization of the system; (iii) development of a mathematical framework; and (iv) model evaluation. In this chapter, we do not consider the details of model development (which have been described previously), but we focus on those aspects of development that are relevant for crop protection models.

2.1 Definition of the Intended Use of the Model

A precise definition of the intended use of a model is crucial for its practical use (Kranz & Royle, 1978; Sutherst, 1993; Wainwright & Mulligan, 2004). The crop protection problem to be addressed should be clearly identified; the involvement of the people who will use the model (e.g., growers, advisors, and policy makers) may help to delineate the problem to be addressed and define its boundaries (Rossi et al., 2010, 2019; Wainwright & Mulligan, 2004).

A main objective for the development of crop protection models is disease prediction (Campbell & Madden, 1990), at both "strategic" and "tactical" levels (Shoemaker, 1981). Strategic use of a model concerns the model's involvement in those decisions that affect the entire life cycle of a crop, i.e., on whether to apply measures for disease prevention and/or suppression (e.g., the choice of the cultivar, the training system, irrigation, and mechanization level). Tactical use of a model involves those decisions limited to the current season, i.e., whether and when to apply plant management and protection actions (Power, 2007; Rossi et al., 2010; 2012). Estimation of the risk that a crop will develop a disease in relation to environmental and agricultural factors is based on the concept of ecological and economic justification (Nutter, 2007). Models able to predict the time of disease outbreak, the risk of infection, and the risk of epidemic development allow users to identify the optimal timing and frequency for the implementation of disease management options (Plant & Mangel, 1987; Rossi et al., 2019).

2.2 Conceptualization of the System

Once the aim of the model has been defined, the second step is conceptualization of the system of interest. In plant pathology, this includes the host crop, the pathogen, the environment, natural enemies or limiters, human interventions (cultural practices), and their multiple interactions. Conceptualization begins with definition of the system boundaries: (i) how many and which elements are involved (e.g., host crop processes, such as growth and development; infection processes; environmental effects on processes; etc.); (ii) the size of the system (e.g., one plant, 1 cm^2, a crop, etc.); and (iii) time characteristics (i.e., time frame and time step).

Conceptualization requires a deep knowledge and understanding of the processes involved in the system. This knowledge can be acquired through a careful review of the literature. Systematic literature reviews, which are well-established in the field of medicine, are useful for identifying, appraising, and synthesizing research-based evidence and for presenting that evidence in an accessible format (Mulrow, 1994). This systematic and explicit approach minimizes bias, facilitates critical appraisal, and provides reliable findings from which conclusions can be drawn (Antman et al., 1992; Oxman & Guyatt, 1993). Model conceptualization helps modelers to organize epidemiological knowledge in order to identify knowledge gaps, to make assumptions or simplifications, and to design specific experiments.

The system structure can finally be organized in a relational diagram through "systems analysis," which was first proposed by Forrester (1961) and was subsequently developed by others (e.g., de Wit, 1993; Rossi et al., 2008; Savary & Willocquet, 2014). Systems analysis involves the following basic elements: state variables, rates, flows, intermediate and auxiliary variables, coefficients, constants, switches, external variables, and driving functions that link external variables to the system. A state variable represents the status of the system, which changes in time and space and determines the dynamics of the system. State variables can be measured in terms of the number of individuals or the amount of matter (i.e., number of insects, number of spores, number of lesions on leaves, etc.). A rate represents the "speed" at which a state variable changes. Rates control flows, which represent an increase (inflow) or a decrease (outflow) of contents of the state variables. Coefficients or parameters are constant or variable, e.g., they may change through model runs or according to the influence of a driving function. External factors are those that are outside the boundaries of the system but that influence the system by affecting driving functions: in plant pathology, external factors are mainly environmental variables such as temperature, rain, relative humidity, etc. Switches have been introduced more recently into systems analysis, and they account for logical operators such as "if condition $=$, $<$, or $>$ x then go to y else go to z" (Rossi et al., 2008; 2010). Computer programs can help to organize the system's structure and the relationships among components of the model. There are several programs and kinds of software, e.g., the STELLA® program (Isee Systems, 2005), that enable developers to encode models and to conceptualize systems.

A basic system structure for plant disease epidemics follows the concepts established by Van der Plank (1963) and Zadoks (1971) and compartmentalizes the main stages of the host-plant tissue during an epidemic, i.e., healthy (H), latent (L), infectious (I), and post-infectious (R) compartments. For the development of a model flowchart that focuses on a polycyclic disease caused by an airborne pathogen that produces lesions on foliage, for example, state variables would be the number of host sites changing from H–L–I–R during an infection cycle (Fig. 1). Because categories of host-plant sites are nonoverlapping, the following equation applies:

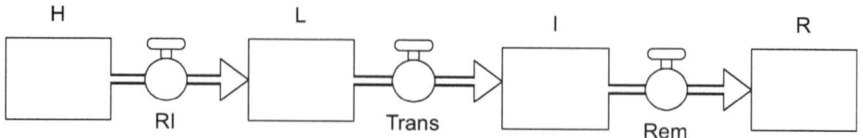

Fig. 1 Example of an H–L–I–R model flowchart drawn using the STELLA® software. Plant sites evolve from healthy (H) to latent (L), infectious (I), and post-infectious (R) categories through rates of infection (RI), transfer (Trans), and removal (Rem)

$$N = H + L + I + R,$$

where N is the total number of sites in the system. A detailed example of an H–L–I–R model that was developed with STELLA® was provided by Savary and Willocquet (2014).

2.3 Development of the Mathematical Framework

Once the model structure has been conceptualized and system size, time frame, and time step have been defined, the model is then mathematically encoded. A framework of equations must be developed by either analytical or numerical integration. The differences between these two approaches are explained in detail by Savary and Willocquet (2014).

As an example of analytical integration of the equation framework, the model in Fig. 1 can be encoded through linked differential equations:

$$dH/dt = -\beta IH$$
$$dL/dt = \beta IH - \theta L$$
$$dI/dt = \theta L - \alpha I$$
$$dR/dt = \alpha I$$

where β is a disease transmission rate and θ and α are constant rates of progress from the latent to infectious and from the infectious to post-infectious compartments, respectively (Madden et al., 2007; Segarra et al., 2001).

The rate of change in the number of infected sites (x) can be written as follows:

$$dx/dt = dL/dt + dI/dt + dR/dt$$

The definition of the mathematical framework of a system, and, specifically, of the quantitative relationships between the components of the model, is based on previously published or specific experimental data. Functions can be used to identify the mathematical relationship that best explains the data. This procedure is called "model fitting." If a general function is known to describe a specific process (e.g., latency or infectiousness), the model equation can be parameterized to achieve the best possible fit to the data (Camase, 1996; Rossi et al., 2010). This procedure is also referred to as "model calibration" (Pascual et al., 2003).

2.4   Model Evaluation

Model evaluation includes model verification and model testing (Teng, 1981). Model verification consists of inspecting the internal consistency of the model by determining whether its structure and equations perform as expected. Model testing includes the following several steps: (i) validation; (ii) sensitivity analysis; and (iii) judgment of utility (Camase, 1996; Rossi et al., 2010). The validation process helps to determine whether the model is useful and relevant with respect to the defined purpose.

For predictive models, validation is mainly carried out by comparing the predictions of the model with observations (i.e., from scouting or experiments). Through this procedure, it is possible to evaluate the *accuracy* of the model as the ratio between the number of predictions and the real observations. Another model characteristic to be evaluated is its *robustness*, i.e., its capability to perform well across the entire range of environmental conditions driving the model (Pascual et al., 2003). For validation of these kinds of models, Bayes' theorem (Yuen & Hughes, 2002) can be useful in evaluating the sensitivity (or true positive rate) and the specificity (or true negative rate) of a predictor.

For simulation models, validation can be performed by comparing the simulated output (e.g., epidemic development) with the actual (a priori) overall pattern of the system.

Sensitivity analysis of a model is conducted to determine whether the model performs as intended and to identify model components that are "sensitive," i.e., that greatly affect model behavior (Savary & Willocquet, 2014). This analysis is usually performed by changing the values of input variables and evaluating the consequent changes in model outputs (Rossi et al., 2010). When a large variation in a parameter value causes only a small change in output, that parameter can be excluded from the model or its measurement can be simplified. Frey and Patil (2002) assigned sensitivity analysis methods to three classes of model: mathematical, statistical, and graphical. Other classifications of sensitivity analysis have also been described (Camase, 1996; Cullen & Frey, 1999; Papastamati & Van Den Bosch, 2007; Saltelli et al., 2000).

As the last step in model evaluation, judgment of utility consists of determining the added value of using the model for the defined purpose compared with the current practice. For instance, judgment of utility might involve comparison of the number of applications of fungicide spray and the disease intensity resulting from

treatments that follow the model vs. (i) treatments that follow the standard grower's
fungicide schedule or (ii) an unsprayed control (Caffi et al., 2010).

3 Models for IPM

Plant disease models have assumed a key role in supporting the decision-making
process in integrated pest management (IPM) (Rossi et al., 2019), which aims to
protect crops against harmful organisms and to reduce the use of chemicals to levels
that are economically and ecologically justified, thereby reducing and minimizing
the risks for humans and the environment (Ehler, 2006). The advantages of using
models in IPM are linked to their ability to process and analyze complex information
and to provide output in support of the decision-making process (Rossi et al., 2019).

3.1 Decision-Making in IPM

Decision-making is a cognitive process resulting in the selection of an action among
several alternative actions: every decision-making process produces a final choice
(March, 1994), which can be strategic, tactical, or operational (Conway, 1984;
McCown, 2002; Rossi et al., 2012).

  Strategic decisions concern the selection of long-term measures, at both the farm
and crop level (e.g., crop rotation and variety sown, respectively), for disease
prevention and/or suppression. Tactical decisions are made day-by-day (or within
a day) and concern whether, when, and which plant protection actions are required at
the crop level. Operational decisions are made by the person responsible for
implementing daily management actions at the crop level; for example, that person
must select spray volume, sprayer speed, and spray timing during the day and must
modify those selections when environmental conditions change (e.g., an increase in
wind speed or the occurrence of unexpected rain during spraying).

  Models can support both strategic and tactical decisions, but operational decisions are usually made in too short a time to benefit from mathematical models.
Nevertheless, there are IT tools that can be used for some operational decisions, such
as determining the optimal setting of the sprayer for each specific intervention
(Anselmi & Legler, 2021).

3.2 Strategic Disease Management

Strategic decisions are focused on cultural practices concerning crop health, hygiene
measures, and the selection of host genotypes that are resistant or tolerant to
pathogens.

They include but are not limited to the following: (i) choice of crop species and cultivars; (ii) choice of crop type (open field or protected cultivation, uniform or in consociation, etc.) and rotational schemes; (iii) elimination of inoculum reservoirs, not only from inside the field but also from the entire farm; (iv) choice of times and types of soil preparation and of planting dates; and (v) choice of cropping methods that may last some years (like the training system and plant density in orchards) or that may require depreciation allowances for equipment (like equipment needed for subsoil or sprinkler irrigation).

Strategic decision-making can be supported by simulation models. González-Domínguez et al. (2020) recently developed a general example of a simulation model that predicts the effect of the main crop management actions (actions commonly recommended in IPM) on plant disease epidemics. Simulation models represent a source of information on the benefits of strategic actions, providing a system to measure their effects, usually measured over a long time scale (e.g., years).

3.2.1   A First Example: Epidemics of Grape Downy Mildew

Bove et al. (2020a) developed a process-based simulation model with the aim of quantitatively synthesizing the literature available for the grapevine (*Vitis vinifera* L.)-downy mildew (*Plasmopara viticola*) pathosystem and of providing a tool to help strategic decisions for disease management. The model (Fig. 2) includes the main processes involved in the epidemics on both leaves and grape clusters, from inoculum mobilization to disease multiplication on foliage. Inoculum mobilization involves the rate of primary infections (RPI) as a function of the onset date (OD), the duration of inoculum mobilization (PD), and the inflow of primary infections (P). Disease multiplication on foliage reflects the rate of infection (RI) of healthy sites (HS) and the rates at which infections on leaves become latent (L), infectious (I), and removed (R). The RI is also affected by the linkage between disease on foliage and disease on clusters; disease on clusters occurs at a rate of cluster infection (RCI), a rate at which healthy clusters (HC) become diseased (DC). The model also includes host dynamics, i.e., crop development (DVS), growth (RG), physiological senescence (RSEN), and disease-induced senescence (RRDS).

The model was used to investigate the behavior of the pathosystem as affected by climate change (Bove et al., 2020b) and resistant varieties (Bove et al., 2021). For the latter purpose, 4 components of partial resistance, i.e., infection efficiency (RRIE), latency period (RRLP), sporulation (RRSP), and infectious period (RRIP), were integrated into the model for 16 grapevine varieties, for which the parameter values for components of partial resistance had been measured previously in laboratory experiments (Bove & Rossi, 2020).

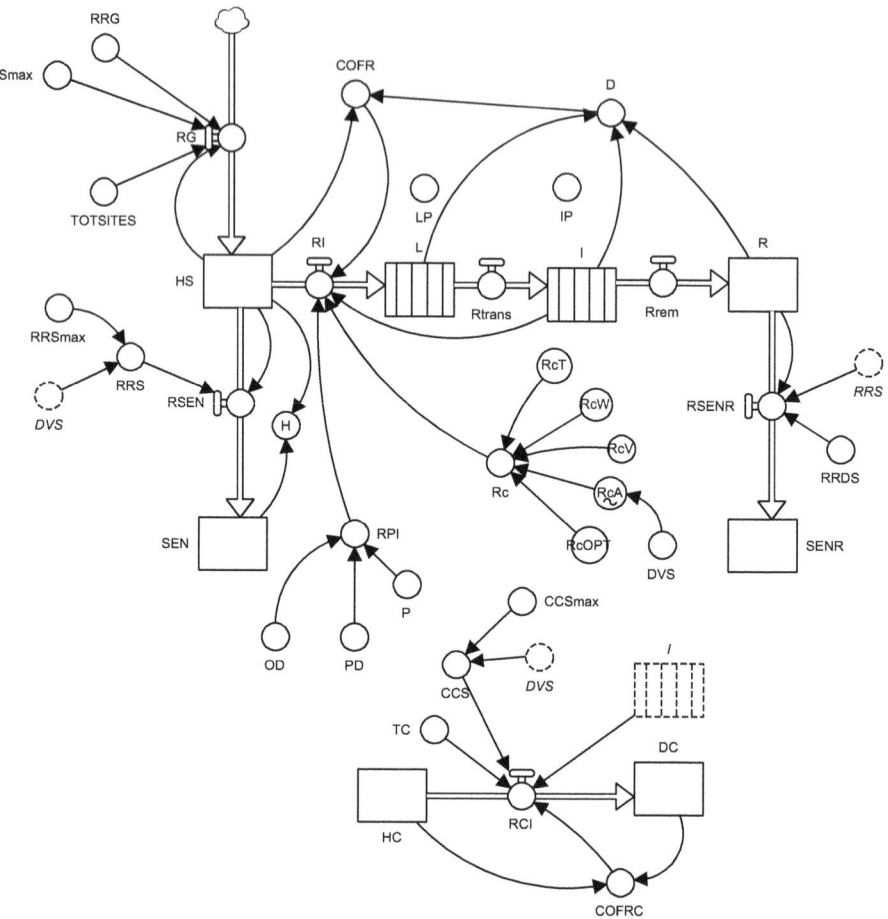

Fig. 2 Flowchart of the grapevine downy mildew (*Plasmopara viticola*) simulation model developed by Bove et al. (2020a). The diagram uses the symbols developed by Forrester (1961). The core of the structure of the model is based on Zadoks (1971), with sites on foliage evolving from healthy (HS) to latent (L), infectious (I), and removed (R). The rate of infection of leaves (RI) depends on primary (P) and secondary infections (Rc and I). Infectious sites on the foliage (I) are the only source of infection of grape clusters, and the rate of cluster infections (RCI) is regulated by the leaf-to-cluster disease transmission coefficient (TC); grape clusters can be healthy (HC) or diseased (DC). The structure incorporates host (canopy) growth (RG), physiological senescence (RSEN), and physiological senescence compounded by disease-induced senescence (RSENR)

3.2.2 A Second Example: Effect of Biocontrol on Epidemic Development

Fedele et al. (2020) developed a model that simulates the effect of any BCA (biocontrol agent) on the daily progress of a foliar disease (Fig. 3). The model considers seven categories of plant tissue: (i) healthy and susceptible to infection (HS); (ii) affected by the pathogen and infectious, i.e., can generate new, secondary

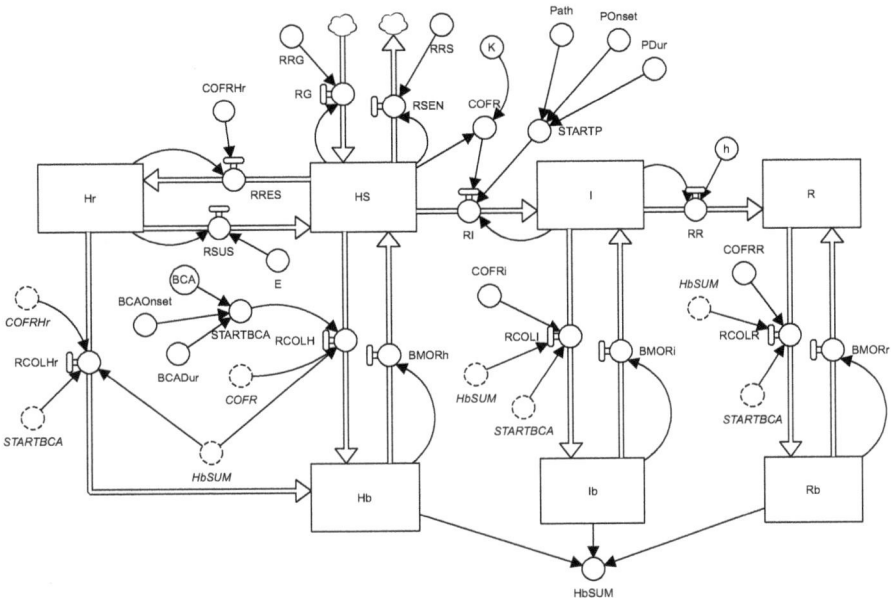

Fig. 3 Simplified flowchart of the model for biocontrol agents (BCAs) developed by Fedele et al. (2020). The core of the model is based on a classic susceptible infected–removed (SIR) model, with tissue evolving from healthy–susceptible (HS) to infectious (I) and removed (R). The rate of infection of tissue (RI) depends on primary (STARTP) and secondary infections (I). The rate of resistance induction by a BCA (RRES) depends on BCA application (STARTB) and the total amount of healthy–susceptible tissue (HS). The rate of BCA colonization (RCOLH, RCOLI, RCOLR, RCOLHr) depends on BCA application (STARTB) and the total amount of tissue colonized by the BCA (BSUM). The structure incorporates host growth (RG) and physiological senescence (RS)

infections (I); (iii) affected by the pathogen and removed, i.e., no longer infectious (R); (iv) healthy and colonized by the BCA and protected from the pathogen through induced host resistance (Hr); (v) healthy and colonized by the BCA and protected from the pathogen by competition or antibiosis (Hb); (vi) infectious and colonized by the BCA, i.e., unable to generate new infections (Ib); and (vii) removed and colonized by the BCA (Rb). The amount of HS increases over time as a consequence of plant growth or decreases as a consequence of senescence (which is relevant for those diseases in which the senescent plant tissue is no longer susceptible to infection). For host-pathogen dynamics, the rate of infection (RI) determines the flow from HS to I, and the rate of removal (RR) determines the flow from I to R. For pathogen-BCA dynamics, introduction of the BCA generates outflows from HS to Hb, a state of the host that cannot be infected, i.e., a state that prevents the flow from H to I. The model considers that this outflow can be caused by BCAs that induce resistance in the host tissue and/or that prevent infection due to competition and/or antibiosis through the following rates: RRES, rate of induced resistance by BCA; RCOLHr, rate of colonization of Hr by BCA; RCOLH, rate of colonization of HS by

BCA; RCOLI, rate of colonization of I by BCA; and RCOLR, rate of colonization of R by BCA. Because induction of resistance in the host tissue is transitory, the model considers that the Hr tissue reverts to HS and becomes susceptible to infection according to the rate at which induced resistance declines (RSUS), or according to the rate of BCA mortality (BMOR), which determines the rate at which plant tissue colonized by BCA reverts to BCA-free tissue.

Rates depend on the characteristics of the pathogen, host plant, and BCA and are influenced by the weather conditions that affect the processes underlying the dynamics of both the pathogen (i.e., infection and infectiousness) and the BCA (i.e., growth and survival capability). Specifically, BCA efficacy is affected by four factors: (i) BCA mechanism of action; (ii) timing of BCA application with respect to timing of pathogen infection (preventative vs. curative); (iii) temperature and moisture requirements for BCA growth; and (iv) BCA survival as influenced by temperature and moisture.

The model can be used to simulate, as a screening tool, the environmental responses and survival capability of BCA candidates during their selection and formulation. In addition, the model can be used to predict the efficacy of BCAs under different scenarios of weather conditions and application timings.

3.3   Tactical Disease Management

Tactical decisions focus on control actions aimed at reducing the progress of epidemics. They include (i) fungicide sprays (application times, products, and dosages) or alternative methods aimed at pathogen control (BCA application or eradication) and (ii) cultural practices that regulate crop growth and canopy density (e.g., fertilization, irrigation times and rates, weed control, etc.) (Rossi et al., 2019). Predictive models provide information that can help growers make tactical decisions.

3.3.1   A First Example: Prediction of Secondary Infections of Grape Downy Mildew

Brischetto et al. (2020) developed a model that produces warnings about the occurrence of secondary infections by *Plasmopara viticola* and that predicts the relative severity of such infection. The model assesses the main biological and epidemiological processes involved in secondary infections by *P. viticola* and is organized into three compartments: (i) sporulation; (ii) dispersal and deposition of sporangia; and (iii) infection.

The model begins when the first seasonal downy mildew (DM) lesions appear in the vineyard; once DM lesions appear, the model assumes that lesions will be present throughout the season and that these lesions may produce sporangia under suitable conditions. The first step in the model consists of "sites with visible DM lesions" (Fig. 4). Sites can be either DM-free or occupied by a DM lesion, and the latter can

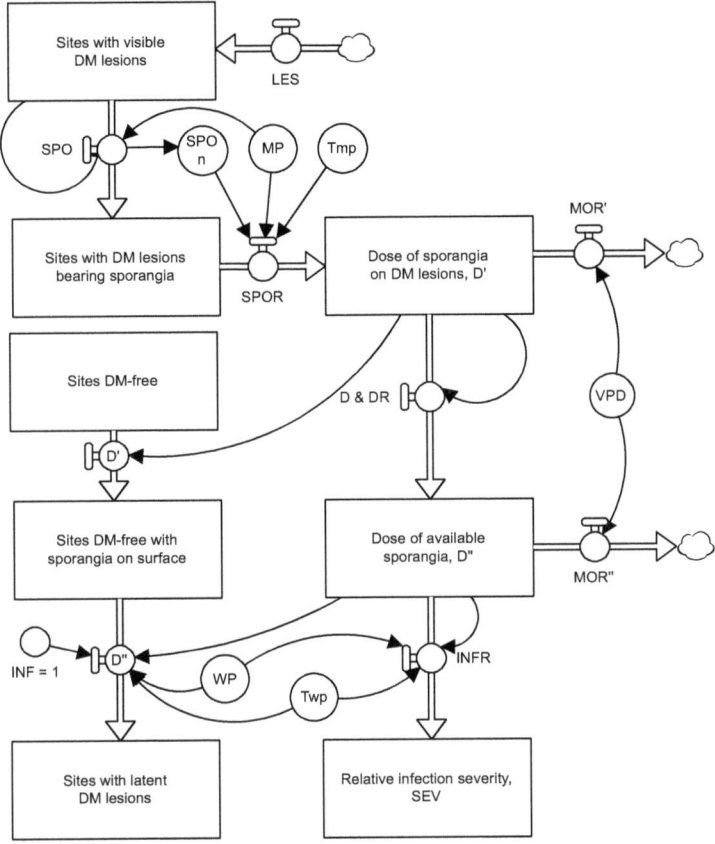

Fig. 4 Relational diagram of the model developed by Brischetto et al. (2020) predicting secondary infection cycles of *Plasmopara viticola*

be either a visible non-sporulating or sporulating lesions. Sites with DM lesions produce sporangiophores and sporangia when weather conditions are favorable as regulated by a switch (SPO) (Fig. 4). If SPO = 1, then lesions advance to the second step in the model, i.e., "sites with DM lesions bearing sporangia"; if SPO = 0, in contrast, lesions remain at the first step (i.e., they remain non-sporulating).

When lesions begin producing sporangia, they continue sporulating for a sporulation period, during which the relative "dose of sporangia on DM lesions" (D′) is calculated through a sporulation rate (SPOR). As the sporulation period ends, the dose of sporangia progressively diminishes because the sporangia still attached to sporangiophores may die according to a mortality rate (MOR'). After being detached from sporangiophores on sporulating DM lesions, sporangia are dispersed into the air, and some of them are deposited on DM-free sites. The model calculates the relative "dose of available sporangia" on DM-free sites (D") through a dispersal rate (D) and a deposition rate (DR). This dose of sporangia, however, progressively diminishes because the sporangia detached from sporangiophores can die under unfavorable conditions (Blaeser & Weltzien, 1979) according to a mortality rate (MOR").

When there are viable sporangia on DM-free sites and conditions are favorable for infection, as regulated by a switch (INF), sites become latently infected. The relative infection severity of each infection period is finally calculated by an infection rate (INFR).

The model was evaluated against observed data (collected over a 3-year period in a vineyard) for its accuracy to predict periods with no sporangia or periods with peaks of sporangia, so that growers can identify periods with no/low risk or high risk. The model was able to identify periods in which the DM risk was nil or very low. It may therefore help growers to avoid fungicide sprays when not needed and lengthen the interval between two sprays, i.e., it will help growers to move from calendar-based to risk-based fungicide schedules for the control of *P. viticola* in vineyards.

3.3.2   A Second Example: Prediction of Stem Rust Infections on Wheat

Salotti (2021) developed a plant-focused model for the epidemics of *Puccinia graminis* f. sp. *tritici*, the causal agent of stem (or black) rust, on wheat (Fig. 5). Although the pathogen goes through five spore stages in its life cycle, the model

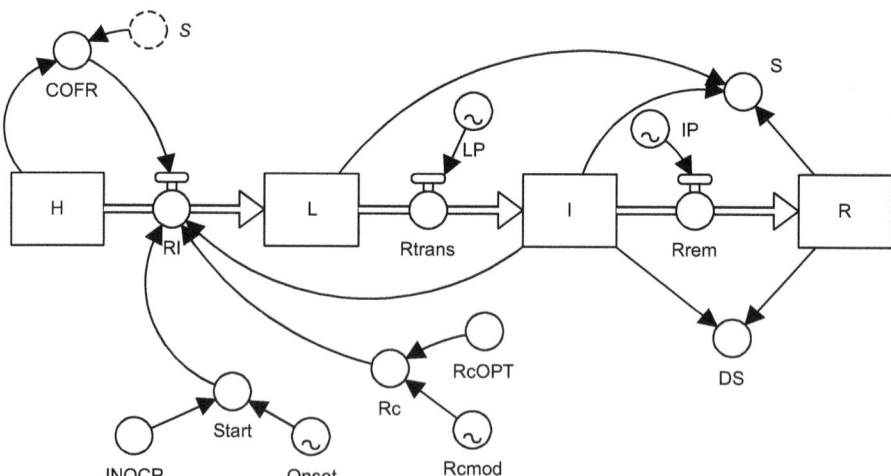

Fig. 5 Flowchart of stem rust of wheat (*Puccinia graminis* f. sp. *tritici*) developed by Salotti (2021). The diagram uses the symbols developed by Forrester (1961). The core of the model structure is based on Zadoks (1971), with sites evolving from healthy (H) to latent (L), infectious (I), and removed (R). The onset of the epidemic depends on primary infection (INOCP, Onset). The progress of the epidemic depends on secondary infections (RI, Rc, and I). The model's core operates with a daily time step

focuses on uredospores, which are the main spores responsible for disease in the major wheat-growing areas (Roelfs, 1985).

The core of the model's structure is based on the epidemiological model developed by Zadoks (1971), in which the crop is considered to consist of a large but finite number of infection sites (i.e., fractions of the host tissues where an infection may take place and where a lesion may develop) that have equal dimensions and equal chances of being infected.

An infection site has one of the following mutually exclusive conditions: (i) healthy (H); (ii) latent (L), where symptoms of stem rust are not visible; (iii) infectious (I), visible lesions producing spores; and (iv) removed (R), lesions that are older and nonsporulating. Sites becoming infected, with a rate of infection (RI), flow from healthy to latent. Latent sites become infectious at the end of a latency period (LP), and infectious sites become removed at the end of an infectious period (IP). The transitions from L to I and from I to R are regulated by the rate of transfer (Rtrans) and the rate of removal (Rrem), respectively. At the beginning of model calculations, the whole crop is healthy (i.e., H = 1), and the model represents the flow from one state to the next as a proportion of H (i.e., on a 0 to 1 scale). The unlimited growth of the epidemic is confined by a correction factor for removals (COFR), which is calculated based on H and diseased sites (S = I + L + R).

Stem rust epidemics are initiated by a single inflow of infections (INOCP), which represents the primary inoculum (uredospores in the atmosphere) that is deposited on host tissues by rain and that cause infection under favorable wetness and temperature conditions. Secondary spread of the disease is governed by RI, which was modelled as a function of I, COFR, and a daily multiplication factor (Rc). The model has a daily time step, with the exception of infection, which is calculated hourly.

The model was evaluated against field data for its ability to predict primary infections (nine epidemics) and disease progress (six epidemics) during the growing season. Overall, the model showed good accuracy and specificity in predicting primary infections and was reliable, accurate, and robust in predicting seasonal dynamics of stem rust epidemics. It follows that the model could help growers schedule fungicide treatments to control *P. graminis* f. p. *tritici* infections.

3.4 Multi-modelling for Decision-Making

Although plant disease models are important tools for IPM, decision-making requires additional information. For example, information is needed on (i) susceptibility of the plant to the disease, (ii) the residual efficacy of previous protection measures, (iii) selection of the most appropriate plant protection product and its dose, and (iv) the suitability of environmental conditions for fungicide application. This multi-criteria decision process requires a multi-modelling approach that includes models for plant disease, for plant growth and development (with consideration of disease resistance), and for the effects of fungicides (Rossi et al., 2012).

An example of a multi-modelling approach to determine whether a fungicide application against *P. viticola* is necessary was provided by Rossi et al. (2014); the approach included models that predict the risk of primary and secondary infections (Caffi et al., 2013; Rossi et al., 2008), a model that predicts grapevine phenology (Cola et al., 2014), and a model that predicts the residual protection provided by the last fungicide application (Caffi & Rossi, 2018).

3.4.1 Modelling the Host Plant

Plant growth is the increase in size and weight of plants or plant parts. The amount of host tissue directly affects disease progress because the quantity of host tissue available for infection determines the "carrying capacity" or maximum potential disease. Host dynamics can be predicted by crop growth models. For instance, the grapevine growth model developed by Cola et al. (2014) calculates daily the plant phenological stages, the formation and unfolding of leaves, leaf area development, total cluster fresh weight (i.e., yield), and the source-sink balance given as the leaf area-to-yield ratio.

In the grapevine growth model (Fig. 6), global solar radiation (GSR) is converted into photosynthetically active radiation (PAR) determined by the canopy light interception rate; PAR is used to estimate the absorbed photosynthetically active radiation (APAR). The canopy light interception rate is a function of canopy dimensions (phenological stage), which depends on air temperature (T) and vineyard-specific characteristics such as geographical position, aspect, slope, and between-row and in-row vine spacing (solar position). The rate of photosynthesis, which is a function of the atmospheric CO_2 concentration, is used to estimate the daily potential gross CO_2 assimilation (GASS) based on the APAR. The potential net assimilation (PNA), defined as dry matter available for plant growth and storage replenishment, is determined by the rate of losses (respiration losses) associated with conversion of primary photosynthates into structural plant matter. The daily final production of dry matter (DMI FIN) is determined by limitation rates to PNA due to thermal and water resources (thermal limitation and water limitation, respectively). The DMI_FIN is partitioned into roots, leaves, stems, and clusters based on phenological stage. Final grape yield, expressed as "cluster fresh weight" is obtained by converting the DMI_FIN inflow into "clusters fresh weight" using a constant water dilution factor (water input).

The model can be used to predict the beginning of grapevine susceptibility to DM, because stomata are not functional in leaves before they unfold (Allègre et al., 2007), and unfolded leaves are therefore not susceptible to *P. viticola* infection through stomata (Gessler et al., 2011). The model can also be used to determine the duration of fungicide activity after treatment, because leaf unfolding and leaf area development directly influence the dilution of the fungicide on and inside the host (see Sect. 3.4.1).

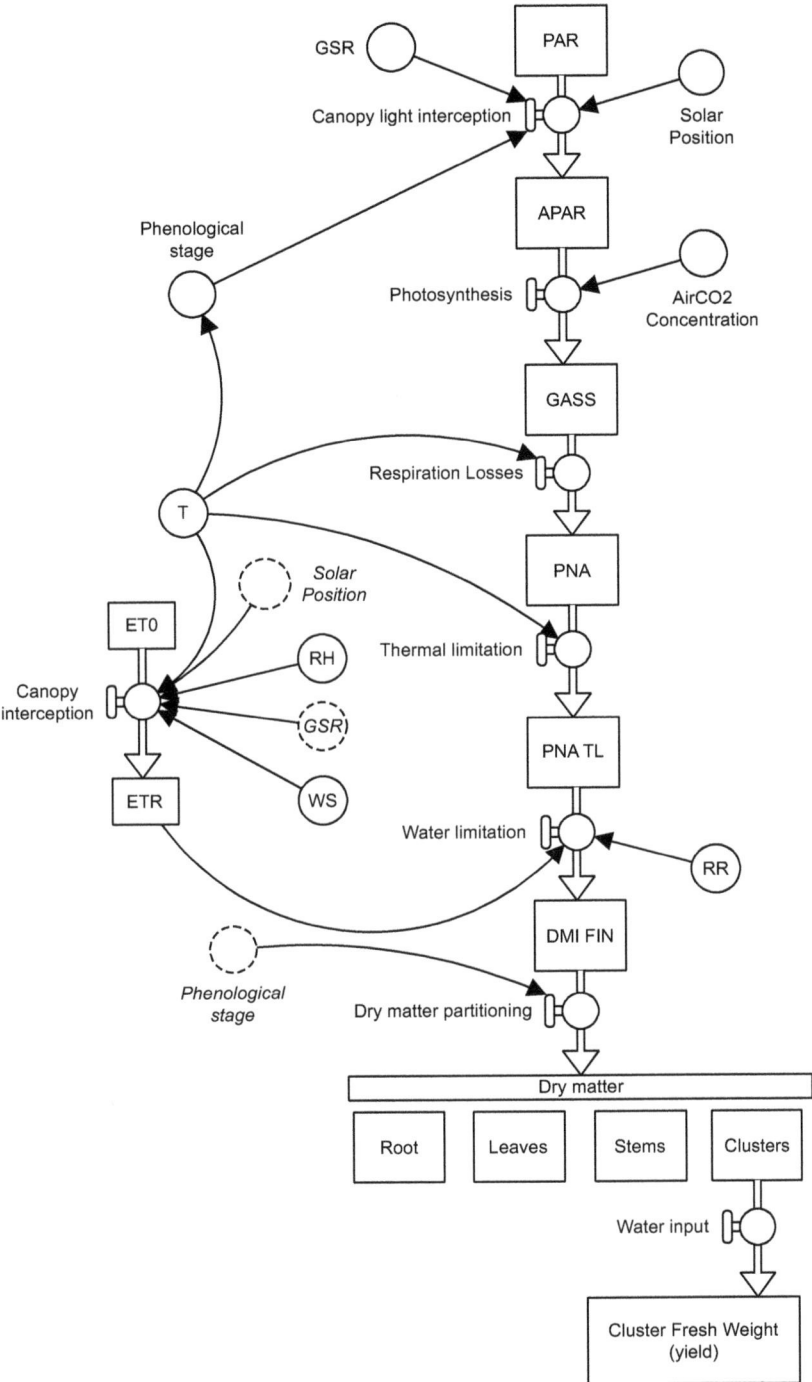

Fig. 6 Simplified version of the model flowchart developed by Cola et al. (2014) describing the cascade of matter triggered by the photosynthetic process. The driving variables inserted in black circles are the weather inputs: global solar radiation (GSR), air temperature (AT), relative humidity (RH), precipitation (RR), and wind speed (WS)

3.4.2 Modelling the Effects of Fungicides

Information on the duration and degree of pre- or post-infection activity (i.e., preventive or curative) or on the eradicant activity of fungicides and how those activities are affected by weather and plant-related variables made it possible to develop models that realistically incorporate the effects of fungicide on the dynamics of plant disease epidemics (Pfender, 2006).

Caffi and Rossi (2018) developed a model that predicts the decline in fungicide efficacy after application based on weather conditions (which affect the drying, rainfastness, and rainfall tenacity of the fungicide) and plant growth and development.

This model incorporates the complexity of the model of Arneson et al. (2002), which includes two compartments: (i) a fungicide concentration compartment, which predicts the dynamics of fungicide residue on the plant tissue on a daily basis after application as influenced by fungicide characteristics, weather conditions, and plant growth, and (ii) a fungicide efficacy compartment, which predicts the effect of the fungicide residue. The daily fungicide residue concentration is estimated based on the following factors: (i) reduction of the initial fungicide concentration, which is a function of the physical mode of fungicide action, i.e., fungicide activity in relation to the host-pathogen interactions (e.g., pre- or post-infection activity) and of fungicide localization on or within the plant (i.e., protectant or penetrant); (ii) rainfall wash-off, which applies only to non-penetrant fungicides and is a function of the amount of precipitation and fungicide susceptibility to wash-off by rainfall; and (iii) dilution due to plant growth, which is a function of plant biomass and is estimated separately for non-penetrant and penetrant fungicides.

3.4.3 Combining Models

Model outputs can be combined by expert systems, i.e., software that is developed for a particular purpose and that provides solutions similar to those that would be provided by human experts (Gonzalez-Dominguez et al., 2016; Patterson, 2004). In crop protection, expert systems are used mainly to help growers to make correct and objective decisions, regardless of their level of expertise.

The fuzzy control system (FCS) developed by Gonzalez-Dominguez et al. (2016) is an example of an expert system for the sustainable management of vineyards. With the information provided by the previously mentioned models that predict grapevine phenology, the risk of *P. viticola* infection, and residual fungicide protection, the FCS was able to reproduce expert reasoning regarding the decision to apply a fungicide against *P. viticola* in a vineyard. The FCS was tested by comparing the scheduling of copper fungicides against *P. viticola* in 18 organic vineyards of Italy determined by a panel of 5 experts vs. the FCS. The FCS was able to reproduce the expert reasoning with an overall accuracy (on a scale from 0 to 1) of 0.992.

4   Conclusions and Perspectives

The rational control of organisms that harm plants is a goal of IPM and is fundamental for ensuring agricultural productivity while maintaining economic and environmental sustainability. The high degree of complexity of making decisions linked
to IPM requires careful evaluation of the benefits and costs, both economic and
environmental, associated with each management action. Population dynamics and
epidemiological models predicting the dynamics of harmful organisms have
assumed a key role in supporting decision-making (at strategic, tactical, and operational levels) in pest management. The advantages of using models in IPM are
linked to their ability to process and analyze complex information and to provide
output supporting the decision-making process.

   Models often concentrate on a single problem (e.g., a single disease), but farmers
need to cope with a broad range of problems (e.g., multiple pests and their interactions with cropping practices) (Magarey et al. 2002). Decision-making for the IPM
of plant disease depends on several risk factors: (i) the risk of disease or infection, as
indicated by plant disease models; (ii) susceptibility of the host to the disease, as
indicated by plant growth models; and (iii) the residual efficacy of previous fungicide applications, as indicated by fungicide models. The use of an FCS (Gonzalez-
Dominguez et al., 2016) allows the integration of all of these risk factors. It uses
information provided by multiple models (including those concerning the host, the
pathogen, and management) to determine whether a fungicide application is needed
to control a disease in the field; it expresses combinations of the risk algorithms as
"if-then" rules, resulting in simple output such as "treatment or no-treatment."

   Once it has been determined that a plant protection product should be applied, and
once the plant protection product is selected, farmers need to know how much and
when the product should be applied. Systems for defining the product dose and when
environmental conditions are suitable for application are therefore needed to complete the decision-making process (Caffi & Rossi, 2018). For these purposes,
methods for crop-adapted spray applications (Gil & Escolà, 2009; Walklate &
Cross, 2010) and for choosing the best moment of the day to apply a treatment
(Bouma, 2003) have been developed.

   A multi-modelling approach and the use of expert systems have been
implemented into decision support systems (DSSs) to help farmers solve complex
problems while reducing the time and resources needed for analyzing the available
information and for selecting the best solution (Narayana Reddy & Rao, 1995).
Advances in information and communication technologies have made it possible to
incorporate models into DSSs and to deliver these systems to farmers. For implementation of IPM, farmers must have access to reliable, user-friendly, accessible,
and up-to-date decision tools that are adaptable to the farmer's specific conditions
(Rossi et al., 2019).

   One example of a widely used DSS is vite.net® (Rossi et al., 2014), which is
intended for vineyard managers. The DSS has two main components: (i) an integrated system for real-time monitoring of vineyard data (air temperature and

humidity, soil temperature and moisture content, plants, pests, and diseases); and (ii) a web-based tool that uses advanced modelling techniques to analyze these data and then provides up-to-date information for vineyard management as alerts and decision supports; the decision about fungicide application remains with the users. This DSS has the following characteristics: (i) holistic treatment of crop management problems (including pests, diseases, fertilization, canopy management, and irrigation); (ii) conversion of complex decision processes into simple and easy-to-understand "decision supports"; (iii) easy and rapid access via the Internet; and (iv) two-way (push-and-pull) communication between users and providers that make it possible to consider context-specific information. The use of the DSS has made it possible to maintain yields and grape quality while reducing pesticide usage by as much as 50%. The DSSs increase farmer profits (Pertot et al., 2017) and enhance the environmental and social sustainability of grape production systems (Metral et al., 2013).

Conflict of Interest The authors declare that there are no conflicts of interest.

References

Allègre, M., Daire, X., Héloir, M. C., Trouvelot, S., Mercier, L., Adrian, M., & Pugin, A. (2007). Stomatal deregulation in *Plasmopara viticola*-infected grapevine leaves. *The New Phytologist*. https://doi.org/10.1111/j.1469-8137.2006.01959.x

Anselmi, A., & Legler, S. E. (2021). Un'app innovative per la taratura delle irroratrici. *Informatore Agrario, 2*, 46.

Antman, E. M., Lau, J., Kupelnick, B., Mosteller, F., & Chalmers, T. C. (1992). A comparison of results of meta-analyses of randomized control trials and recommendations of clinical experts: Treatments for myocardial infarction. *Journal of the American Medical Association*. https://doi.org/10.1001/jama.1992.03490020088036

Arneson, P.A., Oren, T.R., Loria, R., Jenkins, J.R., Goodman, E.D., Cooper, W.E., 2002. Applescab Model Description. www.scientificsocieties.org/APS/AppleScab/Applescab_Manual.pdf.

Baldacci, E. (1947). Epifitie di *Plasmopara viticola* (1941–46) nell'Oltrepó Pavese ed adozione del calendario di incubazione come strumento di lotta. *Atti Ist Bot Lab Crittogam, VIII*, 45–85.

Blaeser, M., & Weltzien, H. C. (1979). Epidemiologische studien an *Plasmopara viticola* zur ve rbesserung der spritzterminbestimmung. Zeitschrift fur Pflanzenkrankheiten und Pflanzenschutz. *Journal of Plant Diseases and Protection, 86*, 489–498.

Bouma, E. (2003). GEWIS, a weather-based decision support system for timing the application of plant protection products. *Bulletin OEPP/EPPO Bulletin, 33*, 483–488.

Bove, F., & Rossi, V. (2020). Components of partial resistance to *Plasmopara viticola* enable complete phenotypic characterization of grapevine varieties. *Scientific Reports*. https://doi.org/10.1038/s41598-020-57482-0

Bove, F., Savary, S., Willocquet, L., & Rossi, V. (2020a). Designing a modelling structure for the grapevine downy mildew pathosystem. *European Journal of Plant Pathology, 157*, 251–268. https://doi.org/10.1007/s10658-020-01974-2

Bove, F., Savary, S., Willocquet, L., & Rossi, V. (2020b). Simulation of potential epidemics of downy mildew of grapevine in different scenarios of disease conduciveness. *European Journal of Plant Pathology, 158*, 599–614. https://doi.org/10.1007/s10658-020-02085-8

Bove, F., Savary, S., Willocquet, L., & Rossi, V. (2021). Modelling the effect of partial resistance on epidemics of downy mildew of grapevine. *European Journal of Plant Pathology*. https://doi.org/10.1007/s10658-021-02367-9

Brischetto, C., Bove, F., Languasco, L., & Rossi, V. (2020). Can spore sampler data be used to predict *Plasmopara viticola* infection in vineyards? *Frontiers in Plant Science, 11*, 1–12. https://doi.org/10.3389/fpls.2020.01187

Butt, D. J., & Jeger, M. J. (1985). The practical implementation of models in crop disease management. In C. A. Gilligan (Ed.), *Mathematical modelling of crop disease* (pp. 207–230). Academic.

Caffi, T., & Rossi, V. (2018). Fungicide models are key components of multiple modelling approaches for decision-making in crop protection. *Phytopathologia Mediterranea, 57*, 153–169. https://doi.org/10.14601/Phytopathol

Caffi, T., Rossi, V., & Bugiani, R. (2010). Evaluation of a warning system for controlling primary infections of grapevine downy mildew. *Plant Disease, 94*, 709–716. https://doi.org/10.1094/PDIS-94-6-0709

Caffi, T., Gilardi, G., Monchiero, M., & Rossi, V. (2013). Production and release of asexual sporangia in *Plasmopara viticola*. *Phytopathology*. https://doi.org/10.1094/PHYTO-04-12-0082-R

Camase, 1996. Register of agro-ecosystems models DLO research institute for agrobiology and soil fertility, .

Campbell, C. L., & Madden, L. V. (1990). *Introduction to plant disease epidemiology*. Wiley.

Christie, M., Cliffe, A., Dawid, P., & Senn, S. (2011). *Simplicity, complexity and modelling, simplicity, complexity and modelling*. https://doi.org/10.1002/9781119951445

Cola, G., Mariani, L., Salinari, F., Civardi, S., Bernizzoni, F., Gatti, M., & Poni, S. (2014). Description and testing of a weather-based model for predicting phenology, canopy development and source-sink balance in *Vitis vinifera* L. cv. Barbera. *Agricultural and Forest Meteorology, 184*, 117–136. https://doi.org/10.1016/j.agrformet.2013.09.008

Conway, G. R. (1984). *Pest and pathogen control: Strategic, tactical, and policy models*. Wiley.

Cullen, A., & Frey, C. (1999). *Probabilistic techniques in exposure assessment, in*. Plenum Press.

De Vallavieille-Pope, C., Giosue, S., Munk, L., Newton, A. C., Niks, R. E., Ostergard, H., Pons-Kuhnemann, J., Rossi, V., & Sache, I. (2000). Assessment of epidemiological parameters and their use in epidemiological and forecasting models of cereal airborne diseases. *Agronomie*. https://doi.org/10.1051/agro:2000171

de Wit, C. T. (1993). Philosophy and terminology. In P. A. Leffelaar (Ed.), *On system analysis and simulation of ecological processes* (pp. 3–9). Kluwer Academic Publishers. https://doi.org/10.1007/978-94-011-4814-6_1

De Wolf, E. D., & Isard, S. A. (2008). Disease cycle approach to plant disease prediction. *Annual Review of Phytopathology, 45*, 203–220. https://doi.org/10.1146/annurev.phyto.44.070505.143329

Edminister, E. D. (1978). *Concepts for using modeling as a research tool* (Technical ed). U.-S. Department of Agriculture.

Ehler, L. E. (2006). Integrated pest management (IPM): Definition, historical development and implementation, and the other IPM. *Pest Management Science*. https://doi.org/10.1002/ps.1247

Fedele, G., Bove, F., González-Domínguez, E., & Rossi, V. (2020). A generic model accounting for the interactions among pathogens, host plants, biocontrol agents, and the environment, with parametrization for *Botrytis cinerea* on grapevines. *Agronomy, 10*. https://doi.org/10.3390/agronomy10020222

Forrester, J. W. (1961). *Industrial dynamics*. M. I. T.Press.

Forrester, J. W. (1997). Industrial dynamics. *The Journal of the Operational Research Society, 48*, 1037–1041. https://doi.org/10.1057/palgrave.jors.2600946

Frey, H. C., & Patil, S. R. (2002). *Identification and review of sensitivity analysis methods, in: Risk analysis*. https://doi.org/10.1111/0272-4332.00039

Fry, W. E., & Fohner, G. R. (1985). Construction of prediction models I: Forecasting disease development. In C. A. Gilligan (Ed.), *Advances in plant pathology* (pp. 161–178). Academic.

Gessler, C., Pertot, I., & Perazzolli, M. (2011). *Plasmopara viticola: A review of knowledge on downy mildew of grapevine and effective disease management*. Phytopathol. https://doi.org/10.14601/Phytopathol_Mediterr-9360

Gil, E., & Escolà, A. (2009). Design of a decision support method to determine volume rate for vineyard spraying. *Applied Engineering in Agriculture, 25,* 145–151.

Gilligan, C. A. (1985). Introduction. In C. A. Gilligan (Ed.), *Advances in plant pathology* (Mathematical modelling of crop disease) (Vol. 3, pp. 1–10). Academic.

Gonzalez-Dominguez, E., Caffi, T., Bodini, A., Galbusera, L., & Rossi, V. (2016). A fuzzy control system for decision-making about fungicide applications against grape downy mildew. *European Journal of Plant Pathology, 144,* 763–772. https://doi.org/10.1007/s10658-015-0781-x

González-Domínguez, E., Fedele, G., Salinari, F., & Rossi, V. (2020). A general model for the effect of crop management on plant disease epidemics at different scales of complexity. *Agronomy, 10.* https://doi.org/10.3390/agronomy10040462

Isee Systems, Inc. (2005). *STELLA. System thinking for education and research.* Available online: https://www.iseesystems.com/

Kranz, J., & Hau, B. (1980). Systems analysis in epidemiology. *Annual Review of Phytopathology.* https://doi.org/10.1146/annurev.py.18.090180.000435

Kranz, J., & Royle, D. (1978). Perspectives in mathematical modelling of plant disease epidemics. In P. Scott & A. Bainbridge (Eds.), *Plant disease epidemiology* (pp. 111–120). Blackwell Scientific Publications.

Krause, R. A., & Massie, L. B. (1975). Predictive systems: Modern approaches to disease control. *Annual Review of Phytopathology, 13,* 31–47.

Lehoczky, J. (1990). Statistical methods. In D. Heyvan & I. M. Sobe (Eds.), *Stochastic models.* Elsevier Science Publishers.

Madden, L. V., Ellis, M.A. (1988). How to develop plant disease forecasters, In: *Experimental techniques in plant disease epidemiology.* https://doi.org/10.1007/978-3-642-95534-1_14

Madden, L. V, Hughes, G., van den Bosch, F., 2007. *The study of plant disease epidemics.* The American Phytopathological Society. https://doi.org/10.1094/9780890545058

Magarey, R. D., Travis, J. W., Russo, J. M., Seem, R. C., & Magarey, P. A. (2002). Decision support systems: Quenching the thirst. *Plant Disease, 86*(1), 4–14.

March, J. G. (1994). *Primer on decision making: How decisions happen.* Simon and Schuster.

McCown, R. L. (2002). Changing systems for supporting farmers' decisions: Problems, paradigms, and prospects. *Agricultural Systems, 74,* 179–220.

Metral, R., Gary, C., Fortino, G., Delière, L., Hoffmann, C., Per- tot, I., Colombini, A., Mugnai, L., Duso, C., Rossi, V., & Caffi, T. (2013). Ex-post assessment of IPM solutions tested in experimental stations and farms and updates of database of alternatives to pesticides and IPM solutions. In *Deliverable 6.1, project PURE "pesticide use-and-risk reduction in European farming systems with integrated Pest management" (FP7–265865).* Available at http://www.pure-ipm.eu/node/319. (accessed on 02/01/17) (p. 85).

Mills, W.D. (1944). Efficient use of sulfur dusts and sprays during rain to control apple scab. *Cornell Extension Bulletin, 630,* 4.

Mulligan, M., & Wainwright, J. (2013). Modelling and model building. In *Environmental modelling: Finding simplicity complex* (2nd ed., pp. 7–26). https://doi.org/10.1002/9781118351475.ch2

Mulrow, C. D. (1994). Systematic reviews: Rationale for systematic reviews. *British Medical Journal, 309,* 597. https://doi.org/10.1136/bmj.309.6954.597

Narayana Reddy, M., & Rao, N. H. (1995). *GIS-based decision support Systems in Agriculture.* National Academy of Agricultural Research Management.

Norton, G. A., Holt, J., & Mumford, J. D. (1993). Introduction to pest models. In G. A. Norton & J. D. Mumford (Eds.), *Decision tools for Pest management* (pp. 89–99). CAB International.

Nutter, F. W. (2007). The role of plant disease epidemiology in developing successful integrated disease management programs. In: *General concepts in Integrated Pest and Disease Management*. https://doi.org/10.1007/978-1-4020-6061-8_3

Oxman, A. D., & Guyatt, G. H. (1993). The science of reviewing research. *Annals of the New York Academy of Sciences*. https://doi.org/10.1111/j.1749-6632.1993.tb26342.x

Papastamati, K., & Van Den Bosch, F. (2007). The sensitivity of the epidemic growth rate to weather variables, with an application to yellow rust on wheat. *Phytopathology*. https://doi.org/10.1094/PHYTO-97-2-0202

Park, E. W., Seem, R. C., Gadoury, D. M., & Pearson, R. C. (1997). DMCAST: A prediction model for grape downy mildew development. *Viticulture and Enology Science, 52*, 182–189.

Pascual, P., Stiber, N., & Sunderland, E. (2003). *Draft guidance on the development, evaluation, and application of regulatory environmental models*. US Environmental Protection Agency.

Patterson, D. W. (2004). *Introduction to artificial intelligence and expert systems*. Prentice-Hall.

Pertot, I., Caffi, T., Rossi, V., Mugnai, L., Hoffmann, C., Grando, M. S., Gary, C., Lafond, D., Duso, C., Thiery, D., et al. (2017). A critical review of plant protection tools for reducing pesticide use on grapevine and new perspectives for the implementation of IPM in viticulture. *Crop Protection, 97*, 70–84.

Pfender, W. F. (2006). Interaction of fungicide physical modes of action and plant phenology in control of stem rust of perennial ryegrass grown for seed. *Plant Disease*. https://doi.org/10.1094/PD-90-1225

Plant, R. E., & Mangel, M. (1987). Modeling and simulation in agricultural pest management. *SIAM Review*. https://doi.org/10.1137/1029043

Power, D. J. (2007). A brief history of decision support systems. *Decision Support Systems*.

Rabbinge, R., & de Wit, C. T. (1989). Systems, model and simulation. *Tetrahedron Letters, 23*, 4461–4464.

Renard, P., Alcolea, A., & Ginsbourger, D. (2013). Stochastic versus deterministic approaches. In *Environmental modelling: Finding simplicity in complexity* (2nd edn). https://doi.org/10.1002/9781118351475.ch8

Roelfs, A. P. (1985). *Wheat and rye stem rust, in: Diseases, distribution, epidemiology, and control* (pp. 3–37). Academic.

Rossi, V., Caffi, T., Giosuè, S., & Bugiani, R. (2008). A mechanistic model simulating primary infections of downy mildew in grapevine. *Ecological Modelling*. https://doi.org/10.1016/j.ecolmodel.2007.10.046

Rossi, V., Giosuè, S., & Caffi, T. (2010). Modelling plant diseases for decision making in crop protection. *Precision Crop Protection – The Challenge and Use of Heterogeneity, 241*. https://doi.org/10.1007/978-90-481-9277-9

Rossi, V., Caffi, T., & Salinari, F. (2012). Helping farmers face the increasing complexity of decision-making for crop protection. *Phytopathologia Mediterranea, 51*, 457–479. https://doi.org/10.14601/Phytopathol_Mediterr-11038

Rossi, V., Salinari, F., Poni, S., Caffi, T., & Bettati, T. (2014). Addressing the implementation problem in agricultural decision support systems: The example of vite.net®. *Computers and Electronics in Agriculture, 100*, 88–99. https://doi.org/10.1016/j.compag.2013.10.011

Rossi, V., Sperandio, G., Caffi, T., Simonetto, A., & Gilioli, G. (2019). Critical success factors for the adoption of decision tools in IPM. *Agronomy*. https://doi.org/10.3390/agronomy9110710

Salotti, I. (2021). *Development and validation of a mechanistic, weather-based model for predicting Puccinia graminis f. sp. tritici infections and stem rust development in wheat*. Doctoral thesis: "Development of epidemiological models for wheat and legumes in crop rotation".

Saltelli, A., Tarantola, S., & Campolongo, F. (2000). Sensitivity analysis as an ingredient of modeling. *Statistical Science*. https://doi.org/10.1214/ss/1009213004

Savary, S., & Willocquet, L. (2014). Simulation modeling in botanical epidemiology and crop loss analysis. *Plant Health Instructor*. https://doi.org/10.1094/phi-a-2014-0314-01

Segarra, J., Jeger, M. J., & Van den Bosch, F. (2001). Epidemic dynamics and patterns of plant diseases. *Phytopathology*. https://doi.org/10.1094/PHYTO.2001.91.10.1001

Shoemaker, R. A. (1981). Changes in taxonomy and nomenclature of important genera of plant pathogens. *Annual Review of Phytopathology*. https://doi.org/10.1146/annurev.py.19.090181. 001501

Sonka, S. T., Bauer, M. E., & Cherry, E. T. (1997). *Precision agriculture in the 21st century: Geospatial and information technologies in crop management.* National Academy Press.

Sutherst, R. (1993). Role of modelling in sustainable pest management. In S. Corey, D. Dall, & W. Milne (Eds.), *Pest control and sustainable agriculture* (pp. 66–71). CSIRO.

Teng, P. S. (1981). Validation of computer models of plant disease epidemics: A review of philosophy and methodology/Zuverlässigkeit von Computermodellen für Epidemien von Pflanzenkrankheiten: Ein Überblick über Grundgedanken und Methodik. *Zeitschrift für Pflanzenkrankheiten und Pflanzenschutz/Journal of Plant Diseases and Protection, 88*, 49.

Teng, P. S. (1985). A comparison of simulation approaches to epidemic modeling. *Annual Review of Phytopathology*. https://doi.org/10.1146/annurev.py.23.090185.002031

Teng, P. S., Blackie, M. J., & Close, R. C. (1980). Simulation of the barley leaf rust epidemic: Structure and validation of BARSIM-I. *Agricultural Systems*. https://doi.org/10.1016/0308-521X(80)90001-3

Van der Plank, J. E. (1960). Analysis of epidemics. In H. JG & C. EB (Eds.), *Plant pathology: An advance treatise* (Vol. 3, pp. 229–289). Academic.

Van der Plank, J. E. (1963). *Plant diseases: Epidemics and control.* Academic.

Van Maanen, A., & Xu, X. M. (2003). Modelling plant disease epidemics. *European Journal of Plant Pathology, 109*, 669–682. https://doi.org/10.1023/A:1026018005613

Waggoner, P. E. (1960). Forecasting epidemics. In H. JG & D. AE (Eds.), *Plant pathology, an advanced treatise* (pp. 291–313III). Academic.

Waggoner, P. E., & Horsfall, J. G. (1969). E.P.I.D.E.M.: A simulator of plant disease written for a computer. *Connecticut Agricultural Experiment Station Bulletin USA, 698*.

Waggoner, P. E., Horsfall, J. G., & Lukens, R. J. (1969). *A simulator of southern corn leaf blight* (Bulletin of the Connecticut agricultural). Wiley.

Wainwright, J., & Mulligan, M. (2004). *Environmental modeling finding simplicity in complexity.* Wiley.

Walklate, P. J., & Cross, J. V. (2010). A webpage calculator for dose rate adjustment of orchard spraying products. *Aspects of Applied Biology, 99*, 359–366.

Xu, X. M., & Ridout, M. S. (1996). Analysis of disease incidence data using a stochastic spatial-temporal simulation model. *Aspects of Applied Biology, 46*, 155–158.

Yuen, J. E., & Hughes, G. (2002). Bayesian analysis of plant disease prediction. *Plant Pathology.* https://doi.org/10.1046/j.0032-0862.2002.00741.x

Zadoks, J. C. (1971). Systems analysis and the dynamics of epidemics. *Phytopathology, 61*, 600–610.

Zadoks, J. C., & Rabbinge, R. (1985). Modelling to a purposed. In C. A. Gilligan (Ed.), *Advances in plant pathology* (Mathematical modelling of crop diseases) (Vol. 3, pp. 231–244). Academic.

Development and Adoption of Model-Based Practices in Precision Agriculture

Jotham Akaka, Aurora García-Gallego, Nikolaos Georgantzis, Clive Rahn, and Jean-Christian Tisserand

Abstract Decision-making in agriculture becomes increasingly challenging as farmers seek to meet agronomic, environmental and compliance goals. There is an increasing amount of data available to help reach these goals, and decision-making can be supported by computer-based DSSs. Unfortunately, not all DSS are adopted or contribute to efficient farming. The reasons for non-adoption are not easy to find, as such failures are often not reported in the literature, hindering the process of learning from our mistakes. In this chapter we attempt to identify some of the causes of failure and propose strategies to overcome them. This involves a mix of many years of experience in model development, improving fertilizer use and more recent studies from social sciences on adoption. It is challenging for natural and social scientists to work together towards understanding the processes in creating model-based solutions in precision farming. In the past, social scientists have concentrated on the behavioural reasons for (non)adoption of model-based practices (MBPs), but it is impossible to improve the usability of even the best scientific model as it is too late after it has been developed. Natural scientists have also developed excellent research models of no use to practitioners. To avoid such mismatches, all stake-holders, particularly potential users, must be involved from the beginning and at all stages in the process. Simultaneously, natural and social scientists must work together to create models suitable for the farmer. In conclusion, co-creation is important and users need to be involved from the start. A model must be designed to take account of stakeholder needs, address relevant questions and provide appropriate answers with minimal data requirements in a user-friendly way. Models must identify and respond to the social and individual-level phenomena which lead to a number of well-documented biases that influence behaviour and attitudes.

J. Akaka (✉) · A. García-Gallego
Department of Economics, Universitat Jaume I, Castelló de la Plana, Spain

N. Georgantzis · J.-C. Tisserand
Burgundy School of Business, Lyon, France

C. Rahn
Crop Centre, School of Life Sciences – University of Warwick, Coventry, UK

© The Author(s), under exclusive license to Springer Nature Switzerland AG 2023
D. Cammarano et al. (eds.), *Precision Agriculture: Modelling*, Progress in Precision
Agriculture, https://doi.org/10.1007/978-3-031-15258-0_4

1   Introduction

Technological progress, efficient use of knowledge and advanced data analysis have led to the development of many informatics applications to assist users in many sectors of society in their everyday decision-making. Examples are GPS (global positioning system)-assisted route choice, leisure choice, consumption and production. However, these tools are not used extensively in agriculture. Explanations sought for this limited use can be classified under two broad categories: solution-specific and user-specific. The existing approaches of identifying factors affecting the adoption of model-based practices (MBPs) have often been one-sided, mainly due to a lack of effective interaction between researchers with a technical focus and, on the other hand, social scientists with a human behaviour focus. The solution-focused approaches have not addressed cases of failure sufficiently, due to a natural preference for promoting the use of successful solutions. User-specific approaches have identified various social and behavioural factors that affect the adoption of MBPs. The probable causes of poor adoption are factors that influence decision-making on a personal level, such as behavioural, social and other idiosyncratic factors. We present a case study demonstrating how user behaviour affects the adoption of decision support systems (DSSs) on the farm.

Several authors have advocated that the development of MBPs can be seen as a viable pathway to provide farm managers with evidence-based recommendations and in the long run increase innovation and reduce wastage (Fountas et al., 2005; McCown et al., 2009; Thorburn et al., 2011). Over the last three decades, several MBPs have been developed, but current usage statistics show they are not as effective as predicted in terms of their use in real-life scenarios. For instance, Aubert et al. (2012) showed that work practices are important factors when it comes to adopting innovations, deviating from mainstream theories of technology acceptance and innovation diffusion theories. Often, developers optimize their MBPs to work with specific types of weeds, crops or scenarios, whereas farmers prefer a more polyvalent MBP. In essence, farmers prefer to have a wider view, whereas the MBPs currently present in the market have a narrow scope in terms of their application (Rossi et al., 2014). In other studies, Van Meensel et al. (2012) claim that some MBPs are overwhelmingly complex. Furthermore, their terminology and functions are not only unsuitable but also irrelevant to the end user. In general, MBPs are developed according to the rate of technological advancement instead of responding to existing demand for the product. This mismatch between what farmers want and what is currently in the market is partly explained by common technology acceptance models which focus on the perceived usefulness, but fail to factor in the environment when evaluating the uptake of MBPs. From the examples above, it is seen that there is an obvious gap between research and implementation.

Poor implementation of MBPs also accounts for the low uptake among farmers. This occurs when a tool is not fully utilized due to technical limitations of the technology and user attitude towards the same. In such cases, the tendency not to adopt is due to farmers' perception that the technology is difficult to use and useless. Behavioural factors such as farm characteristics, farmer's age and education level, high costs (Aubert et al., 2012; Pierpaoli et al., 2013) and a steep learning curve (Kutter et al., 2011) influence farmers' intentions to adopt.

Other research also cites inadequate information on how farmers make decisions in the field that limits the adoption of MBP. Research on decision-making processes are largely based on rational assumptions instead of empirical research from data where the farmer makes decisions in their natural setting (Gray et al., 2009). Moreover, in some cases, input from farmers is not considered in the development and implementation stages. Relevant literature cites that participatory-based approaches in the adoption of sustainable agricultural technologies are more beneficial than cases where farmers are not involved in development and implementation of MBPs. Studies that involve farmers making decisions about their daily farming practices show promising results in the usefulness of MBP as a tool for better decision-making (Lindblom et al., 2013).

Lastly, there is a mismatch between the features of MBPs and what farmers want. In many cases, farmers' needs are not considered when developing MBPs. Parker and Sinclair (2001) claim the degree of involvement of farmers in the design and development of MBP can determine their success or failure. Involving stakeholders and social learning is key for the successful development of agricultural MBPs (Van Meensel et al., 2012). Therefore there is a need to develop MBPs that not only give farmers access to the most relevant information but also involve them in the design and implementation phases of MBPs.

Figure 1 illustrates the factors that potentially affect the process of developing models to support farm practices.

In much of the literature, the problems of model adoption have been analysed at the end-user level. At this stage, it is too late to improve the adoption of a poorly developed model. Good models will engage all stakeholders starting at the design stage, so that the role of the tool is relevant and applicable to the final users. One of the most important stakeholders are the funders. Funders are important in the

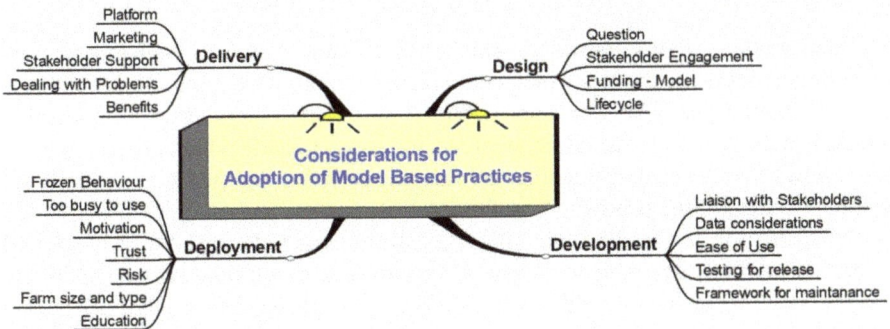

Fig. 1 Relevant factors in every step of developing MBPs. (Rahn, 2004, 2013)

development stage but must also play a part in supporting delivery, deployment and maintenance. Otherwise, the models will have a limited life. This is more challenging in the agricultural sector where the number of users (and thus revenue) is limited compared to wider fields such as accounting packages.

Some models are useful tools to improve scientific understanding where the end users are other scientists. However, the models directed to farmer use must be appropriately designed with end-user involvement at all stages (Woodward et al., 2008) to ensure that their outcomes are relevant and are driven by readily available data. Delivery and deployment must be supported by trusted enthusiasts, such as researchers, and end users that may have been in the original development team but could also be in the cohort of early adopters or initial testers.

This chapter aims to provide insight on factors which affect adoption of MBPs. The aim is to understand which technical and behavioural factors should be considered when developing models for decision support in farming. Box, 1979 said 'all models are wrong but some are useful'. Our aim is to encourage development of useful model-based practices.

In Sect. 2 we give an overview of the literature on MBPs and derive qualitative observations on all phases of the technical development and implementation of MBPs. Likewise, social factors are addressed in Sect. 3. In Sect. 4, behavioural economics is introduced as a method to quantify the influence of various factors that influence adoption or non-adoption of agricultural MBPs. This approach is further described by a case study in Sect. 5. Finally, we draw conclusions in Sect. 6.

2   Design, Development and Delivery of MBPs

2.1   Design

The Problem: Stakeholder Engagement

A model must be well designed to take account of stakeholders' needs to address a relevant question and provide appropriate answers with minimal data requirements in a user-friendly way. Such models need defined objectives, scope and scale.

Models can have diverse purposes which suit different groups of stakeholders. At one extreme there are models which could be used to improve scientific understanding with an end-user group of other scientists and at the other extreme models that would benefit farm practice. Here, we focus on those models which are designed to provide advice on management practices to overcome problems in agricultural activities aimed at the production of food.

Once the problem has been identified, questions need to be asked to refine it, and science needs to be available to address it. In some cases, solutions may be relatively easy to solve by following laid down protocols. Complex problems that depend on varying soil or meteorological factors are more likely to benefit from the development of models. These problems might relate to the control of pests, diseases, weeds and the application of fertilizer. Some models may be on-board farm machinery to make decisions on inputs based on crop sensors as they move across the field. Others may be office-based to provide decisions on farm practices at a field scale. Increasingly, there is potential to use AI (artificial intelligence) and self-learning tools to integrate multiple sources of information to better inform practice (Columbus, 2021).

The decision to devise an MBP could come from a range of stakeholders. Where there are environmental concerns, the sponsoring stakeholder may be the government. For pest, disease and weed control, the sponsors may include government, levy or commercial companies. For any MBP to be effective, end users should be involved at an early stage to ensure their needs and concerns are considered for the greatest chance of adoption. The developers are likely to be researchers in universities or applied research units. Some developers, especially those concerned with software, may be unaware of the practical constraints in agriculture. So, there needs to be a mix of the stakeholders at the design stage to avoid mismatches in development. Developers need to be sure that there is sufficient scientific understanding to support model development. Farmers need to be sure that the relevant question is being asked and that the output decisions are applicable in the field. Mistakes made at this stage are not easy to recover from and could result in poor model uptake.

Scope and Scale: Field/Farm
Models may be used at a field scale or farm scale. Models designed to look at a particular problem such as how much fertilizer should be applied or pest disease or weed control are likely to be at a field scale. Farm-scale models are likely to be used for crop and financial management. Also, they are likely to record and report on farm activities for compliance with food quality and traceability protocols, environmental and ecosystem service audits. Models may be used to decide on immediate field decisions, while others may be run in 'what if' mode to decide on farm management strategies. Additionally there is an increasing demand for farm-scale systems to integrate with blockchain systems involved with traceability and logistics.

The scope of models can be diverse, with some tailored to specific crops, countries or climatic zones. It is important to define the scope of the model at an early stage of the project. The scope of a model may change over time. Williams et al. (2013) commented that the simplifications for pastoral systems in the Overseer model may limit the model's suitability for use in more complex arable and horticultural rotations.

Model Life Cycle: The Funding Model
Funding is vital for the development and sustainability of any DSS. Some policy-based models may be wholly funded by governments. Some large-scale models such as those developed for climate change may get sustained funding, but long-term funding for agricultural models is less likely. Models designed for field- and farm-scale issues may be funded by both government and commercial sources. In all cases, for a model to survive beyond the lifetime of the initial development, thought needs to be given to how the model in use can be supported (Fig. 2)

ALL STEPS NEED FUNDING

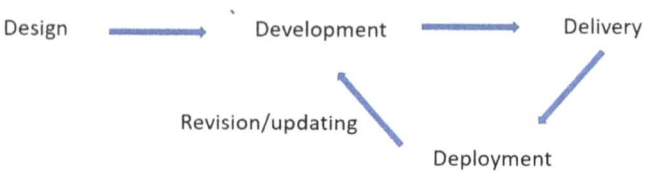

Fig. 2 All stages of the model life cycle need sufficient financial support. (Rahn, 2004, 2013)

Many models have sunk after initial development. Experiences of such failures are rarely published, so there is a danger of repeating the same mistakes. In some cases, models die after key researchers leave full-time employment. Models will need funding to deliver them to the end users as well as support their maintenance as errors in programming or science will generally be found. Science will develop with time, so revisions will need to be funded. Changes in the platform such as the operating system (e.g. Windows, Android or Apple Mac) will also drive the need for revision.

2.2 Development

Throughout the model's life cycle, the project must accommodate stakeholders' inputs (e.g. scientists and potential users) and adapt with developing science to provide timely, effective and efficient management strategies. To achieve this, it is important to identify the main factors for successful development of an agricultural DSS. Good project management brings on board all stakeholders. This sets the stage for developing sound insights into the outputs needed to inform decisions that support timely change in farm practices.

Data
Given the amount of data available, it may be taxing to the modeller to collect, store and analyse this information. The aim is to use the appropriate data to explain a particular situation. Therefore, certain data types, such as field conditions where the crop is grown, soil properties or even meteorological data are required for different decisions. In other cases, the context can be specific to one type of crop. Information on the planting season, pesticide and fertilizer inputs can be collected for certain crops. The multiplicity of farming contexts implies that different models are needed for different scenarios. Perhaps, a significant drawback of having many independent models is the risk that similar data may need to be entered several times to suit the needs of each model. Technologies such as integrated pesticide management (IPM) allow for a combination of several MBPs for different pests across multiple geographical locations. This allows for better management and organization; fast

loading speeds; ease in implementing changes and updates; and better support. The data could also be divided according to the context, such as the field in which the crop is sown, information on soil properties and meteorological data. Other crop-specific data can include sowing dates, pesticide and fertilizer inputs. Conventions such as the EU-funded project IPM Decisions (https://www.ipmdecisions.net/) facilitate better integration of inputs and outputs of models developed to provide decision support. In the future, field-scale models will need to deliver data to other parts of the management system responsible for environmental and ecosystem service audits.

Meteorological data may be sourced from the internet or from localized weather stations. Some DSS tools may need to be supported by weather forecasts. It is important that the meteorological data needed to run the models should be available, preferably from local or government sources. However, the acquisition of data might be costly when it is not available for certain locations. Some models may need crop and soil data from sensors or field visits to operate. The way in which data is relayed to the models must be considered at an early stage to increase their ease of use.

Ease of Use
According to the technology acceptance model (TAM) (Davis, 1989), the perceived ease of use and usefulness have a direct influence on individual behaviour through the intention to adopt technologies. The perceived ease of use determines the perceived usefulness because the easier it is to use a technology, the greater the perceived usefulness. In other words, the perceived ease of use is the subjective evaluation of effort required to use the system. Other research on behavioural decision-making shows that individuals try to minimize the effort used when performing tasks (Payne et al., 1993). Systems that are easy to use are useful and therefore more likely to be adopted.

Past research in the fields of information systems and psychology shows the importance of individual beliefs about computerized systems on the perceived ease of use (Davis, 1989; Doll & Ajzen, 1992). In behavioural decision theory, anchoring and adjustment heuristics influence the perceived ease of use and the adoption. Individuals often use general information (the anchor) when there is no specific knowledge. When new information becomes available, the decision is adjusted accordingly, but they still rely on this anchor. Based on these findings, it is expected that potential users anchor their perception of the ease of use of a novel system on the basis of their beliefs about computers and computerized systems. However, as they become familiar with the system, their perception of the perceived ease is adjusted according to the experience gained from direct use. In addition to improving computer system design characteristics to increase perceived ease of use, it is equally important to consider the importance of individual behaviour, heuristics and experience.

Ideally, all the individual models integrate into a farm system model, which logs activities carried out at the field scale. This data is useful in information audit for regulatory, ecosystem service inputs and feedback on MBPs to improve decision-making. Then, it may be possible for farm management models to support various individual models by integrating them into the whole farm system. Additionally, there is increasing demand for integrating farm and blockchain systems involved with traceability of food and logistics.

Such frameworks allow the addition of new tools into the system as new problems and science develop. It is common for individual companies to have their own in-house styles that are important for their competitive advantage, which may pose a problem. For example, Windows, Android and Apple operating systems are tailored for different market segments and are rarely interoperable.

The design of the interface is critically important. The ability to enter data on a single screen without having to switch between multiple windows is valuable. The ability of users to review input data and correct any errors before model runs is desirable. If the process of data entry involves multiple steps, repetition or contains hidden menus, it is likely to be difficult to use.

Output screens should be easy to read and interpret, possibly having a simpler first screen but also provide further information on the inputs required as an option. Thought should be given on how the outputs are communicated to managers and farm equipment. For instance, disease or pest models must be able to signal warnings on crops that need to be sprayed. Others may need output that can be read by other models or tools on the farm. For example, if a model recommends that a crop be sprayed, it is helpful to identify appropriate pesticides for use. Afterwards, an automatic order can be placed with a supplier. A job note could be raised informing contractors or staff of pending work to do. Finally, after the job has been done, a record is filed for the cropping record on the whole farm system as part of the traceability needed by the consumers of the harvested product.

Simply updating a model does not always mean that it is easier to use. The DOS (disk operating system) version of WELL_N (Rahn et al., 1996), a fertilizer prediction model for field vegetables, was easier to use than the later Windows versions. The latter was easier on the eye but involved data entry into multiple windows, difficult to review and correct before the model ran. In this instance, there was insufficient contact between the programmers and users to rectify this problem. Additionally, the inclusion of hints and context-specific help screens can help the users to operate the model. The provision of contact details for help and assistance from a real person via a helpline or email address is useful. Such contact also provides developers with information for debugging and introducing new features in future revisions.

In any computerized system, it is important to understand and incorporate user needs in the development of the system. This has a long-term impact on the success and continuous use of many computerized systems. Ji et al. (2018) showed that customising the user interface according to the user's cognitive response and behaviour significantly improved usability and satisfaction among 200 users.

If it takes more than a few attempts to run a new model, the motivation to persevere with it and the enthusiasm to use similar technology in the future decline. Again, the involvement of the end users at an early design stage is necessary to develop a user interface that is 'fit for purpose'. Additionally, during development, panels and workshops with potential users can identify differences between what is offered and end user needs.

Testing for Release

Modelling is an intellectually rewarding endeavour. Sharing the outcome of this work will encourage debate and may lead to further development of the model. Models that will be used for decision support must be well-tested before they are used in practice. No model is ever going to be completely free of errors, so when is it good enough to be released? Project deadlines may push one to release a model earlier. However, is it better to release a beta version early, or a fully tested version later, given that 'buggy' software may not be trusted by customers? These errors can arise from the software hosting the model or the model itself.

A trial version of the main DSSs can be made available before it is fully adopted (Doss, 2003). A study by Adesina and Zinnah (1993) identified that the perception of the characteristics of modern rice varieties had a significant effect on the decision to adopt them. It is vital that before the introduction of any new technology to farmers, they should be involved in the evaluation to determine how suitable the technology is to their circumstances (Sinja et al., 2004).

Model-based software may provide incorrect results, broadly speaking, for two reasons. First, the model itself may be an inaccurate representation of the decision problem. Second, the software hosting the model could contain errors. Statistical tests can assess models for robustness and complexity (Smith et al., 1996). However, a model may still be useful even if it fails some of these tests. Absolutely precise models are rare because of limitations in the field. For instance, fertilizer application rates that are smaller than can be practically applied are not useful.

Models can be inaccurate in different ways. For example, they can be biased: simulated values are systematically higher (or lower) than observed values (Fig. 3a). This suggests that the model captures the important processes but has been parameterized incorrectly. Models can be adrift in predicting the timing of certain events (Fig.3b), for example peaks of simulated N mineralization, or pest populations. A pest model that predicts high population density ahead of observed peaks in

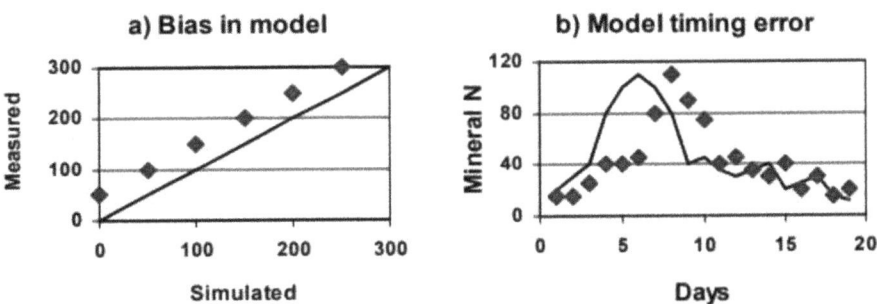

Fig. 3 Examples of failures in model prediction (Rahn, 2004). Simulated = solid line, measured = point (**a**) Bias in a model $r^2 = 1$. (**b**) Model showing timing error $r^2 = 0.25$

population density would be wrong in theory, but such a model would still provide a benefit if correcting actions could be applied in the field to avoid peak infestation.

A DSS should be determined to be suitable, i.e. simple to operate and give advice that is beneficial. The above should be achieved without data requirements that cannot be met by practitioners. The research model underlying WELL_N (Rahn et al., 1996) was circulated and tested in the field by an advisory body ADAS (Agricultural Development and Advisory Service) before the decision support system itself was released. During testing, the DSS was operated by experienced advisors, which allowed errors to be identified and reported to the programming team. A standard dataset was used to test subsequent versions of the model. A wide range of cropping scenarios was generated to identify errors. It was discovered that the biggest problems were related to programming errors rather than to model concepts.

Sensitivity analysis can be used to identify inputs and parameters that have profound effects on model outputs (de Wit, 1982; Rahn et al., 2001). In the case of WELL_N, nitrogen recommendation was identified. Sensitivity analysis is an important element of testing the model described above.

2.3   Delivery

It's pointless developing models that improve farm practice if they are not delivered effectively. At the design stage, having some insight into the delivery strategy is beneficial. Initial stakeholders' and early adopters' support in the delivery of MBPs is vital.

Additionally, the five principles of marketing and other aspects of relational marketing (i.e. the relationship between the developers and users) also apply (Alajoutsijärvi & Tikkanen, 2000). Importantly, it is all about marketing the benefits of using the MBPs and not the practices/tools themselves.

Product: Benefits
A vital question for the stakeholder is whether using the model provides a benefit. In nitrogen modelling, the most obvious benefits to consider are a reduction in the amount (and cost) of fertilizer used as well as a reduction in environmental pollution. Even though such benefits are important to scientists, they might be insignificant to growers. Specifically, does the business become more competitive? In the end, a model will only be used if it adds value to the production process as a whole. The 'value chain' (Porter, 1985) in Fig. 4 shows various areas where value can be added and where a higher profit margin can be achieved. A grower may use a model to demonstrate compliance with good agricultural practice, thereby boost marketing of the produce and thereby increase the profit margin. Using a model can also improve operations and logistics by optimizing purchase and use of fertilizers. Quality and storability (i.e., value) of produce may be increased by targeted fertilizer

Fig. 4 Porter's (1985) value chain – areas where contributions can be made to improve margin

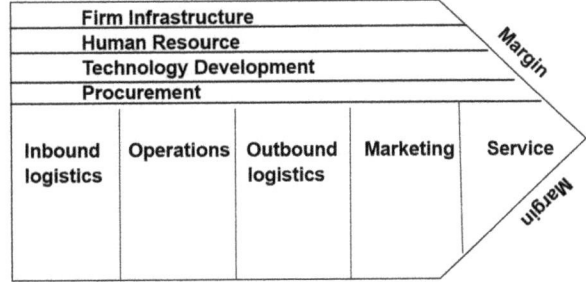

applications. If the benefits of using DSSs can be demonstrated, it is likely that stakeholders will be interested (Rahn, 2004).

Price

There are different strategies of pricing DSSs. Some can be free; some may be provided as part of a package with a commitment to purchase pesticides and fertilizers or sold independently at a premium. However, in specialized industries, such as horticulture, the chances of recovering the cost of development are small. Nonetheless, it is possible to recover the costs of maintenance and revision. Regardless, if the price exceeds the perceived benefits, it is likely the DSS will not be purchased. A subscription model which involves online support may be more beneficial. There are many textbooks that go into this subject area in great detail, but our objective is to highlight how pricing strategies affect adoption and longevity of DSSs.

Placement: Where to Release

Another important question is where the model might be placed. The DSSs can be distributed in many places, for example onboard computers in tractors, phones, on the cloud, etc. Each option has its advantages and disadvantages. In cloud-based distribution, the system is easier to control and upgrade but relies on users having reliable Internet that is more accessible in the office than in the field. Other complex models may need to be run by consultants on behalf of the farmers as part of a farm management contract. Pest and disease models may benefit from centralised systems as they can easily access national monitoring data. The need to use data from yield, soil, moisture and fertility maps dictates the appropriate platform for a particular model.

The location of data has implications on security and run time. Offline systems can run without an Internet connection. However, distribution, upgrading and updating can be difficult. They are also prone to problems on computers that might stop the models from operating and are related to specific machines where user support is important.

Promotion and People: Getting the Message Across

It is important to identify the appropriate mix of methods to promote access and use. Initial stakeholders and early adopters can make a significant contribution to the process. Model champions should be used to spearhead the process. Field-scale demonstration of use and benefits of a model by early adopters will be seen by their peers and neighbours. Publications in the farming press and blogs will all help. Farmers can be invited to workshops and trained to use the models. The mix between group and one-to-one methods will depend on many factors (Rahn, 2013).

An important distinction should be made between advertising/promotion campaigns on the one hand, and informative demonstration sessions on the other. A recent paper by Akaka et al. (2021) has shown that while farmers' trust in a given DSS increases with exposure to demonstration sessions, advertising has the opposite effect, hindering the users' trust.

Key in getting the messages across and helping the users to use the models is contact with the developers by email, phone or blog:

1. To help users run the models.
2. To identify revisions that may need to be made to the model interfaces and core model.

Challenges of Getting the Message Across for Fertilizer Management: An Example in China

China's farming landscape is very different to that in Europe. For example, there are 910,000 small-scale farmers in Beijing alone. Such a large number of clients is hard to reach via individual advisory visits. A project from the School of Oriental and African Studies (SOAS) at the University of London identified a different approach (Smith, 2015). This project assembled a team of social and biological scientists in the UK and China to study what influences farmer behaviour with respect to fertilizer recommendations (Bellarby et al., 2017; Smith et al., 2015). A farmer's neighbours had a greater influence on decisions pertaining to fertilizer use than advisers (including those from fertilizer companies), in *Zea mays*, *Triticum aestivum* and greenhouse vegetable cropping systems. The national recommendation system of China Agriculture University (Zhang et al., 2009) was little used.

As neighbours can be important in the diffusion process, innovators and early adopters can be used to influence small groups of farmers. Figure 5 shows a modified Rogers innovation curve. Innovators and early adopters can be the first growers to test out the new practices, providing demonstrations locally for others to see and follow. On-farm and context-specific demonstrations are crucial for bridging the chasm between early adopters and the early majority.

Early adopters are keen to benefit from a new system and can be persuaded to modify their behaviour by discussion. An advisory service can maintain contact with this small group without undue effort. Early adopters are well aware of the critical constraints to production, and it is advantageous to involve them when designing new recommendation systems. Early adopters also play a role in demonstrating the benefits of a new recommendation system to their neighbours at a meaningful scale. Some practitioners will not be convinced even by such demonstrations and will only use the new system when legislation such as the Nitrate Directive (EU, 1991) in Europe is put in place (Rahn, 2018).

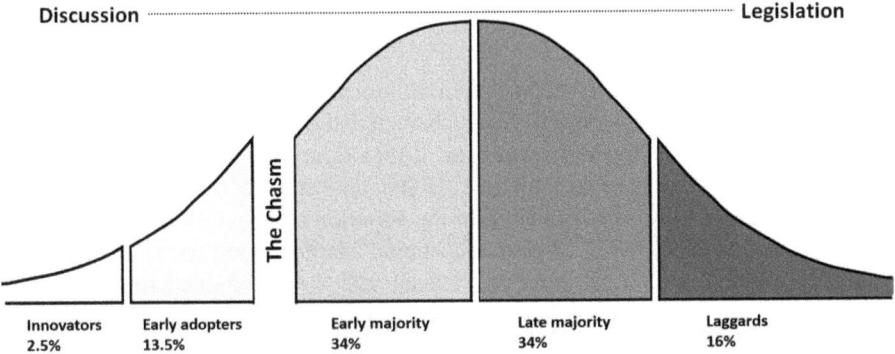

Fig. 5 Modified innovation adoption curve. (Rahn, 2018; Rogers, 1962)

3   Social Factors Affecting Adoption of MBPs

Technical aspects are often blamed for low adoption of MBPs, but it is important to identify how individual attributes contribute to the low adoption of MBPs. Following such concerns, the social determinants of adoption of new farming practices have been extensively documented (Kuehne et al., 2017). This chapter organizes and sorts the adoption factors of farming technology (Feder & Umali, 1993; Kuehne et al., 2017). Conventionally, decision-making process theories identify three main factors: farmer, technological, and institutional characteristics (Meijer et al., 2015). A review by Kumar et al. (2018) named five main factors: the source of information, features of the technology, economic factors, farm characteristics, institutional and sociodemographic factors. We present a revised analytical framework outlining the interactions between end-user characteristics and structural performance features of assessed MBPs in the decision-making process involving technological adoption by users at the farm. Other decision-making processes include a psychological process in addition to the factors identified above. As suggested by Davis (1989) in the technology acceptance model, technology acceptance and usage is determined by the perceived usefulness and the perceived ease of use. Perceived usefulness is utility gained from improving one's work performance due to the use of a certain technology, whereas perceived ease of use is related to the effort required to use a certain technology (Davis, 1989; McDonald et al., 2016).

The decision to adopt new technologies is often seen as an interaction of individual attributes: risk, uncertainty, incomplete information, institutional factors and infrastructure (Uaiene, 2011). Social networks and learning also have an effect on the decision to adopt innovative technologies (Uaiene, 2011). The determinants identified in the literature depend on the technology under investigation and location (Bonabana-Wabbi et al., 2002). In this study, these factors are categorized into demographic, societal and farm characteristics.

3.1 Demographic Factors

Individual characteristics and the social environment have been identified to have a significant influence on adoption. Such characteristics include education, age and gender (Keelan et al., 2009; Mignouna et al., 2011). Education has a positive influence on the decision to adopt new technologies. Higher education levels increase the ability to obtain, process and use information in the adoption of new technology (Lavison, 2013; Mignouna et al., 2011). Higher educational attainment influences attitudes and thoughts, making users open-minded, rational and well-positioned to analyse the benefits of a new technology (Waller et al., 1998). Therefore, innovations can be introduced and adopted easily by such users (Adebiyi & Okunlola, 2013). However, another study by Uematsu and Mishra (2010) reported a negative influence of formal education towards adopting genetically modified crops.

Age has also been identified as a determinant in the adoption of new technologies in agriculture. Research shows that older farmers are more knowledgeable and experienced when evaluating technologies compared to younger farmers (Kariyasa & Dewi, 2013; Mignouna et al., 2011). At the same time, other studies report a negative relationship between age and the adoption of technology. According to Mauceri et al. (2007) and Adesina and Zinnah (1993), older farmers are more risk-averse and avoid long-term investments. However, younger farmers are often risk-seeking and more inclined to experiment with new technologies. In a study by Alexander and Van Mellor (2005) in the adoption of genetically modified maize varieties, adoption increased with age for younger farmers as they gained experience and increased their stock of human capital, but declined with age for farmers closer to retirement. Gender roles have also been extensively investigated, and researchers report mixed results on the role of gender in technology adoption (Bonabana-Wabbi et al., 2002). A study by Doss and Morris (2000) on the impact of gender in technology adoption did not identify any significant effects of gender on the probability of adopting improved maize varieties in Ghana. Rather, adoption depends on access to land, labour and other resources. However, if males had greater access to these resources, then the benefits of the technologies will benefit women to a lesser degree.

3.2 Societal Factors

Being part of a social group facilitates the accumulation of social capital. Consequently, trust and information exchange grow in such situations (Mignouna et al., 2011). In these groups, there are opinion leaders who can influence how social groups behave. Successful adoption of technologies in the presence of such individuals hinges on gaining their trust and support by providing them with information about new environmental management approaches to disseminate in their groups.

Farmers in the same social networks learn about new technologies from each other. According to Uaiene (2011), social network effects are important for individual decisions. In the context of agricultural innovations, it is common for farmers to share data and learn from each other. The amount of data or knowledge accessible to each group, a linear combination of individual input (static externalities) and available knowledge (learning externalities), is a key driver of productivity. Even though many researchers report that social groups have a positive influence on adoption, others report that they can be detrimental to technology adoption evidenced by free-riding behaviour because some members tend to benefit from positive learning externalities at a minimal cost.

In the same generation agricultural networks, seasoned farmers contribute more towards the accumulation of available knowledge than their amateur counterparts. New and inexperienced farmers are more likely to benefit from the accumulated knowledge in the social group in comparison to their contribution, leading to free-riding behaviour. However, under the same assumptions listed above, today's generation of farmers are more productive than previous generations because they can access knowledge faster. Thus, older farmers tend to free ride in mixed generation networks.

A study by Foster and Rosenzweig (1995) on the adoption of Green Revolution technologies in India reported that not only learning externalities (e.g. diffusion of information, social learning, spillovers) within social networks increased the profitability of adoption, but also there was evidence of free-riding by other farmers on their neighbours' experimentation with the new technology. Bandiera and Rasul (2006) report that learning externalities generate opposite effects. The greater the experimentation with new technologies, the more advantageous it is to join such social networks. At the same time, it was easier to free-ride others' experimentation as farmers can rely on other members to gain experience and later increase the usage as it becomes profitable to them. Stemming from the outcome of these contradictory effects, network effects are positive at low rates of adoption but negative at high rates of adoption.

Acquisition of information about a new technology enables farmers to learn about the existence, effective use and, eventually, its adoption. Farmers will only adopt the technology they are aware of. As farmers accumulate information about a technology's performance, their assessment changes over time from subjective to objective (Bonabana-Wabbi et al., 2002; Caswell et al., 2001). The accumulated information does not automatically lead to adoption because farmers may perceive and evaluate the technology differently from scientists (Uaiene, 2011).

Access to extension services provides the farmer with information about the existence, use and benefit of new technologies through extension agents. They connect developers to farmers and reduce transaction costs incurred when passing on specialized information about the new technology.

According to Mohamed and Temu (2008), access to finance eases adoption by increasing the capability to cope with risk (Simtowe & Zeller, 2006), avoid suboptimal strategies and take on riskier investments with larger rewards. In some countries, female-headed households have reduced access to credit facilities compared to their male counterparts. In the long term, gender effects lead to low adoption due to reduced ability of female households to acquire these technologies (Muzari et al., 2012), which must be addressed through appropriate policies (Simtowe & Zeller, 2006).

3.3 Farm Size

Farm size affects and in turn is affected by other factors influencing adoption (Lavison, 2013). For instance, some DSSs can be scale-dependent because of the size of the farm, i.e. some are suited for big or small farms, but not both (Bonabana-Wabbi et al., 2002). Other studies have reported a positive correlation between farm size and the adoption of agricultural technologies (Mignouna et al., 2011; Uaiene, 2011). Larger farms can dedicate a part of their land to experiment with new farming technologies compared to smaller farms (Uaiene, 2011). More so, heavy machinery needs big spaces to reach economies of scale and maintain profitability (Feder et al., 1995). On the contrary, other studies report a negative correlation between farm size and the adoption of agricultural technologies. Smaller farms are better suited to adopt technologies that aim to minimize input, land and labour. Greenhouse technologies and zero-grazing substitutes increased agricultural output for minimal input (Harper et al., 1990; Yaron et al., 1992). After all, large-scale farming is riskier because of the high cost of equipment, capital and inputs. Other studies have reported a neutral correlation between farm size and the adoption of new farming technologies. Samiee et al. (2009) reported that farm size did not influence the adoption of IPM.

Other studies have also reported that farm size does not influence IPM adoption (Bonabana-Wabbi et al., 2002; Grieshop et al., 1988; Ridgley & Brush, 1992; Samiee et al., 2009; Waller et al., 1998). Dissemination may take place independently of farming scale. Kariyasa and Dewi (2013) report that larger farms had no significant effect on the probability of adopting Integrated Crop Management Farmer Field School. Perhaps, it is best to consider extent instead of total farm size. Total farm area facilitates comparison of productivity realized from using new technology (Lowenberg-DeBoer, 2000). This makes it easier to compare productivity gains among farmers with different farm sizes.

4 Behavioural Approaches to the Adoption of Models for Precision Farming

Novel tools and methods may not be adopted by farmers due to a variety of reasons leading to a mismatch between the solutions offered and the needs these are supposed to satisfy. Technical aspects of the tools and solutions are often blamed, assuming that the decision-making process of the potential adopter is flawless. However, behavioural economics provides strong evidence that humans are affected by social- and individual-level phenomena which lead to a number of well-documented biases. In fact, non-adoption of innovative agricultural methods has been observed even in successful novel solutions (Abedullah & Qaim, 2015; Villano et al., 2015). It is, therefore, interesting to connect the literature on individual and social features of human decision-making with the biases that influence the decision to adopt a given method or a model-based practice.

In what follows we provide a list of idiosyncratic characteristics and individual or collective behavioural biases as determinants of the match or mismatch between farmers' needs and the solutions offered to them.

4.1 Risk Attitudes

Risk attitudes are among the most prominent features of individual behaviour, identified in early literature (Feder, 1979) as a possible idiosyncratic barrier to the adoption of new technologies. Despite a long series of paradigm changes and developments since the seminal theory of expected utility theory by von Neumann and Morgenstern (1944), the term 'risk attitudes' is still often confused with or limited to the property of 'risk aversion', namely, decision-makers' preference for options with less dispersion of possible outcomes, other things being equal. Several authors (Barham et al., 2014; Barham et al., 2015) have hypothesized that risk aversion makes a farmer hesitant to adopt new methods or innovative decision support tools. Using the risk aversion parameters estimated for farmers, the reverse path has also been explored, determining the optimal levels of inputs in herbicide management problems (Pannell, 1991, 1995). However, from a theoretical point of view, risk aversion could be expected to create incentives both towards and against adoption decisions (a 'Jekyll and Hyde' effect) when the new solutions or methods considered could provide more safety towards climate or yield uncertainty. Therefore, the existing findings reported on the effects of risk aversion on adoption may have been less salient than they really are, given that the aforementioned second (adoption-enhancing) effect of risk aversion has not been modelled so far. A general theoretical model which considers both roles of risk aversion on adoption decisions does not exist. This constitutes an important gap in the existing approaches.

Following evidence against the expected utility theory and critique by Allais (1953), the attribution of risk attitudes and, thus, risk aversion to the curvature of the decision-maker's utility function was shown to be insufficient to accommodate a plethora of other phenomena observed in decision-making under uncertainty. Kahneman and Tversky's (1979) prospect theory opened the way to a whole range of new terms and phenomena, leading people to evaluate uncertain prospects in very different ways from that suggested by the simple mathematical expectation of the 'lottery' of possible outcomes. Among such terms, reference points, loss aversion and probability weighting were shown to play an important role in decision-making under uncertainty.

Probability weighting denotes the tendency of decision-makers to distort probabilities in a way that leads to overestimation of low probabilities and underestimation of high ones. *Loss aversion* is the behavioural pattern leading to a greater decrease from a loss than the utility increase from a gain of the same size. Finally, determining a *reference point* (current earnings, aspiration level, subsistence level, etc.) makes it possible for all outcomes to be classified as gains and losses, leaving a further degree of freedom to prospect theory by allowing risk aversion or risk seeking to vary across the two domains. A consequence of the freedom to consider different reference points has revealed the importance of *framing* decisions as gains or losses, depending on the reference point adopted. The plethora of studies on risk aversion contrasts with the scarcity of studies accounting for these phenomena, which when neglected can lead to severe underestimation of the role of risk attitudes in farming decisions. Recently, Bontemps et al. (2021) presented a study on French farmers, showing that the use of a more complete specification of prospect theory (cumulative prospect theory) with a joint consideration of reference points, loss aversion and probability weighting raised the percentage of pesticide use decisions explained through risk attitudes from 4% to 19%. In other words, the whole spectrum of aspects captured by recent theories of risky decision-making is necessary for a reliable explanation of farmers' adoption decisions.

Spatial dependence. When considering the determinants of adoption decisions, socioeconomic factors have already obtained some attention, but most approaches have focused on a small subset of isolated drivers and barriers. While the term 'socioeconomic' refers to both individual and collective characteristics of the decision-makers and their social environment, adoption of probability models focuses mostly on farmers' individual characteristics and to a much lesser extent no community features and norms. Areal et al. (2012) and Ambali et al. (2019, 2021) present a joint consideration of individual and collective (geographically clustered) risk attitudes which partly determine Nigerian farmers' decisions to adopt new rice varieties. The evidence provided there suggests that risk attitudes may have evolved in similar ways among populations exposed to similar climatic conditions. It has also been argued that farmers' risk attitudes could have evolved differently from other groups (Iyer et al., 2020). Therefore, the farmers' natural environment can be a determinant of risk attitudes, which in their turn determine their decisions. The interplay among different behavioural drivers and the co-evolution of farming and societal characteristics needs to be jointly and exhaustively taken into account when explaining the role of the human decision-making processes in tool and technique adoption decisions. Clustering of decisions by geographical location could simply be an effect of imitation of others in a community, leading to *herding behaviour*, but this is a more general aspect of human behaviour beyond risky decisions, which are discussed below.

4.2 Behavioural Factors Beyond Risk Attitudes

The growing field of experimental economics and economic psychology has produced a long list of behavioural biases which systematically affect decision-making in all domains of human life, including agriculture. Some of these are discussed below.

Anchoring refers to the choice of the default or status quo option more frequently than would be optimal. Holst et al. (2015) prove that farmers' decisions do suffer from this behavioural bias. Imitation of successful others and *herding* makes an individual choose according to the choice of the majority. While herding in financial markets, including those for agricultural commodities (Apergis et al., 2020), there is little systematic evidence on this phenomenon, which seems central to the way farmers face new methods and new MBPs. This behavioural bias increases the inertia of a community, making it less prone to adopting new methods and decision-support tools. This brings us to the importance of social norms. Hillenbrand and Miruka (2019) defined these as collective beliefs or expectations one has about the behaviour of others in a reference group. Such norms moderate the beliefs of a farmer regarding what others consider correct, important or successful, including the transmission of information and advice among others in the community. An important aspect of social norms regards the dimension of gender which should be dealt with well beyond the usual 'demographic approach'. As Hillenbrand and Miruka (2019) observe, gender and gender norms determine, to a large extent, the possibility of a fair share of the benefits from agriculture and farming innovations, as well as the changing context of labour relations in agriculture.

Finally, other demographic variables have been shown to affect a farmer's propensity to adopt model-based practices in precision farming. Among such variables, the most common in the literature are age and education. Also, the farmers' access and type of exposure (advertising, demonstrations) to a given solution matter, as reported by Akaka et al. (2021). Their dataset is partly used in the next section to estimate the model discussed to give a simple demonstration of the quantitative methodology that can be adopted to study farmers' adoption decisions as a function of their individual characteristics.

5   An Empirical Approach to Research on Adoption of Models

The growing literature on behavioural economics has pointed out that the complexity of human behaviour can only be studied by taking into account all the aforementioned behavioural and socioeconomic factors jointly. While useful for pilot results and further design of more systematic approaches, qualitative methods expecting users to guess their own motivations through experience and introspection have failed to provide a complete picture of the underlying relations that hinder or facilitate the match between solutions and needs. Quantitative analysis of extensive surveys may reveal unexpected barriers and drivers of adoption decisions, beyond what is detected by each user from their own experience.

In what follows, we discuss the results from a subsample of an ongoing survey in the framework of IPM Decisions whose analysis offers insights on the drivers and barriers in a user's decision to adopt a DSS. At the same time, the approach is used as

Table 1 Percentage of responses related with the variables under study

Age		Type of farm	
18 to 25	7.38%	Conventional	40.74%
25 to 35	25.50%	Integrated	29.63%
35 to 50	27.52%	Organic	9.26%
50 to 60	29.53%	Biodynamic	0.67%
60 or more	10.07%	Not indicated	20.37%
Education		Farm size	
Vocational	40.27%	Under 5 ha	10.07%
Bachelor	27.52%	5 to 10 ha	2.68%
Masters	17.45%	10 to 30 ha	18.12%
PhD	4.70%	30 to 60 ha	14.09%
Others	6.71%	Over 60 ha	54.36%
Access to high-speed Internet			
Under 20 K	21.48%	No	10.00%
20 to 25 K	20.81%	In the office only	38.00%
35 to 50 K	20.13%	Also in the fields	52.00%
50 to 75 K	12.08%		
Over 75 K	22.15%		
Gender		Already tried using a DSS	
Male	80.00%	Yes	60.00%
Female	19.00%	No	24.00%
Not indicated	1.00%	Not indicated	16.00%

Akasska et al. (2021)

an example of behavioural models whose role is crucial in the diagnosis of barriers to adoption, especially focusing on the user's point of view. We present first the data collected to analyse the variables that shape the probability for a farm manager to use a DSS. We conducted a questionnaire survey that was administered in 11 different European countries, eliciting numerous personal and farm-specific characteristics relevant to the adoption of DSSs. A total of 107 European farmers answered the questionnaire. The questionnaires were collected during a period of 6 months which ended in March 2020. Table 1 shows the main descriptive statistics of the sample regmontharding the profile of the farmers and the characteristics of the farm.

Econometric Analysis

To identify potential avenues for improving DSS, we did an econometric analysis of the data collected. The purpose of this analysis was to identify the appropriate statistical model allowing us to measure the impact of the different variables that we have collected on the farmers' adoption rate of DSSs. We propose a model where we hypothesize that the characteristics of the farmer as well as of the farm and farm equipment influence the decision to use DSSs. Considering the binary nature of our dependent variable (i.e. using or not using DSSs), we use a logit model to perform this analysis. The model is specified as follows:

$$P(i) = \frac{1}{1 + \exp\left(-\beta_0 - X_{1,i}\beta_1 - X_{2,i}\beta_2 - \dots - X_{15,i}\beta_{15}\right)}, \tag{1}$$

where $P(i)$ is the probability for farmer i to use DSSs, $X_{j,\,i}$ is the value of variable j for subject i and β_j is the parameter to be estimated that captures the specific impact of variable j on the probability for a farmer to use a DSS. Table 2 presents the explanatory variables of the model, while Table 3 presents the results of the model.

In the first column, the farm size and the willingness to pay to use a DSS both show a positive and significant effect on the rate of DSS use ($p < 0.01$). Managers of the largest farms (i.e. total surface area) are more likely to use DSSs, all other parameters remaining equal. Also, the probability of using a DSS increases with the willingness to pay for one.

Other personal characteristics of the manager of the farm do not significantly affect the adoption rate. The level of education only shows a positive and significant effect ($p < 0.1$). Regarding the characteristics of the farm, vegetable producers are significantly more likely to use DSS than other types of producers ($p < 0.05$). This might be explained by the fragile nature of some vegetables faced with diseases. On the other hand, flower growers are significantly less likely to use DSSs than other types of producers.

Column 3 (see Table 3) shows that some farm characteristics influence managers' trust in DSS. Managers of integrated and biodynamic farms tend to show a

Table 2 Explanatory variables of the model

Personal characteristics	Variable symbol	Explanation of variable
Age	X_1	Age of farmer in years
Education	X_2	Highest degree obtained among five categories
Gender	X_3	Dummy variable = 1 if male
Income	X_4	Net income in euros
Need IT teaching	X_5	Dummy variable = 1 if teaching needed
Farm characteristics		
		Dummy variable = 1 if animals in the farm
Animal	X_6	
Arable crops	X_7	Dummy variable = 1 if arable crops in the farm
Flowers	X_8	Dummy variable = 1 if flowers in the farm
Orchards	X_9	Dummy variable = 1 if orchards in the farm
Vegetables	X_{10}	Dummy variable = 1 if vegetables in the farm
Vineyard	X_{11}	Dummy variable = 1 if vineyard in the farm
Farm size	X_{12}	Size of the farm in hectares
Biodynamic	X_{13}	Dummy variable = 1 if biodynamic farm
Conventional farm	X_{14}	Dummy variable = 1 if conventional farm
High-speed internet	X_{15}	Dummy variable = 1 if high-speed internet in the office and in the field

Akaka et al. (2021)

Table 3 Binomial logit and ordered logit estimates with DSS adoption rate (first column) and trust (second column) as endogenous variables in two separate regression analyses

	DSS use	Marginal effects	Trust in DSS
Age	0.1530(0.2638)	0.0292(0.0503)	−0.0534(0.1978)
Education	0.5880*(0.3015)	0.1120*(0.0574)	0.0534(0.2225)
Gender (male)			−0.0174(0.9761)
Farm size	**1.0590***(0.3518)**	**0.2020***(0.0678)**	0.3910*(0.2273)
Income	−0.1850(0.2403)	−0.0353(0.0458)	−0.1760(0.1630)
Arable crops	−0.2580(0.6789)	−0.0492(0.1295)	−0.1720(0.5059)
Animals	−1.1200(1.0769)	−0.2130(0.2029)	−0.2730(0.9750)
Orchards	0.3820(0.9550)	0.0729(0.1823)	−0.8370(0.6805)
Vegetables	**3.0470**(1.1949)**	**0.5810***(0.2184)**	0.1980(0.6600)
Flowers	−2.5950*(0.1.5727)	−0.4940*(0.2923)	1.0010(1.1375)
Vineyards			0.5760(0.7783)
Integrated farm			0.7430*(0.4060)
Biodynamic farm			**3.8190**(1.7438)**
Legislative requirement	0.5610(0.7013)	0.1070(0.1338)	−0.1890(0.5400)
High Speed Internet	1.1570*(0.5995)	**0.2200**(0.1095)**	
Need IT teaching	−0.8950(0.5967)	−0.1710(0.1103)	0.4480(1.0182)
Willingness to pay	**1.9550***(0.6517)**	**0.3730***(0.1184)**	
Already used DSS			−0.2740(0.4152)
Exposed to marketing			**−2.6730**(1.1000)**
Exposed to demonstrations			1.7610*(1.0298)
_cons	**−6.1200***(1.9185)**		**4.063***(1.2981)**
N	107	107	106

Akaka et al. (2021)
Standard errors to the nearest 4 dp in parenthesis. Significance level: $*p < 0.1$, $**p < 0.05$ and $***p < 0.001$

significantly higher level of trust towards DSSs (respectively, $p < 0.1$ and $p < 0.05$). Also, managers of larger farms tend to trust DSS more ($p < 0.1$). Interestingly, results regarding the type of communication managers have been exposed to show how DSS is advertised matters. While farm managers that have been exposed to simple advertising show a significantly lower level of trust toward DSS ($p < 0.1$), farm managers who attended regular demonstrations trust DSS significantly more ($p < 0.05$).

The characteristics of the farm manager have a very limited impact on the decision to use DSSs. The age, gender, income, and need of IT teachings do not influence it. Only the degree of the farm manager shows a significant impact ($p < 0.1$), with a higher degree associated with a higher probability to use DSSs.

Regarding the farm characteristics, the size of the farm as well as the type of cultivated product are relevant in explaining the probable use of DSSs. Results show that a larger farm size is associated with a significantly higher probability of using DSSs ($p < 0.01$) and that farms cultivating vegetables are significantly more likely to use DSSs ($p < 0.05$). It is also interesting to note that farmers managing conventional farms compared to those managing organic, biodynamic and integrated farms are less likely to use DSSs ($p < 0.1$).

The analysis above demonstrates how behavioural research can be used to identify statistically significant drivers and barriers of adoption of a given DSSs, using econometric modelling of trust and the decision to adopt any given DSS as a function of various personal characteristics, farm-related features and how farmers learn about DSS through demonstration and advertising. For behavioural and social scientists interested in identifying the determinants of adoption of DSSs, all the significant findings identified above would be difficult if not impossible to identify using other qualitative methods such as in-depth interviews. Users may not be aware of the effect of each of these factors in their own adoption decisions. Hypothetical bias and misjudgement of one's own preferences contribute to higher levels of bias that call for calibration in future research. In particular, studies designed to compare statistically identified effects based on qualitative methods and self-reported perceptions collected through qualitative methods must pay attention to possible sources of bias and put measures in place to minimize them.

6 Discussion

In our analysis, MBPs can boost farm productivity by supporting decision-making at the farm. We addressed the issue of adoption at various stages of the model's development, coupled with an overview of the relevant socio-behavioural factors influencing farmers' decision to adopt MBPs. This multidisciplinary approach emphasises the need for a duplicitous perspective not only for addressing the problem of adoption but also contributing to ongoing research on agricultural DSSs. Furthermore, it demonstrates the importance of involving various stakeholders at all stages of the MBP's lifecycle with a focus of understanding the interaction between the technical and farmer behaviour in the adoption of DSSs.

In our study, we acknowledge that this chasm can be caused by either technical or end-user limitations, or both. Sometimes, a well-designed model is not adopted because there was no co-creation between various stakeholders and end users. It is possible that failure to adopt is due to technical reasons. For instance, models may be inaccurate because of incorrect representation of the problem, or they are out of synchronization with the decision-making process. In agriculture, such complexity necessitates collaboration between social and natural scientists to identify causes of failure. More importantly, it is necessary to know how to avoid such problems in the first place. Unfortunately, we cannot learn from mistakes made in the past as cases of failure are rarely published. Failure can occur at any point in the design, development or deployment process. If poor decision support is provided by poorly developed models, they will fail. Paradoxically, a good solution can fail if it is unappealing, useless and unsupported after it is released.

To address these problems, we must shift our perspective on the roles of key stakeholders at different stages of development. Indeed, farmers are experts in their own right, and scientists must capture different aspects of the farmer and incorporate it into various phases of the model life cycle. MBPs may become more appealing as a result of this shift in perspective, which focuses on co-creation among various stakeholders. Finally, because (1) the individual farmer makes the decision to adopt and (2) decision-making in agriculture is complex and context-dependent, factors influencing adoption must be examined from a social, behavioural and technical standpoint to increase uptake.

Even though co-creation has the potential to simultaneously facilitate interaction between multiple scientific fields and involve a wide range of stakeholders, it is far from a panacea for increasing MBP adoption. Coordinating stakeholders with differing viewpoints and levels of expertise can be difficult. Similarly, at various stages of the co-creation process, it is difficult to reach a consensus on whose input has more weight at each phase. In this case, the best course of action is to devise a formula for determining the best input for each stage and to resolve any conflicts that arise along the way. Experimentation can help farmers deal with risk and uncertainty in the long run, as well as contribute to increased adoption of MBPs.

Conflict of Interest   The authors declare that there are no conflicts of interest.

References

Abedullah, S. K., & Qaim, M. (2015). Bt cotton, pesticide use and environmental efficiency in Pakistan. *Journal of Agricultural Economics, 66*, 66–86.

Adebiyi, S., & Okunlola, J. O. (2013). Factors affecting adoption of cocoa farm rehabilitation techniques in Oyo state of Nigeria. *World Journal of Agricultural Sciences, 9*(3), 258–265.

Adesina, A., & Zinnah, M. (1993). Technology characteristics, farmers' perceptions and adoption decisions: A Tobit model application in sierra Leona. *Agricultural Economics, 9*(4), 297–311.

Akaka, J., Tisserand, J-C., García-Gallego, A. & Georgantzis, N. (2021). *Decision support systems adoption in pesticide management* (Working Paper 2021/8). Economics Department, Universitat Jaume I, Castellón.

Alajoutsijärvi, K., & Tikkanen, H. (2000). Customer relationships and the small software firm: A framework for understanding challenges faced in marketing. *Information & Management, 37*(3), 153–159.

Alexander, C., & Van Mellor, T. (2005). Determinants of corn rootworm resistant corn adoption in Indiana. *AgBioforum, 8*(4), 197–204.

Allais, M. (1953). Le comportement de l'homme rationnel devant le risque: critique des postulats et axiomes de l'école Américaine. *Econometrica, 21*, 503–546.

Ambali, O.I., Areal, F. J. & Georgantzis, N. (2019). *The role of spatial dependence in risk preference: Evidence from Nigeria*. Sixth International Conference of the African Association of Agricultural Economists (AAAE), September 23–26, 2019, Abuja, Nigeria 295871.

Ambali, O. I., Areal, F. J., & Georgantzis, N. (2021). Roles of risk preferences and spatial dependence in decisions to adopt improved rice technology. *Sustainability, 13*, 5943.

Apergis, N., Christou, C., Hayat, T., & Saeed, T. (2020). U.S. Monetary policy and herding: Evidence from commodity markets. *Atlantic Economic Journal, 48*(3), 355–374.

Areal, F. J., Balcombe, K., & Tiffin, R. (2012). Integrating spatial dependence into stochastic frontier analysis. *Australian Journal of Agricultural and Resource Economics, 56*, 521–541.

Aubert, B. A., Schroeder, A., & Grimaudo, J. (2012). IT as an enabler of sustainable farming: An empirical analysis of farmers' adoption decision of precision agriculture technology. *Decision Support Systems, 54*, 510–520.

Bandiera, O., & Rasul, I. (2006). Social networks and technology adoption in Northern Mozambique. *The Economic Journal, 16*(514), 869–902.

Barham, B. L., Chavas, J. P., Fitz, D., Ríos-Salas, V., & Schechter, L. (2014). The roles of risk and ambiguity in technology adoption. *Journal of Economic Behaviour and Organization, 97*, 204–218.

Barham, B. L., Chavas, J. P., Fitz, D., Ríos-Salas, V., & Schechter, L. (2015). Risk, learning, and technology adoption. *Agricultural Economics, 46*, 11–24.

Bellarby, J., Siciliano, G., Smith, L. E. D., Xin, L., Zhou, J., Liu, K., Jie, L., Meng, F., Inman, A., Rahn, C., Surridge, B., & Haygarth, P. M. (2017). Strategies for sustainable nutrient management: Insights from a mixed natural and social science analysis of Chinese crop production systems. *Environmental Development, 21*, 52–65.

Bonabana-Wabbi, J., Taylor, D. B., Bertelsen, M., & Mcguirk, A. (2002). *Assessing factors affecting adoption of agricultural technologies: The case of Integrated pest Management (IPM) in Kumi District.* Virginia Tech.

Bontemps, C., Bougherara, D., & Nauges, C. (2021). Do risk preferences really matter? The case of pesticide use in agriculture. *Environmental Modeling and Assessment, Forthcoming, 26*(4), 609–630.

Box, G. E. P. (1979). Robustness in the strategy of scientific model building. In R. L. Launer & G. N. Wilkinson (Eds.), *Robustness in statistics* (pp. 201–236). Academic.

Caswell, M., K. Fuglie, Ingram, C., Jans, S., & Kascak, C. (2001). *Adoption of agricultural production practices: lessons learned from the U.S. department of agriculture area studies project.* (Resource Economics Division, Economic Research Service, Agriculture Economic Report No.792). U.S. Department of Agriculture, Washington, DC.

Columbus L. (2021). *10 ways AI has the potential to improve agriculture.* In: 2021 https://www.forbes.com/sites/louiscolumbus/2021/02/17/10-ways-ai-has-the-potential-to-improve-agriculture-in-2021/

Davis, F. D. (1989). Perceived usefulness, perceived ease of use, and user acceptance of information technology. *MIS Quarterly, 13*(3), 319–339.

de Wit, C. T. (1982). Simulation of living systems. In F. W. T. Penning de Vries & H. H. van Laar (Eds.), *Simulation of plant growth and crop production* (Vol. 86, pp. 3–8).

Doll, J., & Ajzen, I. (1992). Accessibility and stability of predictors in the theory of planned behavior. *Journal of Personality and Social Psychology, 63*(5), 754–765.

Doss, C. R. (2003). *Understanding farm-level technology adoption: Lessons learned from CIMMYT micro surveys in Eastern Africa.* CIMMYT.

Doss, C. R., & Morris, M. L. (2000). How does gender affect the adoption of agricultural innovations? *Agricultural Economics, 25*(1), 27–39.

EU. (1991). *Directive concerning the protection of waters against pollution caused by nitrates from agricultural sources (91/676/EEC).* The Council of The European Communities Brussels.

Feder, G. (1979). Pesticides, information, and pest management under uncertainty. *American Journal of Agricultural Economics, 61*(1), 97–103.

Feder, G., & Umali, D. L. (1993). The adoption of agricultural innovations. A review. *Technological Forecasting and Social Change, 43*(3–4), 215–239.

Feder, G., Just, R. E., & Zilberman, D. (1995). Adoption of agricultural innovations in developing countries: A survey. *Economic Development and Cultural Change, 33*(2), 255–298.

Foster, A. D., & Rosenzweig, M. R. (1995). Learning by doing and learning from others: Human capital and technical change in agriculture. *Journal of Political Economy, 103*(6), 1176–1209.

Fountas, S., Blackmore, S., Ess, D., Hawkins, S., Blumhoff, G., Lowenberg-Deboer, J., & Sorensen, C. G. (2005). Farmer experience with precision agriculture in Denmark and the US Eastern Corn Belt. *Precision Agriculture, 6*, 121–141.

Gray, D. I., Parker, W. J., & Kemp, E. (2009). Farm management research: A discussion of some of the important issues. *Journal of International Farm Management, 5*(1), 1–24.

Grieshop, J. I., Zalom, F. G., & Miyao, G. (1988). Adoption and diffusion of integrated pest management innovations in agriculture. *Bulletin of the Entomological Society of America, 34*(2), 72–79.

Harper, J. K., Rister, M. E., Mjelde, J. W., Drees, B. M., & Way, M. O. (1990). Factors influencing the adoption of insect management technology. *American Journal of Agricultural Economics, 72*(4), 997–1005.

Hillenbrand, E. & Miruka, M. (2019). Gender and social norms in agriculture: A review. In A. R. Quisumbing, R. S. Meinzen-Dick, & J. Njuki (Eds.), *2019 Annual trends and outlook report: Gender equality in rural Africa: From commitments to outcomes* (Chapter 2, pp. 11–31).

Holst, S., Hermann, D., & Musshoff, O. (2015). Anchoring effects in an experimental auction – Are farmers anchored? *Journal of Economic Psychology, 48*, 106–117.

Iyer, P., Bozzola, M., Hirsch, S., Meraner, M., & Finger, R. (2020). Measuring farmer risk preferences in Europe: A systematic review. *Journal of Agricultural Economics, 71*(1), 3–26.

Ji, H., Yun, Y., Lee, S., Kim, K., & Lim, H. (2018). An adaptable UI/UX considering user's cognitive and behavior information in a distributed environment. *Cluster Computing, 21*, 1045–1058.

Kahneman, D., & Tversky, A. (1979). Prospect theory: An analysis of decision under risk. *Econometrica, 47*, 263–291.

Kariyasa, K., & Dewi, Y. A. (2013). Analysis of factors affecting adoption of integrated crop management farmer field school (ICM-FFS) in swampy areas. *International Journal of Food and Agricultural Economics, 1*(2), 29–38.

Keelan, C., Thorne, F., Flanagan, P., Newman, C., & Mullins, E. (2009). Predicted willingness of Irish farmers to adopt GM technology. *AgBioforum, 12*(3–4), 394–403.

Kuehne, G., Llewellyn, R., Pannell, D. J., Wilkinson, R., Dolling, P., Ouzman, J., & Ewing, M. (2017). Predicting farmer uptake of new agricultural practices: A tool for research, extension and policy. *Agricultural Systems, 156*, 115–125.

Kumar, G., Engle, C., & Tucker, C. (2018). Factors driving aquaculture technology adoption. *Journal of the World Aquaculture Society, 49*(3), 447–476.

Kutter, T., Tiemann, S., Siebert, R., & Fountas, S. (2011). The role of communication and co-operation in the adoption of precision farming. *Precision Agriculture, 12*(1), 2–17.

Lavison, K. R., (2013). Factors influencing the adoption of organic fertilizers in vegetable production in Accra. Thesis submitted for the Master of Philosophy Degree in Agribusiness, University of Ghana, Legon.

Lindblom, J., Rambusch, J., Ljung, M., & Lundstrom, C. (2013). *Decision-making in agriculture— Farmers' lifeworld in theory and practice.* 21st European Seminar on Extension Education (ESEE13).

Lowenberg-DeBoer, J. (2000). Comment on site-specific crop management: Adoption patterns and incentives. *Review of Agricultural Economics, 22*(1), 245–247.

Mauceri, M., Alwang, J., Norton, G., & Barrera, V. (2007). Effectiveness of integrated pest management dissemination techniques: A case study of potato farmers in Carchi, Ecuador. *Journal of Agricultural and Applied Economics, 39*(3), 765–780.

McCown, R. L., Carberry, P. S., Hochman, Z., Dalgliesh, N. P., & Foale, M. A. (2009). Re-inventing model based decision support with Australian dryland farmers: Changing intervention concepts during 17 years of action research. *Crop and Pasture Science, 60*(11), 1017–1030.

McDonald, R., Heanue, K., Pierce, K., & Horan, B. (2016). Factors influencing new entrant dairy farmer's decision-factors influencing new entrant dairy farmer's decision-making process around technology adoption. *Journal of Agricultural Education and Extension, 22*, 163–177.

Meijer, S. S., Catacutan, D., Ajayi, O. C., & Sileshi, G. W. (2015). The role of knowledge, attitudes and perceptions in the uptake of agricultural and agroforestry innovations among smallholder farmers in sub-Saharan Africa. *International Journal of Agricultural Sustainability, 13*(1), 40–54.

Mignouna, D. B., Manyong, V. M., Rusike, J., Mutabazi, K. D. S., & Senkondo, E. M. M. (2011). Determinants of imazapyr-resistant maize technologies and its impact on household income in Western Kenya. *AgBioforum, 14*(3), 158–163.

Mohamed, K. S., & Temu, A. E. (2008). Access to credit and its effect on the adoption of agricultural technologies: The case of Zanzibar. *Savings and Development, 32*, 45–89.

Muzari, W., Gatsi, W., & Muvhunzi, S. (2012). The impacts of technology adoption on smallholder agricultural productivity in sub-saharan Africa: A review. *Journal of Sustainable Development, 5*(8), 69–77.

Pannell, D. J. (1991). Pests and pesticides, risk and risk aversion. *Agricultural Economics, 5*(4), 361–383.

Pannell, D. J. (1995). Optimal herbicide strategies for weed control under risk aversion. *Review of Agricultural Economics, 17*(3), 337–350.

Parker, C., & Sinclair, M. (2001). User-centred design does make a difference. The case of decision support systems in crop production. *Behaviour and Information Technology, 20*(6), 449–460.

Payne, J. W., Bettman, J., & Johnson, E. J. (1993). *The adaptive decision maker*. Cambridge University Press.

Pierpaoli, E., Carli, G., Pignatti, E., & Canavari, M. (2013). Drivers of precision agriculture technologies adoption: A literature review. *Procedia Technology, 8*, 61–69.

Porter, M. E. (1985). *Competitive advantage* (p. 37). New York Free Press.

Rahn, C. R. (2004). The use of models to optimise production of field vegetable crops with minimal impact on the environment. *Acta Horticulturae, 654*, 81–88.

Rahn, C. R. (2013). The challenges of knowledge transfer in the implementation of the nitrates directive. In D'Haene, K., Vandecasteele, B., De Vis, R., Crappé, S., Callens, D., Mechant, E., Hofman, G., De Neve, S. (Eds.), *NUTRIHORT (nutrient management, innovative techniques and nutrient legislation in intensive horticulture for an improved water quality) proceedings*, Ghent, pp. 9–15.

Rahn, C. R. (2018). Challenges of devising nitrogen recommendation systems for open field vegetables. Proceedings V international symposium on ecologically sound fertilization strategies for field vegetable production, Beijing May 2015. *Acta Horticulturae, 1192*, 11–20.

Rahn, C. R., Greenwood, D. J., & Draycott, A. (1996). Prediction of nitrogen fertiliser requirement with HRI WELL_N computer model. In *Proceedings of 8th nitrogen workshop, Ghent, 5-8 September 1994* (Progress in nitrogen cycling studies) (pp. 255–258). Kluwer.

Rahn, C. R., Mead, A., Draycott, A., Lillywhite, R., & Salo, T. (2001). A sensitivity analysis of the prediction of the nitrogen fertiliser requirement of cauliflower crops using the HRI WELL_N computer model. *The Journal of Agricultural Science, 137*, 55–69.

Ridgley, A. M., & Brush, S. (1992). Social factors and selective technology adoption: The case of integrated pest management. *Human Organization, 51*(4), 367–378.

Rogers, E. M. (1962). *Diffusion of innovations* (1st ed.). Free Press.

Rossi, V., Salinari, F., Poni, S., Caffi, T., & Bettati, T. (2014). Addressing the implementation problem in agricultural decision support systems. *Computers and Electronics in Agriculture, 100*, 88–99.

Samiee, A., Rezvanfar, A., & Faham, E. (2009). Factors influencing the adoption of Integrated Pest Management (IPM) by wheat growers in Varamin county, Iran. *African Journal of Agricultural Research, 4*(5), 491–497.

Simtowe, F., & Zeller, M. (2006). The impact of access to credit on the adoption of hybrid maize in Malawi: An empirical test of an agricultural household model under credit market failure. *2007 second international conference*, August 20–22, 2007, Accra, Ghana 52076, African Association of Agricultural Economists (AAAE).

Sinja, J., Karugia, J., Baltenweck, I., Waithaka, M.M., Miano, M.D., Nyikal, R., & Romney, D., (2004). Farmer perception of technology and its impact on technology uptake: The case of fodder legume in Central Kenya Highlands. *2004 Inaugural Symposium*, December 6–8, 2004, Nairobi, Kenya 9543, African Association of Agricultural Economists (AAAE).

Smith, L. (2015). *PPM-Nutrients: Policy and practice for management of nutrients.* Available online at https://www.soas.ac.uk/cedep/research/ppm-nutrients/. Accessed 14 Dec 2021.

Smith, J., Smith, P., & Addiscott, T. (1996). Quantitative methods to evaluate and compare soil organic matter (SOM) models. In P. Powlson, P. Smith and J. Smith (Eds.), *Evaluation of soil organic matter models* (NATO ASI, 1(38): 181–200).

Smith, L., Siciliano, G., Inman, A., Rahn, C., Bellarby, J., Surridge, B., Haygarth, P., Xin, L., Guilong, Z., Ji, L., Zhou, J., Meng, F. & Burke, S., (2015). *Delivering improved nutrient stewardship in China: The knowledge, attitudes and practices of farmers and advisers* (SAIN policy brief 13). UK-China Sustainable Agricultural Innovation Network (SAIN).

Thorburn, P. J., Jakku, E., Webster, A. J., & Everingham, Y. L. (2011). Agricultural decision support system facilitating co-learning. *International Journal of Agricultural Sustainability, 9*(2), 322–333.

Uaiene, R. N. (2011). Determinants of technology adoption in Mozambique. In *10th African crop science conference proceedings*, Maputo, Mozambique, 10–13 October 2011 (p. 375) ref. 10.

Uematsu, H., & Mishra, A. K. (2010). *Can education be a barrier to technology adoption?* Agricultural & Applied Economics Association 2010 AAEA 38.

Van Meensel, J., Lauwers, L., Dempen, I., Dessein, J., & Van Huylenbroeck, G. (2012). Effect of a participatory approach on the successful development of agricultural decision support systems: The case of Pigs2win. *Decision Support Systems, 54*(1), 164–172.

Villano, R., Bravo-Ureta, B., Solis, D., & Fleming, E. (2015). Modern rice technologies and productivity in the Philippines: Disentangling technology from managerial gaps'. *Journal of Agricultural Economics, 66*, 129–154.

Von Neumann, J., & Morgenstern, O. (1944). *Theory of games and economic behavior.* Princeton University Press.

Waller, B. E., Casey, W., Hoy, J. L., Henderson, B. S., & Welty, C. (1998). Matching innovations with potential users, a case study of potato IPM practices. *Agriculture, Ecosystems and Environment, 70*(2–3), 203–215.

Williams, R., Brown, H., Dunbier, M., Edmeades, D., Hill, R., Metherell A., Rahn C., & Thorburn P. (2013). *A critical examination of the role of Overseer in modelling nitrate losses from arable crops.* Available online at http://flrc.massey.ac.nz/workshops/13/Manuscripts/Paper_Wiliams_2013.pdf. Accessed 14 Dec 2021.

Woodward, S. J. R., Romera, A. J., Beskow, W. B., & Lovatt, S. J. (2008). Better simulation modelling to support farming systems innovation: Review and synthesis. *New Zealand Journal of Agricultural Research, 51*(3), 235–252. https://doi.org/10.1080/00288230809510452

Yaron, D., Dinar, A., & Voet, H. (1992). Innovations on family farms: The Nazareth region in Israel. *American Journal of Agricultural Economics, 74*(2), 361–370.

Zhang, F. S., Chen, X. P., & Chen, Q. (2009). *Fertilizer application guideline for main crops of China.* Chinese Agricultural University press.

Part II
State of the Art

Process-Based Models and Simulation of Nitrogen Dynamics

Davide Cammarano ⓘ**, Fernando E. Miguez, and Laila Puntel**

Abstract This chapter illustrates how some process-based models simulate the nitrogen dynamics. There are many crop simulation models available in the literature and some of the simulated nitrogen processes included in the crop growth models might differ. The aim of this chapter is to give readers who are less familiar with the simulation of nitrogen processes an overview on how crop models handle N balance. In addition, two case studies are presented to illustrate the application of crop simulation models for N management in different agro-environmental conditions.

1 Introduction

Food supply needs to be increased by about 70% by 2050 in order to meet global food demand. This has to be achieved by considering the following constrains: (i) producing on the same (or less) amount of agricultural land; (ii) reducing the negative environmental footprint associated with agronomic practices in terms of groundwater pollution and greenhouse gas emissions while keeping farming economically viable; and (iii) producing in a changing climate (Cammarano et al., 2016; Davis et al., 2021; Tian et al., 2021).

Nitrogen (N) is one of the most important factors affecting crop production and the environment. Since the green revolution, grain yield for cereal crops has increased with a combination of better agronomic management and improved genetics and N inputs from synthetic fertilizers (Raun & Johnson, 1999). In terms

D. Cammarano (✉)
Department of Agroecology, iClimate, Centre for Circular Bioeconomy (CBIO),
Aarhus University, Tjele, Denmark
e-mail: davide.cammarano@agro.au.dk

F. E. Miguez
Department of Agronomy, Iowa State University, Ames, USA

L. Puntel
Department of Agronomy and Horticulture, University of Nebraska, Lincoln, USA

of N fertilization, deciding on the appropriate rate of N application is challenging because this decision is made at the beginning of the growing season. The amount depends on future crop needs and soil N mineralization rates, which are, in turn, controlled by the weather and edaphic conditions and are unknown at the time (Mandrini et al., 2021). In addition, the optimal amount of N varies within a field due to the spatial variation caused by crop-soil interactions and among years due to the effects of rainfall and temperature (Basso et al., 2011b). Given the dynamics of cropping systems, on-farm N use efficiency tends to be low (30–40%) (Cassman & Dobermann, 2022), which naturally results in 60–70% of N fertilizer not being used by the crop (Angus, 2001; Coskun et al., 2017). The N not used by crop is not all lost because some is stabilized in soil organic matter or forms complexes with soil minerals. The rest of this N is effectively lost and causes known environmental problems, such as N leaching from the rootzone and greenhouse gas emissions in the form of nitrous oxide (N_2O), which impairs water quality and contributes to GHG emissions, respectively. Increasing sustainability of agriculture, and in particular of N fertilization, is a priority of many governmental agendas (Dalgaard et al., 2014; EU, 1991; Mandrini et al., 2021).

The interactions between crop genotype and agronomic management affect the spatial agronomic management of N, which is complicated by the impacts of climate variability, extreme events, and climate change. In addition, for some crops, the trade-off between environmental and economic sustainability is complicated by the need to achieve grain quality production standards. For example, Cammarano et al. (2021) demonstrated that optimizing the two-way trade-off (economic and environmental) of spring barley requires considerations of grain quality, not just yield. These quality standards are in place to ensure its end-use in the malting and distilling industry.

Different approaches have been proposed in the development of N management recommendations (Morris et al., 2018). Among the main ones are: the yield goal, soil nitrate test, the maximum return with respect to N fertilization, defining management zones within the field, proximal and remote sensing, N-rich strip treatment, and checkerboards of N spatially distributed within a field (Bundy & Andraski, 1995; Cammarano et al., 2021; Franzen et al., 2016; Jin et al., 2019; Puntel et al., 2018; Raun et al., 2005; Solie et al., 2012; Stanford, 1973).

Results of crop yield response to site-specific N management have been contradictory, which highlights the challenge of considering the spatial and temporal interactions of soil-plant-atmosphere-management interactions. In this context, yield maps from yield monitors are analyzed and interpreted for making agronomic management decisions based on the observed variation. However, most studies that aim at understanding the optimal N application are conducted over a limited time span (e.g., 3–4 years or less) and therefore cannot fully capture the temporal and spatial interactions of the soil-plant-atmosphere system. In fact, Basso, Ritchie, et al. (2011b) stated that while the spatial maps of soil and crop properties are easy to derive with modern tools, temporal variation has not received enough attention. Mandrini et al. (2021) showed how the N management strategy made by considering the variation in interannual weather with a combination of field data and crop simulation modelling achieved the best results. In their study, they demonstrated that such a methodology would reduce the mean N-leaching by 12.7% with no changes to profit with respect to the typical N management done by local services. Measuring soil N has been among the most important agronomic practice and cannot be easily replaced by other types of data. A potential for replacing in-situ measurements is the use of crop simulation models (described in chapter "Process Based Modelling of Soil–Crop Interactions for Site Specific Decision Support in Crop Management") which are becoming a common modelling tool in precision agriculture research related to N management.

Despite their limitations, crop models offer the chance to integrate field data (limited in time and in the quantity of variables) with temporal simulations of the soil-plant-atmosphere-management interactions. And to conduct "what-if" simulations to explore alternative management scenarios or evaluate the sensitivity of a cropping system to climatic events. Given that the modelling of water dynamics has been described in chapter "Modelling Soil Water Dynamics", in this chapter, the focus will be on the simulation of N dynamics. There are many crop simulation models available in the literature (Asseng et al., 2013), and some of the N processes included in the crop growth models might differ. However, the aim of this chapter is to give readers who are less familiar with the simulation of N processes an overview on how crop models handle N balance. In addition, two case studies are presented to illustrate the application of crop simulation models for N management in different agro-environmental conditions.

2 Modelling Nitrogen

In this section, we provide an overview of generic modelling of N dynamic simulation in the soil-plant system. It is a non-exhaustive review, and its aim is to simplify and illustrate the concepts beyond some of the main modelling approaches adopted by most crop models. An important point of clarification is that N simulations in crop models are not designed to work in isolation (e.g., as a stand-alone model), but they are components of a more complex relationship involving simulating water and energy balances, crop development, and growth processes as affected by weather, soil, genotype, and agronomic management.

2.1 Nitrogen Dynamics in the Soil (Supply)

Crop plants use only inorganic N, which regardless of the source is considered as inorganic N. Nitrogen mineralization, which is a microbiological process involving the conversion of organic to inorganic forms (Fig. 1), is independent of plant needs. There are several environmental and edaphic factors affecting the rates of mineralization, as well as root exudates and plant uptake. When N is not taken up by crops, it

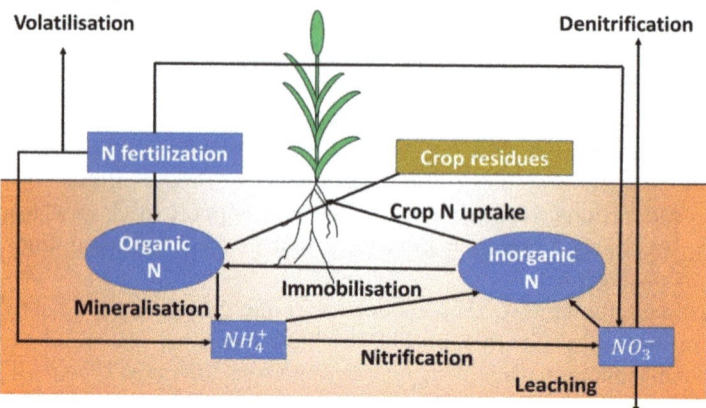

Fig. 1 Simplification of the dynamic soil processes that are simulated in most crop simulation models

can be lost through several processes—mainly leaching, denitrification, and volatilization (Fig. 1). Nitrification, the oxidation of ammonium (NH_4^+) to nitrates (NO_3^-) happens under aerobic conditions, and the main limiting factors are the substrate NH_4^+, oxygen, soil pH, and temperature. Denitrification is the reduction of NO_3^- to gaseous products such as NO, N_2O, and N_2 and is a microbial process occurring under anaerobic conditions. It is mainly affected by soil organic carbon content, soil aeration, temperature, and pH.

The main factors affecting N supply to crop plants considered by most modelling approaches are: (i) concentration of mineral N at the root depth (in terms of nitrate-N and ammonium-N); (ii) root length and density; (iii) the maximum N uptake per unit root length; and (iv) available soil water content (Godwin & Singh, 1998).

Crop simulation models have generally been coded and built by accounting for soil N and carbon (C) cycling in a simple but effective way. The simplicity of soil N models is related to the fact that measuring or calibrating for too many parameters is difficult; therefore, models of soil N that require less input are generally preferred. The required inputs also have to be easy to derive and measure. The models of soil N dynamics are generic enough that, given the right inputs, they can work on a diversity of soils and yet provide acceptable simulations of the underlying processes.

The DSSAT (Decision Support System for Agrotechnology Transfer) crop model (Hoogenboom et al., 2019) utilizes two sub-routines for simulating soil organic matter dynamics: the Godwin-Singh model which was based on the PAPRAN model (Seligman & Van Keulen, 1981) and the CENTURY model (Parton et al., 1988). Godwin and Singh (1998) described how the PAPRAN model was adapted within the DSSAT framework, and the link with the CENTURY model and its underlying assumptions/equations are described in Gijsman et al. (2002) and Porter et al. (2010).

The movement of nitrates (NO_3^-) to different soil layers is related to the flow of water. The water flow in DSSAT is based on the tipping bucket approach, and the volume of water moving from a given layer (L) to the one below is calculated as a flux (FLUX). The NO_3^- lost from a given layer (Nout) is calculated as function of the water that is drained and the one that is left in the current layer as:

$$Nout = SNO3(L) * FLUX(L)/[SW(L) * DLAYR(L) + FLUX(L)], \quad (1)$$

where SNO3(L) is the content of NO_3^- in the current layer N (expressed as kg $N \, ha^{-1}$), SW(L) is the volumetric soil water content of layer L, and DLAYR(L) is the depth of the given layer (Fig. 2a). The leaching from one layer to the one below will continue until, in a given layer L, the NO_3^- concentration falls below 1 µg $NO_3^- \, g^{-1}$ soil.

The model also simulates the upward movements of NO_3^- and urea as water evaporates from surface layers, but this movement is mainly simulated for the upper soil layers and is very small overall due to the small volume of moving water (Godwin & Singh, 1998). The upward movement of N (Nup) as a function of the upward flux (FLOW) is calculated as:

$$Nup = SON3(L) * FLOW(L)/[SW(L) * DLAYR(L) * FLOW(L)]. \quad (2)$$

The N fertilizer is partitioned between urea, NO_3^-, and ammonium (NH_4^+) depending on the type of fertilizer used. While the mineralization process involves the release of mineral N from the decay of organic matter, the immobilization involves the transformation of inorganic compounds into organic forms due to microbial processes (Porter et al., 2010). Overall, the model considers such processes and a balance between them; for example, if crop residues are added, and have a large C:N ratio, then there could be net immobilization for a given period of time, and only when soil carbon is used for respiration does the net mineralization resume (Godwin & Singh, 1998). This way of simulating N dynamics is not too dissimilar from the Agricultural Production Systems sIMulator (APSIM) soil N sub-routine (Holzworth et al., 2014).

The nitrification subroutine and partition of organic matter pools were modified in the PAPRAN model implemented in DSSAT. Additional modifications were made

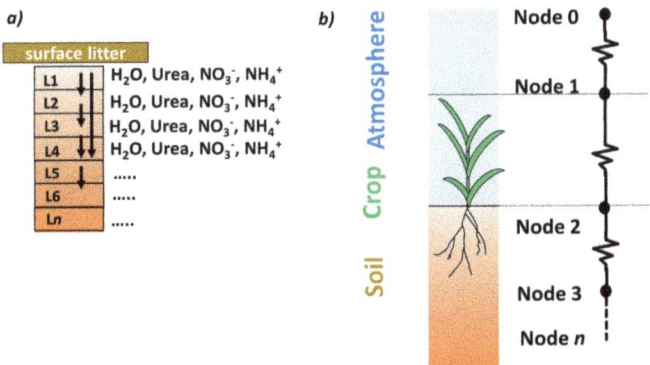

Fig. 2 Schematic representation of how the plant nutrient fluxes are simulated for (**a**) DSSAT and (**b**) CropSyst

to temperature and water subroutines in order to adapt it to the DSSAT framework (Hoogenboom et al., 2019). In this model, the mineralization and immobilization simulate the decay of two types of organic matter: (i) fresh organic matter (FOM) which consists of crop residues (and/or green manure), and it is divided into three pools such as carbohydrate (20%), cellulose (70%), and lignin (10%), with values that can be changed by the user, and (ii) a stable or humic fraction (HUM). Since the model considers the effects of crop residues, the user should state the amount of crop residues and its C:N ratio, as well as an estimate of root residue from the previous crop. The other input is the soil organic carbon for each layer (OC) because it is used to calculate HUM and the N associated with the HUM (HUMN) pool. Regarding the FOM, the three fractions have decay constants determined under nonlimiting conditions of 0.8, 0.05, and 0.0095, respectively. Given that effects of soil temperature (TF), water content (WF), and residue compositions (RF) will impact such non-limiting conditions in the soil, the model includes some modifying factors (0–1) to take this into account. Detailed information on these multipliers (TF, WF, and RF) are given in Godwin and Singh (1998). Briefly, the WF is determined from the SW(L) relative to its lower limit (LL) and drain upper limit (DUL). These will affect N processes such as ammonification, nitrification, and denitrification (Fig. 3a).

This model simulates the potential rate of nitrification and the 0 to 1 factor that reduces this potential rate. The potential rate of nitrification is a Michealis-Menten kinetic function that depends on NH_4^+ concentration and not on soil type. The nitrification capacity index is calculated with the aim of considering the lag phase on nitrification in conditions during the previous two days which were not optimal for nitrification. The Michealis-Menten equation is used to calculate the following:

$$RNTF = \frac{[A * 40 * NH_4(L)]}{[NH_4(L) + 90]} * SNH_4(L), \qquad (3)$$

where RNTF is the rate of nitrification (kg N ha^{-1} d^{-1}), $NH_4(L)$ is the ammonium-N concentration in a given layer (μg N g soil^{-1}), and $SNH_4(L)$ is the NH_4^+ content in

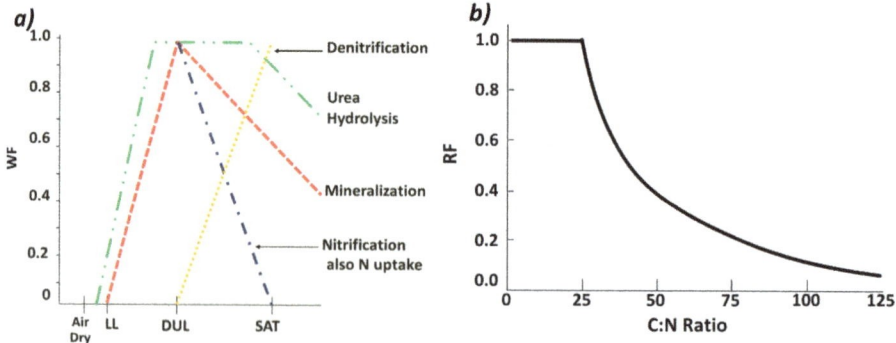

Fig. 3 The two coefficients, water content (WF) and residue compositions (RF), as considered within DSSAT. (Adapted from Godwin & Singh, 1998)

the layer L (kg N ha^{-1}). The A is an indicator that considers the factors discussed above and is calculated as

$$A = \min (RP2, WF, TF, PHN), \tag{4}$$

where RP2 is a 0 to 1 index for the nitrification potential based on the immediate past (Godwin & Singh, 1998), WF is the index discussed above (Fig. 3a), TF is the temperature factor for the rate of nitrification, and PHN is the factor for the effect of soil pH on rates of nitrification.

The DSSAT–CENTURY subroutine, on the other hand, contains three soil organic matter (SOM) pools (SOM1: microbial or active material; SOM2: recalcitrant substance, decomposed SOM1; SOM3: inert matter, and two FOM (metabolic litter which is easily decomposed, and structural litter which is recalcitrant fresh residue). The model estimates the fraction of soil SOM3 (stable fraction of C) in three different ways described in Porter et al. (2010), but they essentially comprise (i) direct input from the user; (ii) a field history which is based on a procedure developed by Basso et al. (2011a) to represent the effects of given management scenarios on the stable C fraction; and (iii) using a regression equation to estimate the stable C as:

$$StableC = 0.0015 * (Clay + Silt) + 0.069, \tag{5}$$

where stable C (SOM3), clay, and silt are expressed as g soil component/100 g soil. In this approach the ratre of decomposition for each organic matter pool depends first on order rate constants adjusted for temperature, soil water, agronomic management, and soil texture; it is calculated as follows:

$$\frac{dC}{dt} = -k_t * C, \tag{6}$$

$$k_t = k_0 * TF * WF * MF * TexF, \tag{7}$$

where C is organic C in the decomposing pool (kg ha^{-1}), t is the daily time step, k_0, k_t are the base and the modified rate constants (day^{-1}), TF is the temperature factor (Fig. 4a), WF is the water factor (Fig. 4b), MF is the management factor, and TexF is the texture factor.

In APSIM the organic matter is divided into the BIOM pool which is the labile fraction, a HUM pool, which is similar to the "slow" pool in the CENTURY model, and the rest is considered to be an inert pool. Overall, the flow among pools is modelled as carbon, and following the C:N ratio of the receiving pool, the N flows are derived. The user-defined input of the C:N ratio for the organic carbon can vary for each soil layer, and the C:N ratio of the FOM can also be modified. Like DSSAT, the decomposition of these pools follows a first-order calculation with rate constants and modifier (factors from 0 to 1) as a function of temperature and water for a given layer. The underlying assumptions for mineralization and immobilization are not too

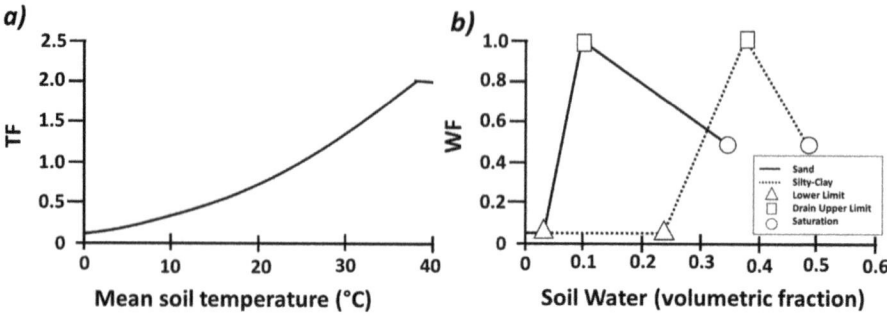

Fig. 4 The two coefficients, temperature factor (TF) and water content (WF) as considered within DSSAT–CENTURY model. (Adapted from Porter et al., 2010)

dissimilar to the CERES model. In APSIM the flows are represented in terms of the proportion of carbon retained from the system (as efficiency coefficients) and the amount of retained carbon that is synthesized in the BIOM pool. At the beginning of the simulation, the HUM and BIOM carbon in each soil layer are calculated from the user-defined inputs (Organic Carbon, Finert, Fbiom). For example, the inert pool is simply calculated as:

$$\text{Intert}_c = \text{Finert} * \text{Organic C.} \tag{8}$$

Decomposition of the fresh organic matter (FOM) is calculated as:

$$\text{FOM dec} = \text{Fpool} * \text{decay rate} * \text{WF} * \text{TF} * \text{CNF}, \tag{9}$$

where the Fpool is the fraction consisting of carbohydrate, cellulose, and lignin; decay rate is the decay rate of each fraction; and WF, TF, and CNF are the water, temperature, and C:N ratio factors, respectively. They range between 0 and 1 and like DSSAT they are calculated in a similar manner.

The rates of nitrification rates are calculated using the Michealis-Menten approach with a slight modification from the CERES approach:

$$\text{Potential Rate} = \text{Nitrification}_{\text{pot}} * \frac{\text{NH}_4(\text{L})}{\left(\text{NH}_4(\text{L}) + \text{NH}_4(\text{L})_{\text{half}_{\text{pot}}}\right)}. \tag{10}$$

Like the CERES approach, soil pH, temperature, and water reduce the potential rate as in Eq. (4).

In other modelling approaches such as CropSyst (Stöckle et al., 2003), soil mineral N is simulated as separate budgets for NO_3^- and NH_4^+. In the model, the simulated N processes include transformations, ammonium sorption, symbiotic N fixation, crop N demand, and crop N uptake (Stöckle et al., 2003). Soil N transformations (net mineralization, nitrification, and denitrification) and ammonium sorption are derived from the methodology developed by Stockle and Campbell (1989) where soil layers are divided into nodes where there is a transfer of mass and energy through them like an electric analog (Fig. 2b). The first node (Node 0) is the height at which weather data are measured; Node 1 is the zero-plane displacement height within the crop canopy; Node 2 is the soil surface, while the other nodes are the different soil depths. The transfer of heat and water through this system will not be described here, and interested readers might refer to Stockle and Campbell (1989). The N changes due to microbial activities are assumed to happen within the upper 30 cm of the soil; in this particular subroutine, they include net mineralization, nitrification, and denitrification. These are calculated as irreversible first-order kinetics (Stockle & Campbell, 1989) in which the change of N as s function of time is calculated as:

$$dN/_{dt} = -KN, \qquad (11)$$

where dt is a given time interval and K is a rate constant. The net mineralization of organic matter to NH_4^+, defined as the difference between mineralization and immobilization, is calculated as:

$$M = \left[M_0\left(1 - \exp^{(-K_M \Delta t)}\right)\right] F(f_{ps}) \qquad (12)$$

where M is the mineralized N (kg N m^{-2}) at time Δt (s) under optimum water content, M_0 (kg N m^{-2}) is the potentially mineralizable N at the beginning of Δt, $F(f_{ps})$ is a factor that is function of soil water content, and K_M is the rate constant of the mineralization (s^{-1}) and is calculated as function of the absolute soil temperature as:

$$K_M = \exp\left[17.753 - \frac{6350.5}{T_s + 273}\right]/(604,\, 800), \qquad (13)$$

where T_s is the soil temperature (°C) and a soil temperature of 35 °C is considered an upper threshold. Soil nitrification (N_{nit}), which is the amount of NH_4^+ converted into NO_3^- at a time Δt, is calculated as follows:

$$N_{nit} = \left[NH_{4ai}^+\left(1 - \exp^{(-K_N \Delta t)}\right)\right] F(f_{ps}), \qquad (14)$$

where NH_{4ai}^+ is the available amount of NH_4^+ that undergoes conversion at the beginning of Δt. The water factor ($F(f_{ps})$) uses the same concept illustrated above, but the model's parameters for its calculation differ as shown in Stockle and Campbell (1989). Finally, denitrification (N_{den}; $kg\ NO_3 m^{-2}$) is calculated as:

$$N_{den} = N_{den,0}\left[1 - \exp^{(-K_D \Delta t)}\right], \qquad (15)$$

where K_D (s^{-1}) is the denitrification rate constant and like the others is then adjusted for temperature and water effects and $N_{\text{den}, 0}$ is the available amount of NO$_3^-$ for denitrification at the beginning of Δt. The N balance at each node for NH$_4^+$ and NO$_3^-$ is derived as:

$$dN_{\text{nit}}/dt = \text{Flux IN} - \text{Flux OUT} + \text{Mineralization} - \text{Nitrification} - \text{Transport to Rizosphere} \tag{16}$$

$$dN_{\text{den}}/dt = \text{Flux IN} - \text{Flux OUT} + \text{Nitrification} - \text{Denitrification} - \text{Transport to Rizosphere} \tag{17}$$

2.2 N Dynamic in Plants (Demand)

Nitrogen in plants is found in many organs such as leaves, stems, grain, and roots. In the leaves, N is needed because of its role in the metabolic and structural functions in plants. In the plant tissue, N regulates growth (and to some extent development), but in the stem there is a given amount of N that is used as structural N. In the grain, N is stored as protein, and for some crops (e.g., barley, wheat), it has implications regarding its subsequent industrial use. The roots have N stored in proteins and enzymes to facilitate the water and N update from the soil to the plant.

The main component of N simulation in plants is the calculation of N demand, its accumulation, distribution among organs, translocation to grain, and the calculation of N factors affecting photosynthesis and growth processes.

Crop N demand is driven by changes in biomass for the various plant components and the N concentration in different plant components. To estimate N concentration, crop simulation models use the concept of minimum (N_{min}), critical (N_{crt}), and maximum (N_{max}). The N_{min} is structural N and cannot be re-translocated; N_{crt} is the minimum N that a plant tissue needs to maintain it and it drives N demand, and N_{max} is the maximum N concentration that allows a given tissue to accumulate more N than needed up to a given maximum threshold (Gregory et al., 1979). There are some slight changes in the way different models define those limits, but as an overall concept, it is rather conservative. For example, Stockle and Debaeke (1997) pointed out how four main crop simulation models use the relationship between plant N% as function of crop growth stage, fraction of the growth cycle, thermal time, and aboveground biomass to create the functional relationship of the N curves. These concentrations have given thresholds that vary with the different growth stages and different organs of the plants (Fig. 5).

Typically, it is assumed that N has little or no effect on crop phenology and only under very extreme conditions is phenology affected by N stress. Crop models do not account for the effects of N stress on phenology, although the APSIM model (Holzworth et al., 2014) has included an N stress factor that can be used in the future to consider N effects on phenology. This stress factor on phenology (f_{ph}) is

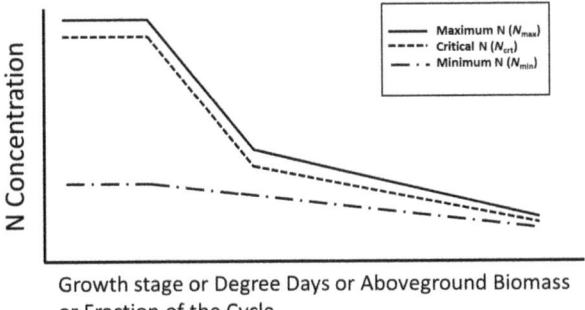

Fig. 5 Conceptualization of the relationship between N concentration at different growth stages for the maximum N (N_{max}; full line), critical N (N_{crt}; dashed line), and minimal N (N_{min}; dot-dash line). This relationship will be different for (i) different X-axis variables and (ii) for different plant organs (stem, leaf, grain)

calculated as the difference between the actual N concentration of an organ (N_{act}) and its N_{min} and N_{crt} as:

$$f_{ph} = N_{phe} \sum_{leaf} \frac{N_{act} - N_{min}}{N_{crt} * f_{Nstem} - N_{min}}, \qquad (18)$$

where N_{phe} is a multiplier for considering the effects of N deficit on phenology and it is a parameter with a default value of 1.5 and f_{Nstem} is a factor equal to 1 for stem and it is a relationship between leaf N_{crt} and the CO_2 concentration of the leaf. Currently, it is set to 1 as it has no effect. The f_{ph} can affect the calculation of thermal time and the reaching of a particular growth stage hence impacting phenology.

Leaves are the organs that require the most N, and any N deficit will limit the leaf area expansive growth processes. Under N stress, leaf area development is affected, causing an increase in specific leaf weight because leaf dry matter accumulation is less affected (Grindlay, 1997). The amount of N that is translocated in seed during the grain filling period is a function of the N accumulated in leaves and stem prior to this growth stage.

One modelling approach (mostly used in CERES-type of models) simulates the effects of N shortages through the calculation of some N reduction factors (N_f) that affect growth, and they are calculated from the ratio of actual to optimum N content (Godwin & Singh, 1998). This, in turn, will limit the leaf expansive growth processes, accelerate leaf senescence and the rate of photosynthesis, and is calculated as follows:

$$N_f = 1 - \frac{(N_{crt} - N_{act})}{(N_{crt} - N_{min})}. \qquad (19)$$

Several modelling approaches use different methods to calculate such stressors, but in the end they all affect biomass accumulation, the rate of leaf appearance and expansion, and grain filling. For example, in APSIM these stressors are calculated using the same functions as Eq. (1) (except for grain N), but where the first multiplicator (N_{phe}) differs among them.

In the CERES-like approach, the N demand comprises two components: (i) the deficiency N demand (DND) which is the N required to reach N_{crt} in the current actual aboveground biomass and (ii) growth N demand (GND) which is the N required to reach N_{crt} in the current potential aboveground biomass (ABG_{pot}). The GND is calculated as follows:

$$\text{GND} = \text{ABG}_{\text{pot}} * N_{\text{crt}}. \tag{20}$$

The potential N supply (N_{sup}) is calculated for each soil layer and is a function of root length density, soil water content, and root distribution. The total plant N demand (N_{dem}) of the crop takes into account (1) shoot growth demand (SDem); (2) shoot deficiency N demand (SDef); (3) root growth N demand (RDem); and (4) root deficiency N demand (RDef) so that:

$$N_{\text{dem}} = \text{SDem} + \text{SDef} + \text{RDem} + \text{RDef}. \tag{21}$$

The demand for N for each plant organ (e.g., leaf, stem) attempts at maintaining a level of organ N within the N_{crt}. Such demand is the sum of the demands from previous biomass that needs to reach N_{crt} in addition to the N required to keep N_{crt} for the day on which biomass is produced. This is calculated for each organ of the plant. For example, in APSIM for given organs (e.g., leaves, stem), the N demand (N_{dem}) is implemented as follows:

$$N_{\text{dem,crt}} = \frac{(\Delta Q_o N_{\text{crt}})}{f_{pht}} + f_n(N_{\text{crt}} - N_{\text{act}}) \text{ if } N_{\text{crt}} > N_{\text{crt,o}} \text{ and } Q_o > 0, \tag{22}$$

$$ND_{\text{dem, max}} = \frac{(\Delta Q_o N_{\text{max}})}{f_{pht}} + f_n(N_{\text{max}} - N_{\text{act}}) \text{ if } N_{\text{max}} > N_{\text{crt,o}} \text{ and } Q_o > 0, \tag{23}$$

where ΔQ_o is the dry matter growth of a given organ (e.g., leaves), Q_o is the green dry matter of the organ, f_{pht} is the soil water stress factor affecting biomass, and f_n is a parameter set at 0.0001.

The actual N uptake (N_{act}) is calculated by considering supply and demand simulated separately and then using the minimum of N_{dem} and N_{sup} in both APSIM and CERES models.

Jamieson and Semenov (2000) developed a method where N demand depends on the need to keep a given N concentration in new leaves (on an area base)—this will cause leaves to become thicker if there is not enough N for expansion; this method was also implemented in Soltani and Sinclair (2012). The stems act as reserves for extra N between N_{min} and N_{crt}. During the vegetative stage, the daily N uptake (N content) is calculated as a function of the needs of leaves and stems. The N_{dem} is calculated as follows:

$$N_{\text{dem}} = (\text{GLAI} * N_{\text{crt,leaf}}) + (\text{GST} * N_{\text{crt,stem}}), \qquad (24)$$

where GLA is the daily increase in leaf area index (LAI) and GST is the daily increase of stem dry matter. A factor accounting for N deficiencies in development at the early stages was added by Soltani and Sinclair (2012).

Grain N is simulated in several ways depending on the modelling approaches as reviewed in Rezaei et al. (2022). The simpler approach employs a harvest index approach in which it is assumed that the N harvest index increases linearly with thermal time from zero at the beginning of grain filling to a maximum N rate as a function of thermal accumulation. The other approach is a source-sink type of model. Overall, grain N is the result of the accumulation of N and carbohydrates in the grain. These are accumulated independently, and they are controlled by the potential rate of kernel growth and actual supply and are affected by air temperature.

3   Case Studies

3.1   Case Study 1: A Mediterranean Environment

Mediterranean environments are characterized by erratic rainfall patterns which affect the amount of water stored in the soil, crop production, and N fertilization strategies (Angus & Herwaarden, 2001). There is a balance between the water stored in the soil prior sowing, the fallow rainfall, and the growing season rainfall. All are important components for wheat production, but mismanagement of sowing time and N fertilization can alter their importance in determining optimal yield levels (Basso et al., 2011b). Nitrogen demand by wheat crops growing in such environments is often high in the late winter/spring, a period when expansive growth processes are high and when it difficult to match crop N demand with soil N supply (Sadras & Angus, 2006).

This case study shows how a calibrated crop simulation model can predict the spatial variation of durum wheat (*Triticum turgidum*, var. Durum; cultivar Duilio) yield in a field and how it is used to predict the weather-agronomy-soil interactions.

The study was carried out on a 12-ha field in Foggia, Italy (41° 27′ 47″ N, 15° 30′ 24″ E; 80 m a.s.l.), during 5 years of wheat monoculture (2005/2006; 2006/2007; 2007/2008; 2008/2009; 2009/2010). The N fertilization consisted of one application at sowing of 25 kg N ha^{-1} as diammonium phosphate and a second one at tillering with 65 kg N ha^{-1} as urea. Long-term weather data were recorded at an on-site station from 1953 to 2018 for daily solar radiation, minimum and maximum air temperature, and rainfall. The crop model DSSAT CERES-wheat (Hoogenboom et al., 2019) was calibrated to use with the dataset of Basso et al. (2010) from a long-term N fertilization trial with 0 and 90 kg N ha^{-1}. Once calibrated, the DSSAT-CERES model wheat was evaluated on an independent dataset; these were the data from the 12-ha field above mentioned (Fig. 6).

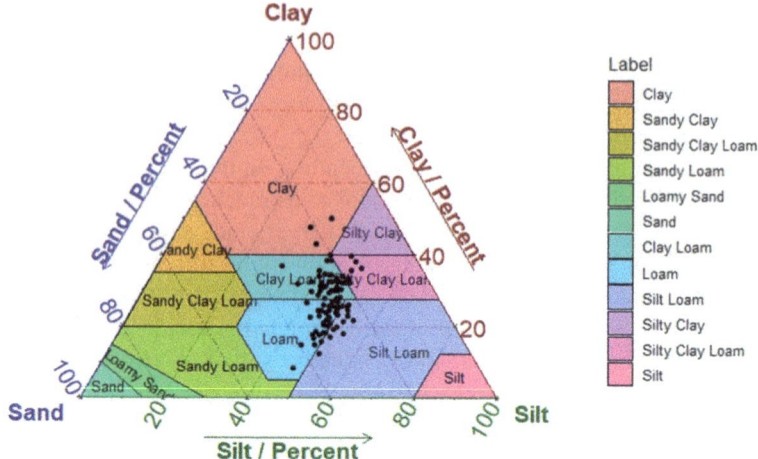

Fig. 6 Soil texture of the 100 points located on the 12-ha field in Foggia (Italy)

The long-term rainfall is shown in Fig. 7. Overall, the monthly cumulative rainfall showed a decrease in August, but the main trend was steady with a slight increase from 1953 to 2018 for January, February, and March. These 3 months are also the key periods for durum wheat growing in this region because the second N application is made in February/March. The results of the model's calibration and evaluation using the N trial experiment is shown in Table 1. The calibration of the DSSAT-wheat model showed RMSE values were in accord with ones reported in the literature; in addition, this calibration dataset had information on plant N% and on soil water content, which is information not often available to modelers.

Evaluation of the crop simulation model showed that the root mean square error (RMSE) was relatively small as a field average and was spatially and inter-annually variable, as shown in Fig. 8. Three growing seasons (2006, 2009, and 2010) had RMSEs above 1000 kg DM ha^{-1}. The simulated wheat yield is shown in Fig. 8b. The long-term wheat mean was 4000 kg DM ha^{-1} with evident spatial patterns across the field, with the top and bottom portions of the field showing higher values and the central section with lower values. A previously published study showed the map of soil resistivity for the field and identified the middle of the field having compacted soil and lower yields (Basso et al., 2012), which accords with the temporal patterns of simulated yield for that field section.

3.2 Case Study 2

The second case study is from a farmer's field in Scotland (UK) with spring barley (*Hordeum vulgare* L.; Cultivar Concerto) as the main cash crop. In such environments, spring barley is important for the distilling and brewing industry because it

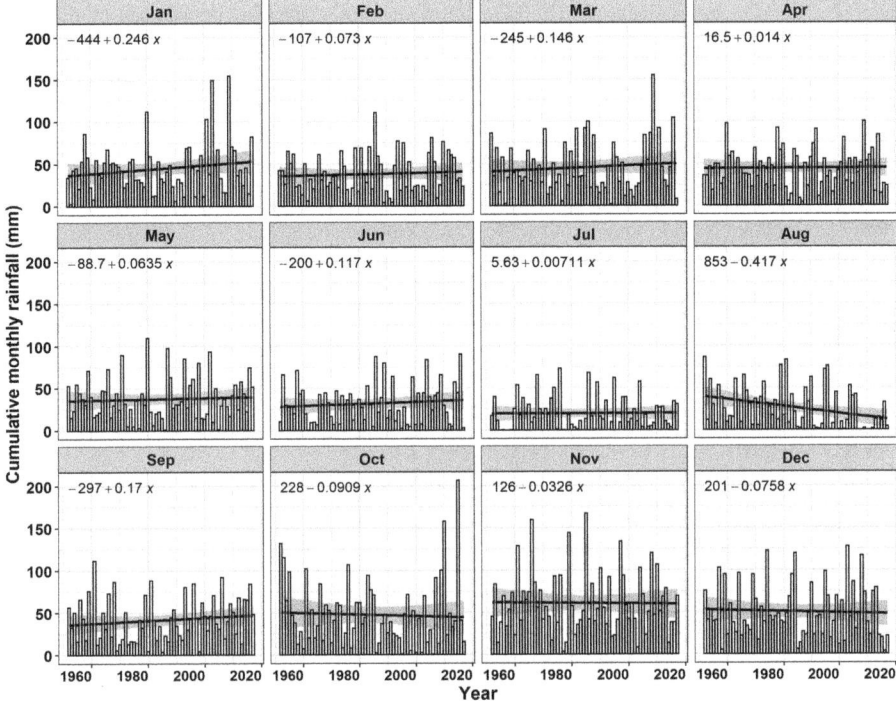

Fig. 7 Monthly cumulative rainfall from 1953 to 2017 for Foggia (Italy)

can be used for whisky or beer production (Cammarano et al., 2021). The grain N% requested by the industry is rather strict as it is tied with the requirement for malting; for example, an optimal grain N% range in the UK is between 1.45 and 1.85% (a range in which farmers are also paid a premium) (UK Malt, 2021).

The optimal grain N% range is rather narrow and can be adjusted by modifying sowing dates, genotypes, plant density, and fertilizer rates. For such a type of use, spring barley is usually sown in March/April and fertilized around sowing/emergence although it has been found that N rate has more effect on quality than timing (Overthrow, 2005).

To optimize the N fertilization strategy that maximize this three-way trade-off (quality-profit-environmental sustainability) long-term experiments are needed to account for the weather-soil-agronomy interactions.

Historical yield maps for this farm were available for 6 years and were used to define management zones using the algorithms developed by Maestrini and Basso (2018). Additional details on how management zones were defined are reported in Maestrini and Basso (2018) and Cammarano et al. (2021). Four yield stability zones were defined in the field as high (HYZ)-, medium (MYZ)-, low (LYZ)-, and unstable (UYZ)-yield zones. Figure 9 shows the spatial yields from overlaying the 6 years of yield maps (Fig. 9a), the temporal variation (Fig. 9b), the four yield stability zones

Table 1 Results of the model's calibration using the dataset from Basso et al. (2010) from Foggia, Italy

Treatment	Variable	Units	Simulated	Observed	Standard deviation observation	RMSE[a]
0 N	Biomass	(kg DM ha^{-1})	872	1045	108.7	826
0 N	Biomass		4135	2876	389.0	
90 N	Biomass		1426	1912	187.4	
90 N	Biomass		5805	4868	140.4	
0 N	Plant N	(%)	2.68	3.01	0.17	0.42
0 N	Plant N		1.52	1.84	0.29	
90 N	Plant N		3.11	2.71	0.17	
90 N	Plant N		1.79	2.37	0.51	
0 N	LAI	–	0.50	0.74	–	0.47
0 N	LAI		1.33	1.07		
0 N	LAI		1.57	1.79		
90 N	LAI		0.77	1.42		
90 N	LAI		2.17	2.70		
90 N	LAI		2.44	3.14		
0 N	Soil water	(%)	0.27	0.26	0.01	0.006
0 N	Soil water		0.22	0.22	0.01	
0 N	Soil water		0.17	0.17	0.01	
0 N	Soil water		0.07	0.07	0.01	
90 N	Soil water		0.28	0.28	0.02	
90 N	Soil water		0.23	0.23	0.02	
90 N	Soil water		0.17	0.15	0.02	
90 N	Soil water		0.07	0.06	0.01	
0 N	Yield	(kg DM ha^{-1})	1314	1173	161	99
90 N	Yield		1981	1986	190	

[a]Root mean square error

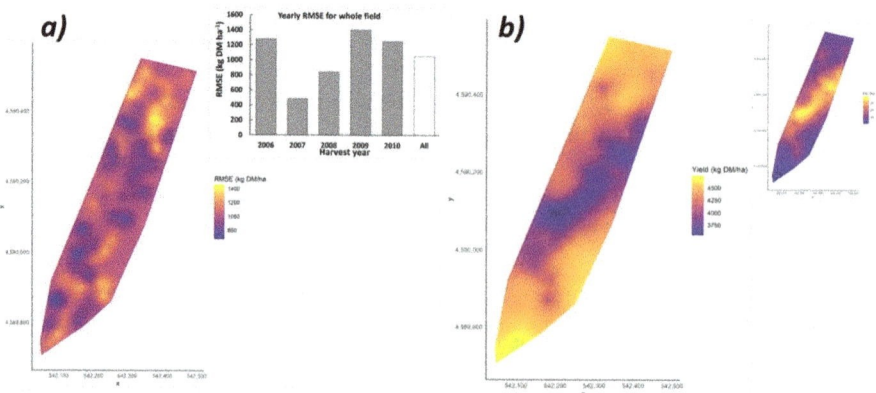

Fig. 8 Results of the (**a**) spatial evaluation of the DSSAT CERES-wheat model for the 5 years of observed spatial yield data; in the inset there are the field average values of the model comparison; and (**b**) mean simulated wheat dry yield for the 1953–2018 continuous simulations and top-right (insert) the coefficient of variation of simulated yield

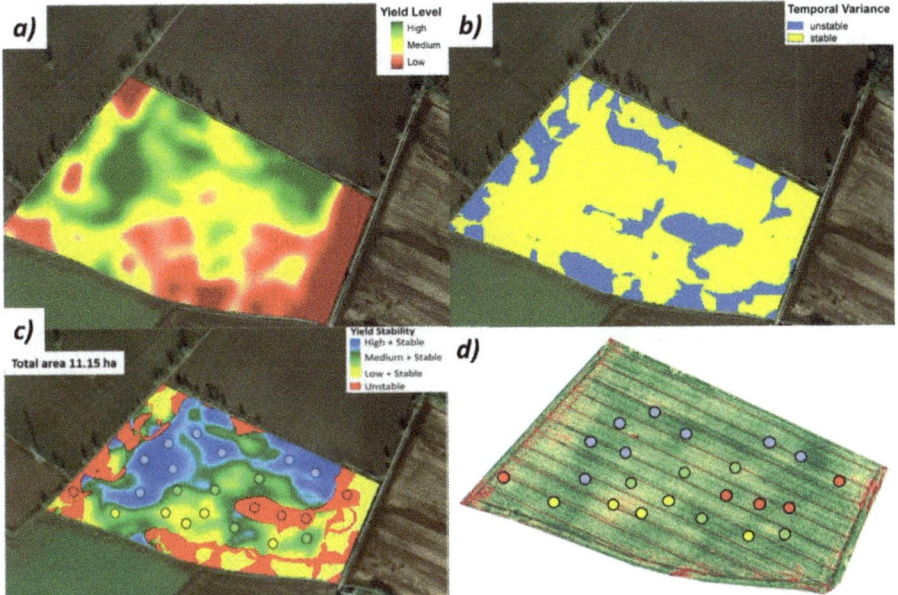

Fig. 9 Spatial distribution of the (**a**) yield and (**b**) temporal variation obtained from overlaying 6 years of spatial yield maps; (**c**) the yield stability zones obtained using the Maestrini and Basso (2018) methods and the corresponding points where soil and plant samples were taken (adapted from Cammarano et al. (2020)); and (**d**) the sampling points overlaid on the NDVI map taken 3 weeks after *N* fertilization (high-yield zone: blue dots; medium-yield zone: green dots; low-yield zone: yellow dots; unstable-yield zone: red dots)

(Fig. 9c), and the overlay of the sampling points for each zone with the NDVI map taken 3 weeks after *N* fertilization (Fig. 9d). At each point soil samples (0–90 cm) and plant samples were collected during the growing season. Soil physical and chemical properties, as well as soil initial conditions in terms of water and N, were recorded 1 month prior sowing; additional details are reported in Cammarano et al. (2021).

The DSSAT-barley crop model was used as described in Cammarano et al. (2019) and Cammarano et al. (2021). The overall evaluation is summarized in Table 2 for the main properties.

Results of soil texture show how the soil physical structure differs within the field and at depth for the different points. The soil has a loamy texture for the first 30 cm and then varies between clay and sandy loam for other points in the field at 60 cm. For the lowest soil layer, examine the soil type ranged between c lay and sandy loam with a wider spread of textural class among the points which is also reflected in the changes in soil bulk density as illustrated in Cammarano et al. (2020).

For the growing season of 2018, rainfall was very low for this type of environment; this is becoming increasingly common and could have practical implications for *N* management and grain *N* quality in the future (Cammarano et al., 2019). The growing season rainfall (Fig. 10) was low or negligible during the vegetative stage.

Table 2 Regression equation parameters between observed and simulated data and root mean square error (RMSE) used for evaluating the goodness of fit of the simulations

Treatment	Unit	a^*	b^+	R^{2**}	n^{++}	$RMSE^-$
SWC 0–0.3 m	$(m^3\ m^{-3})$	−0.08	1.30	0.68	115	0.04
SWC 0.3–0.6 m	$(m^3\ m^{-3})$	0.03	0.94	0.64	115	0.03
Total soil N	$(kg\ N\ ha^{-1})$	56.36	0.67	0.49	115	58.06
Plant N	(%)	104.57	−16.13	0.79	47	0.31
Plant aboveground biomass	$(kg\ DM\ ha^{-1})$	−235.19	0.91	0.69	115	1229.11
Grain yield	$(kg\ DM\ ha^{-1})$	1765.38	0.44	0.60	23	600.7
Grain N	(%)	0.56	0.55	0.35	4	0.06

*Intercept; $^+$slope; **coefficient of determination; $^{++}$number of values; $^-$root mean square error

The interaction between N uptake from the soil and the amount of water held in the soil profile affected the growth of spring barley for that particular growing season. By flowering the LYZ and UYZ did not take up the residual soil N, which was later lost due to more frequent rainfall (Fig. 10).

To identify the optimal N management scenario a series of "what-ifs" N management scenarios were set up with the crop model using a long-term weather series (1984–2017) with the weather data obtained from the UKCP09: Gridded observation datasets (MetOffice, 2019). For this setup, a series of N fertilization amounts was selected based on previous experimental evidence on spring barley in Scotland (Overthrow, 2005). Ten N fertilizer amounts were simulated starting from $20\ kg\ N\ ha^{-1}$ to $200\ kg\ N\ ha^{-1}$ with a $20\ kg\ N\ ha^{-1}$ increase. The rationale for setting up such long-term simulations was because we wanted to quantify the effects of weather on soil-agronomy interactions and highlight the response of the three-way trade-off to different N management strategies. The simulation setup done in this way allows us to account for the probability of weather events that can happen in the near future (e.g., a dry year like 2018 was also observed in the past and might occur again in the near future).

A detailed account of the calculation and demonstration of the three-way trade-off is described elsewhere for each of the management zones (Cammarano et al., 2021). Figure 11 showed that in Year 2 the conditions of high N leaching and low profit were simulated for the field in which $100\ kg\ N\ ha^{-1}$ would optimize the trade-off.

4 Conclusions

In conclusion, there are several approaches for simulating soil and plant N dynamics, and there is not a universal one. Several attempts are being made by the global crop modelling community to compare and improve modelling routines (e.g., see AgMIP). Conceptually, with a crop model that can be applied in a precision agriculture context, the compromise is to have a model that (i) requires a few inputs; (ii) the inputs are easy to obtained by the user; (iii) can work on a range of soil types;

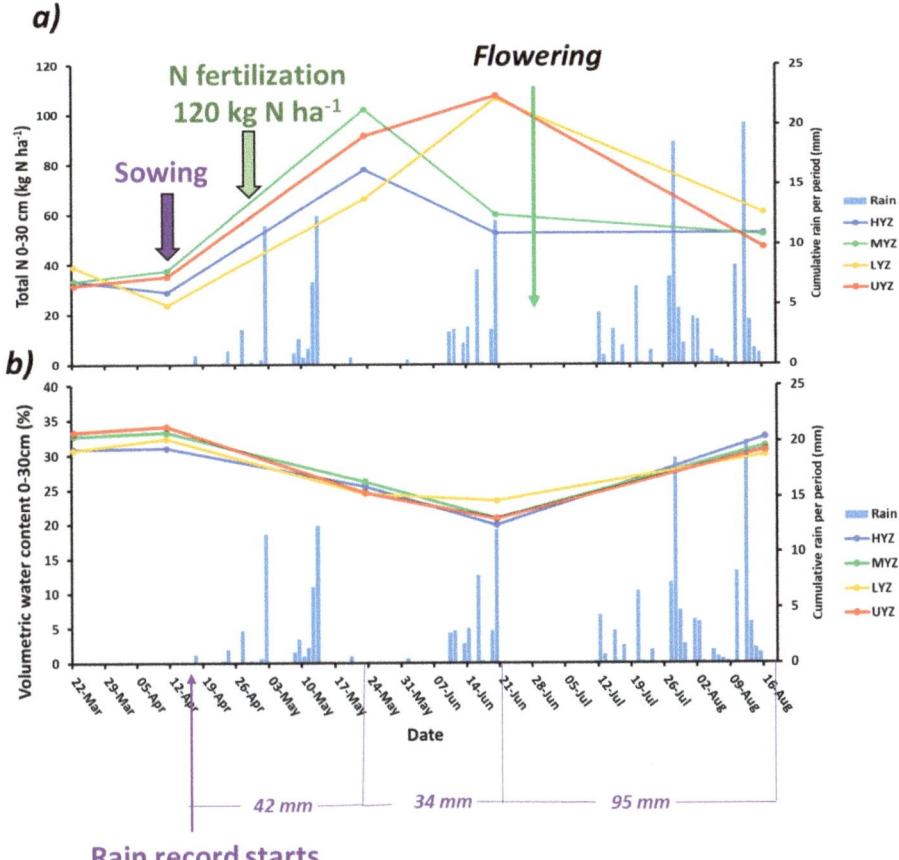

Fig. 10 Experimental observations of (**a**) total soil N (0–30 cm) for the four different yield stability zones (high-yield zone: blue line and dots; medium-yield zone: green line and dots; low-yield zone: yellow line and dots; unstable-yield zone: red line and dots) and (**b**) volumetric soil water content at 0–30 cm. For both figures the secondary Y-axis represents the cumulative monthly rainfall (light blue bars)

and (iv) can simulate the fate of residues of different compositions. Ultimately, the desired crop model requires reliable simulations of the processes involved. Significant breakthroughs in improving the simulation of soil N dynamics are still lacking, and that is due to the following:

(a) The lack of interdisciplinarity in studying the overall system. For example, the crop modelling temperature routines were compared and improved by working with researchers, breeders, crop physiologists, and crop modelers. This leads to tangible results from which the scientific community benefited (Alderman et al., 2013; Asseng et al., 2015; Maiorano et al., 2016; Wang et al., 2017).

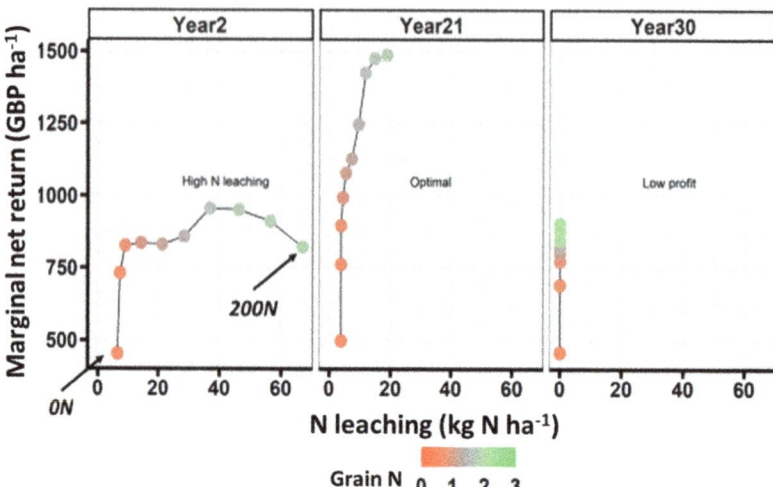

Fig. 11 Relationship between the mean marginal net return (GBP ha^{-1}) and soil N leaching (kg N ha^{-1}). The intensity of the color represents the different levels of grain N% with the grey color being the optimum grain N % content of 1.5%. Each dot represents a given N fertilizer level

(b) Testing it in limited time frames (usually 2–4 years) is not enough to elucidate detailed mechanisms.

Conflict of Interest The authors declare that there are no conflicts of interest.

References

Alderman, P. D., Qulligan, E., Asseng, S., Ewert, F., & Reynolds, M. P. (2013). *Proceedings of the workshop modelling wheat response to high temperature*. CIMMYT.

Angus, J. F. (2001). Nitrogen supply and demand in Australian agriculture. *Australian Journal of Experimental Agriculture, 41*, 277–288. https://doi.org/10.1071/EA00141

Angus, J. F., & van Herwaarden, A. F. (2001). Increasing water use and water use efficiency in dryland wheat. *Agronomy Journal, 93*(2), 290–298. https://doi.org/10.2134/agronj2001.932290x

Asseng, S., et al. (2015). Rising temperatures reduce global wheat production. *Nature Climate Change, 5*, 5. https://doi.org/10.1038/nclimate2470

Asseng, S., et al. (2013). Uncertainty in simulating wheat yields under climate change. *Nature Climate Change, 3*, 827. https://doi.org/10.1038/nclimate1916. https://www.nature.com/articles/nclimate1916#supplementary-information

Basso, B., Cammarano, D., Troccoli, A., Chen, D., & Ritchie, J. T. (2010). Long-term wheat response to nitrogen in a rainfed Mediterranean environment: Field data and simulation analysis. *European Journal of Agronomy, 33*, 132–138. https://doi.org/10.1016/j.eja.2010.04.004

Basso, B., Fiorentino, C., Cammarano, D., Cafiero, G., & Dardanelli, J. (2012). Analysis of rainfall distribution on spatial and temporal patterns of wheat yield in Mediterranean environment. *European Journal of Agronomy, 41*, 52–65. https://doi.org/10.1016/j.eja.2012.03.007

Basso, B., Gargiulo, O., Paustian, K., Robertson, G. P., Porter, C., Grace, P. R., & Jones, J. W. (2011a). Procedures for initializing soil organic carbon pools in the DSSAT-CENTURY model for agricultural systems. *Soil Science Society of America Journal, 75*, 69–78. https://doi.org/10. 2136/sssaj2010.0115

Basso, B., Ritchie, J. T., Cammarano, D., & Sartori, L. (2011b). A strategic and tactical management approach to select optimal *N* fertilizer rates for wheat in a spatially variable field. *European Journal of Agronomy, 35*, 215–222. https://doi.org/10.1016/j.eja.2011.06.004

Bundy, L. G., & Andraski, T. W. (1995). Soil yield potential effects on performance of soil nitrate tests. *Journal of Production Agriculture, 8*, 561–568. https://doi.org/10.2134/jpa1995.0561

Cammarano, D., Basso, B., Holland, J., Gianinetti, A., Baronchelli, M., & Ronga, D. (2021). Modeling spatial and temporal optimal *N* fertilizer rates to reduce nitrate leaching while improving grain yield and quality in malting barley. *Computers and Electronics in Agriculture, 182*, 105997. https://doi.org/10.1016/j.compag.2021.105997

Cammarano, D., et al. (2019). Rainfall and temperature impacts on barley (Hordeum vulgare L.) yield and malting quality in Scotland. *CropM, ft_macsur, 241*, 107559. https://doi.org/10.1016/ j.fcr.2019.107559

Cammarano, D., Holland, J., & Ronga, D. (2020). Spatial and temporal variability of spring barley yield and quality quantified by crop simulation model. *Agronomy, 10*, 393. https://doi.org/10. 3390/agronomy10030393

Cammarano, D., et al. (2016). Uncertainty of wheat water use: Simulated patterns and sensitivity to temperature and CO2. *CropM, ft_macsur, 198*, 80–92. https://doi.org/10.1016/j.fcr.2016. 08.015

Cassman, K. G., & Dobermann, A. (2022). Nitrogen and the future of agriculture: 20 years on. *Ambio, 51*, 17–24. https://doi.org/10.1007/s13280-021-01526-w

Coskun, D., Britto, D. T., Shi, W., & Kronzucker, H. J. (2017). Nitrogen transformations in modern agriculture and the role of biological nitrification inhibition. *Nature Plants, 3*, 17074. https://doi. org/10.1038/nplants.2017.74

Dalgaard, T., et al. (2014). *Farming systems models for regional scale impact assessment in Europe – Case studies of N-losses and greenhouse gas emissions.* Aarhus University.

Davis, K. F., Downs, S., & Gephart, J. A. (2021). Towards food supply chain resilience to environmental shocks. *Nature Food, 2*, 54–65. https://doi.org/10.1038/s43016-020-00196-3

EU. (1991). *Council Directive 91/676/EEC of 12 December 1991 concerning the protection of waters against pollution caused by nitrates from agricultural sources.*

Franzen, D., Kitchen, N., Holland, K., Schepers, J., & Raun, W. (2016). Algorithms for in-season nutrient management in cereals. *Agronomy Journal, 108*, 1775–1781. https://doi.org/10.2134/ agronj2016.01.0041

Gijsman, A. J., Hoogenboom, G., Parton, W. J., & Kerridge, P. C. (2002). Modifying DSSAT crop models for low-input agricultural systems using a soil organic matter–residue module from Century. *Agronomy Journal, 94*, 462–474. https://doi.org/10.2134/agronj2002.4620

Godwin, D. C., & Singh, U. (1998). Nitrogen balance and crop response to nitrogen in upland and lowland cropping systems. In G. Y. Tsuji, G. Hoogenboom, & P. K. Thornton (Eds.), *Understanding options for agricultural production.* Kluwer Academic Publisher.

Gregory, P. J., Crawford, D. V., & McGowan, M. (1979). Nutrient relations of winter wheat: 1. Accumulation and distribution of Na, K, Ca, Mg, P, S and N. *The Journal of Agricultural Science, 93*, 485–494. https://doi.org/10.1017/S0021859600038181

Grindlay, D. J. C. (1997). REVIEW towards an explanation of crop nitrogen demand based on the optimization of leaf nitrogen per unit leaf area. *The Journal of Agricultural Science, 128*, 377–396. https://doi.org/10.1017/S0021859697004310

Holzworth, D. P., et al. (2014). APSIM – Evolution towards a new generation of agricultural systems simulation. *CropM, 62*, 327–350. https://doi.org/10.1016/j.envsoft.2014.07.009

Hoogenboom, G., et al. (2019). The DSSAT crop modeling ecosystem. In K. J. Boote (Ed.), *Advances in crop modeling for a sustainable agriculture.* Burleigh Dodds Science Publishing. https://doi.org/10.19103/AS.2019.0061.10

Jamieson, P. D., & Semenov, M. A. (2000). Modelling nitrogen uptake and redistribution in wheat. *CropM, ft_macsur, 68*, 21–29. https://doi.org/10.1016/S0378-4290(00)00103-9

Jin, Z., Archontoulis, S. V., & Lobell, D. B. (2019). How much will precision nitrogen management pay off? An evaluation based on simulating thousands of corn fields over the US Corn-Belt. *CropM, ft_macsur, 240*, 12–22. https://doi.org/10.1016/j.fcr.2019.04.013

Maestrini, B., & Basso, B. (2018). Predicting spatial patterns of within-field crop yield variability. *CropM, ft_macsur, 219*, 106–112. https://doi.org/10.1016/j.fcr.2018.01.028

Maiorano, A., et al. (2016). Crop model improvement reduces the uncertainty of the response to temperature of multi-model ensembles. *CropM, ft_macsur, 202*, 5–20.

Mandrini, G., Bullock, D. S., & Martin, N. F. (2021). Modeling the economic and environmental effects of corn nitrogen management strategies in Illinois. *CropM, ft_macsur, 261*, 108000. https://doi.org/10.1016/j.fcr.2020.108000

MetOffice, U. (2019) *UKCP09 gridded observation datasets*. https://www.metoffice.gov.uk/climate/uk/data/ukcp09. Accessed 24 May 2019.

Morris, T. F., et al. (2018). Strengths and limitations of nitrogen rate recommendations for corn and opportunities for improvement. *Agronomy Journal, 110*, 1–37. https://doi.org/10.2134/agronj2017.02.0112

Overthrow, R. (2005). *Nitrogen management in spring malting barley for optimum yield and quality* (Report #367. Home Grown Cereals Authority (HGCA)). Daglingworth Cirencester.

Parton, W. J., Stewart, J. W. B., & Cole, C. V. (1988). Dynamics of C, N, P and S in grassland soils: A model. *Biogeochemistry, 5*, 109–131. https://doi.org/10.1007/BF02180320

Porter, C. H., Jones, J. W., Adiku, S., Gijsman, A. J., Gargiulo, O., & Naab, J. B. (2010). Modeling organic carbon and carbon-mediated soil processes in DSSAT v4.5. *Operational Research, 10*, 247–278. https://doi.org/10.1007/s12351-009-0059-1

Puntel, L. A., et al. (2018). A systems modeling approach to forecast corn economic optimum nitrogen rate. *Frontiers in Plant Science, 9*. https://doi.org/10.3389/fpls.2018.00436

Raun, W. R., & Johnson, G. V. (1999). Improving nitrogen use efficiency for cereal production. *Agronomy Journal, 91*, 357–363. https://doi.org/10.2134/agronj1999.00021962009100030001x

Raun, W. R., et al. (2005). Optical sensor-based algorithm for crop nitrogen fertilization. *Communications in Soil Science and Plant Analysis, 36*, 2759–2781. https://doi.org/10.1080/00103620500303988

Rezaei, E. E., Rojas, L. V., Zhu, W., & Cammarano, D. (2022). The potential of crop models in simulation of barley quality traits under changing climates: A review. *Field Crop Research, 286*. https://doi.org/10.1016/j.fcr.2022.108624

Sadras, V. O., & Angus, J. F., (2006). Benchmarking water-use efficiency of rainfed wheat in dry environments. *Australian Journal of Agricultural Research, 57*, 847–856

Seligman, N. G., & Van Keulen, H. (1981). PAPRAN: A simulation model of annual pasture production limited by rainfall and nitrogen. In M. J. Frissel & J. A. Van Veen (Eds.), *Simulation of nitrogen behaviour of soil-plant systems* (pp. 192–220). Pudoc Wageningen.

Solie, J. B., Monroe, A. D., Raun, W. R., & Stone, M. L. (2012). Generalized algorithm for variable-rate nitrogen application in cereal grains. *Agronomy Journal, 104*, 378–387. https://doi.org/10.2134/agronj2011.0249

Soltani, A., & Sinclair, T. R. (2012). *Modeling physiology of crop development, growth and yield*. CABI. https://doi.org/10.1079/9781845939700.0000

Stanford, G. (1973). Rationale for optimum nitrogen fertilization in corn production. *Journal of Environmental Quality, 2*, 159–166. https://doi.org/10.2134/jeq1973.00472425000200020001x

Stockle, C. O., & Campbell, G. S. (1989). Simulation of crop response to water and nitrogen: An example using spring wheat. *Transactions of the ASAE, 32*, 66–0074. https://doi.org/10.13031/2013.30964

Stockle, C. O., & Debaeke, P. (1997). Modeling crop nitrogen requirements: A critical analysis. *European Journal of Agronomy, 7*, 161–169. https://doi.org/10.1016/S1161-0301(97)00038-5

Stöckle, C. O., Donatelli, M., & Nelson, R. (2003). CropSyst, a cropping systems simulation model. *European Journal of Agronomy, 18*, 289–307. https://doi.org/10.1016/S1161-0301(02)00109-0

Tian, X., Engel, B. A., Qian, H., Hua, E., Sun, S., & Wang, Y. (2021). Will reaching the maximum achievable yield potential meet future global food demand? *Journal of Cleaner Production, 294*, 126285. https://doi.org/10.1016/j.jclepro.2021.126285

UK Malt. (2021). *The Maltsers' Association of Great Britain.* http://www.ukmalt.com. Accessed Jul 2021.

Wang, E., et al. (2017). The uncertainty of crop yield projections is reduced by improved temperature response functions. *Nature Plants, 3*, 17102. https://doi.org/10.1038/nplants.2017.102

Modelling Soil Water Dynamics

Marius Heinen

1 Introduction

The International Society of Precision Agriculture defined precision agriculture as follows: 'Precision Agriculture is a management strategy that gathers, processes and analyzes temporal, spatial and individual data and combines it with other information to support management decisions according to estimated variability for improved resource use efficiency, productivity, quality, profitability and sustainability of agricultural production' (www.ispag.org). Bouma (2007) stated more concisely that precision agriculture aims at adjusting and fine-tuning land and crop management to the needs of plants within heterogeneous fields. Besides a reactive approach, using yield maps and sensors, Bouma (2007) also promotes the proactive approach in which simulation models for plant growth and soil processes can be used to predict optimal timing of management practices. The attention required for crop modelling is evident, as high yields of good quality mean a good profit for the farmers. This profit can be increased if the use of resources, such as fertilizers, land management practices and irrigation water, is minimized. Minimizing the use of resources means that we are at the cutting edge. For example, we are balancing between adding too much or too little water and fertilizers. This requires good knowledge on the interaction of soil–water–plant–atmosphere processes. Crop modelling is discussed in chapter "Process Based Modelling of Soil–Crop Interactions for Site Specific Decision Support in Crop Management" of this book, and here we will focus on modelling the dynamics of water in soils. Modelling nutrient dynamics in soils is discussed in chapter "Process-Based Models and Simulation of Nitrogen Dynamics" of this book; see also Heinen et al. (2020a).

M. Heinen (✉)
Wageningen Environmental Research, Wageningen, The Netherlands
e-mail: marius.heinen@wur.nl

© The Author(s), under exclusive license to Springer Nature Switzerland AG 2023   129
D. Cammarano et al. (eds.), *Precision Agriculture: Modelling*, Progress in Precision
Agriculture, https://doi.org/10.1007/978-3-031-15258-0_6

2 Water Movement in Soils

2.1 Law of Conservation of Matter

Water movement in soil always obeys the law of conservation of matter or the law of continuity, here written as (Fig. 1)

$$\frac{\partial \theta}{\partial t} = -\nabla \cdot (\theta \mathbf{v}) - S_w = -\nabla \cdot \mathbf{q} - S_w, \qquad (1)$$

where θ is the volumetric water content ($L^3 L^{-3}$), t is time (T), $\nabla \cdot$ is the divergence operator (L^{-1}), $\mathbf{v}$ is the velocity of the water (LT^{-1}), $\mathbf{q} = \theta \mathbf{v}$ is the volumetric water flux density ($L^3 L^{-2} T^{-1}$) and S_w is the volumetric sink strength for water ($L^3 L^{-3} T^{-1}$), for example root water uptake. In general, all dependent variables in this section, such as θ, $\mathbf{q}$ and S_w, are functions of the independent variables x, y, z and t, where x and y are the horizontal coordinates (L) and z is the vertical coordinate (L), taken positive downwards. For convenience, the spatial and temporal functional dependency is omitted from the notation. Note that here we consider only a sink term, but sources could also be introduced in Eq. (1).

2.2 Darcy–Buckingham

For saturated porous media, Darcy (1856) experimentally obtained that $\mathbf{q}$ is proportional to the gradient in water potential. Buckingham (1907) extended Darcy's law to partially saturated porous media. With the water potential expressed as a head equivalent, the Darcy–Buckingham equation is given by

Fig. 1 The change in water content is due to the difference between incoming and outgoing water fluxes and a sink term

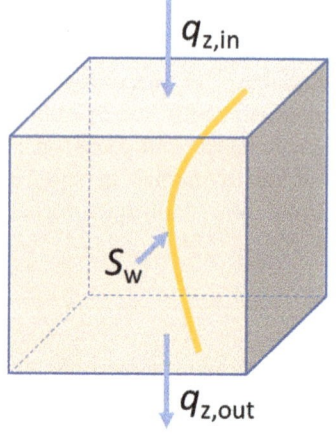

$$q = -K(\theta)\nabla H = -K(\theta)\nabla(h - z_g), \tag{2}$$

where K is the hydraulic conductivity (LT^{-1}) of the porous medium, ∇ is the gradient operator (L^{-1}) and H is the hydraulic head (L) being the sum of the pressure head h (L) and the gravitational head z_g (L) (the minus sign in front of z_g results from z taken positive downwards). The direction of flow is opposite to the gradient in total head, which explains the minus sign in Eq. (2). The dependent variables K, H and h are functions of x, y, z and t as defined above.

2.3 Richards Equation

Richards (1931) combined the Darcy–Buckingham equation and the continuity equation to obtain the governing partial differential flow equation for water in porous media. For variably saturated, heterogeneous, isotropic, rigid, isothermal soils and incompressible water, the Richards equation is

$$\frac{\partial \theta}{\partial t} = \nabla \cdot (K(\theta)\nabla h(\theta)) - \frac{\partial K(\theta)}{\partial z} - S_w. \tag{3}$$

Note that the Richards equation contains three dependent variables (θ, h, K) which are assumed to be mutually related. This will be described further in Sect. 4. In fact, such closed-form relationships must be known in order to solve Eq. (3). Since the mutual (θ, h, K) relationships appear to be (highly) non-linear, the Richards equation is a non-linear partial differential equation and has to be solved numerically. In soil–crop modelling for situations where water movement in the soil is dominantly in the vertical direction only, the Richards equation in one dimension is given by

$$\frac{\partial \theta}{\partial t} = \frac{\partial}{\partial z}\left(K(\theta)\frac{h(\theta)}{\partial z}\right) - \frac{\partial K(\theta)}{\partial z} - S_w, \tag{4}$$

where the first term on the right-hand side describes the variation of the vertical water flux density with depth, the second term refers to the contribution of gravity and the third term is the sink term. One-dimensional approaches are no longer valid when typical multidimensional situations need to be studied, such as studying the extent of the wetted soil underneath a dripper in drip-irrigated crops or simulating flow patterns around a drainage pipe.

2.4 Difference with Tipping-Bucket

In some crop–soil models, water in soil is modelled by the so-called tipping-bucket approach. This approach assumes that the water content in each soil layer varies between two typical values, and when its maximum value is reached, the excess water is added to the next soil layer. Such an approach has the disadvantage that upward movement of water is not possible (although some workarounds have been proposed) and near-saturated conditions are not simulated (unless its maximum water-holding capacity is set at near saturation). Buttler and Riha (1992) and Verburg (1996) showed that both the tipping-bucket approach and the Richards approach can predict soil water contents satisfactorily. However, this does not guarantee that all other water balance components are predicted equally well (Verburg, 1996). For example, Shelia et al. (2017) showed that subsequently calculated crop growth differed when based on tipping-bucket or Richards soil water calculations, especially under water-limited conditions. Boote (2008) stated that a tipping-bucket model works satisfactorily when the soil water-holding properties are well-known and when the root distribution and rooting depth are predicted accurately. However, Scanlon et al. (2002) said that this approach is inherently more approximate than soil water movement simulations based on the Richards equation. In general, the soil water fluxes calculated by tipping-bucket models are not accurate enough to quantify solute transport (Van Dam et al., 2004). Because of its limitations, the tipping-bucket approach is not advocated when studying or modelling situations for precision agriculture. Each of the two approaches has its own advantages and disadvantages, which, according to Shelia et al. (2017), should be tested in future studies. In cases where not enough data are available to parameterize the Richards approach, and data needed for the tipping-bucket approach are present, one could decide to use the tipping-bucket model.

2.5 How to Solve Richards Equation?

As for ordinary differential equations, also for non-linear partial differential equations, the initial and boundary conditions need to be known to solve the problem and in our case also the constitutive relationships between θ–h–K. These conditions and relationships will be elaborated in the next sections. Even with known initial and boundary conditions, in most cases, solution of the Richards equation cannot be obtained analytically. Therefore, the governing flow equation for water in soils must be done with numerical techniques. It goes beyond the scope of this chapter to describe these techniques in detail. In general, the Richards equation can be solved by a finite difference, a finite element or a control volume approach (Wang & Anderson, 1982; Patankar, 1980). All these methods subdivide the soil in small layers or elements, and for each of these, the Richards equation is set up. This results in a set of N equations for N layers/elements. These equations contain the unknown values of the pressure head in space and time, coefficients and known quantities. This system can be represented by the following matrix notation:

$$\mathbf{Ax} = \mathbf{b}, \tag{5}$$

where $\mathbf{A}$ is the coefficient matrix, $\mathbf{x}$ is the vector containing the unknown values for h and $\mathbf{b}$ is the vector containing known quantities. For typical one-dimensional models where the soil is divided into horizontal layers, $\mathbf{A}$ is a typical tri-diagonal matrix, so that Eq. (5) can be solved easily using a tri-diagonal solver (Press et al., 1992). In other cases, alternative solvers are needed, for example an incomplete Cholesky conjugate gradient method (Meijerink & Van Der Vorst, 1977). One important issue is the conservation of mass during the numerical solution procedure. Celia et al. (1990) proposed an iterative scheme for the mixed θ–h Richards equation (Eq. 4) which explicitly takes into account that mass is conserved.

3   Boundary and Initial Conditions

Solving the Richards equation requires starting values for the unknown pressure head (initial condition) and requires information on what applies at the boundaries of the isolated system that is being modelled. The initial condition is given by

$$h(z; t = 0) = h_0(z). \tag{6}$$

Note that for simplicity we consider in this section a one-dimensional situation, but this can be easily extended to two or three dimensions.

Mathematically, we can consider three types of boundary conditions (e.g. McCord, 1991): the first type or Dirichlet condition refers to a known status (pressure head) at the boundary:

$$h(z = z_\Gamma; t > 0) = h_\Gamma; \tag{7}$$

the second type or Neumann condition refers to a known flux perpendicular to the boundary:

$$\frac{\partial (h - z)}{\partial z} (z = z_\Gamma; t > 0) = q_\Gamma; \tag{8}$$

and the third type or Cauchy boundary conditions is a combination of the former two. The subscript Γ refers to a boundary. In practice, soil models make use of the first two types, which may be either constant or variable in time. For example, a Dirichlet bottom boundary condition can be used if the groundwater level is used as input. It may also be used temporarily at the soil surface if water is ponding there. The Neumann or flux condition typically refers to input and output fluxes at the soil surface resulting from rainfall, irrigation and soil evaporation. Flux boundary conditions at the bottom of a soil column can be used if seepage fluxes are known, or in case free drainage is considered (i.e. $q = K$). For one-dimensional models, there are no boundary conditions at the vertical planes along the soil layers (no-flow boundary). For two- or three-dimensional models, boundary conditions must also be supplied at all vertical boundaries to fully define the problem. In two- and three-dimensional studies, often lines (planes) of symmetry are considered to minimize the total volume of soil to be considered, and no-flow (Neumann) conditions apply at these lines of symmetry.

Some one-dimensional models mimic the exchange with lateral elements, such as drains and ditches with an additional sink term in the Richards equation. In this way, these models approximate the true two- or three-dimensional character of water movement from and towards these structures by this sink term. However, when one is interested to determine flow paths around drains and ditches, one must use multidimensional simulations models.

4   Constitutive Hydraulic Properties of Soils

Another requirement in order to solve the Richards equation is to provide information on the constitutive relationships between the main state variables: volumetric water content θ, pressure head h and hydraulic conductivity K. The relationship $\theta(h)$ is known as the water retention characteristic, and $K(h)$ or $K(\theta)$ is the hydraulic conductivity characteristic. These can be measured on intact soil samples in the laboratory (e.g. Dane & Topp, 2002; their Sections 3.3–3.6). From those measurements it appears that both relationships are highly non-linear. Over the years many analytical functional relationships have been proposed to describe such measured relationships. Leij et al. (1997) compared several closed-form relationships. Some relationships appear to be special cases of a more universal expression (Raats, 1992; Heinen & Bakker, 2016). There are two expressions that are used widely in scientific literature. These are the water retention characteristic described by Van Genuchten (1980):

$$S_e(h) = \frac{\theta(h) - \theta_r}{\theta_s - \theta_r} = \frac{1}{\left(1 + |\alpha h|^n\right)^m} \tag{9}$$

and the hydraulic conductivity characteristic described by Mualem (1976; Van Genuchten, 1980):

$$K(S_e) = K_s S_e \left(1 - \left(1 - S_e^{1/m}\right)^m\right)^2, \tag{10}$$

where S_e is the effective degree of saturation (dimensionless), θ_r is the(asymptotic) residual water content ($L^3 L^{-3}$), θ_s is the water content at saturation ($L^3 L^{-3}$), K_s is the hydraulic conductivity at saturation (LT^{-1}), α is a curve shape parameter (L^{-1}) and n, m and λ are dimensionless curve shape parameters. The Mualem equation as used here is only valid under the restriction $m = 1-1/n$. With the help of the RETC model, the parameters can be easily obtained based on measured data (Van Genuchten et al.,

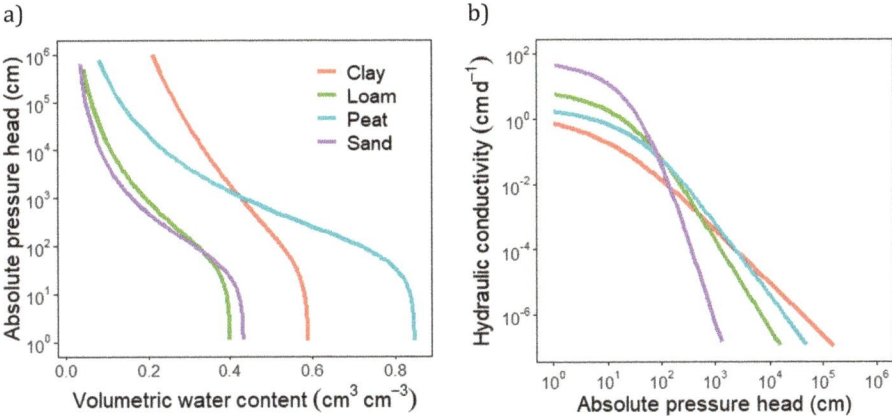

Fig. 2 Examples of (**a**) water retention and (**b**) hydraulic conductivity characteristics for four different soil types (weak loamy, fine sand, light loam, moderately heavy clay and peat; soils B2, B7, B11, O17 from Heinen et al., 2020b)

1991). As these two models are often used, databases exist where the parameters for different soils (soil layers) can be obtained. This can be either parameters obtained for individual soil samples, or averaged parameters for multiple soil samples of the same type, or estimates based on pedo-transfer functional relationships. Examples of such databases are: UNSODA (Nemes et al., 2001), ROSETTA (Schaap et al., 2001), HYPRES (Wösten et al., 1999) and Staring series (Heinen et al., 2020b). Figure 2 provides examples of water retention and hydraulic conductivity curves according to the Mualem–Van Genuchten (MvG) models for four soils.

5 Example 1: Simulated Infiltration Versus Analytical Solution

One of the crucial aspects in modelling soil water movement is the proper handling of the water input from rainfall or irrigation. The conditions in the soil will determine whether or not the water flux density at the soil surface can handle the rate of arrival of rain or irrigation water. Once the soil is unable to handle this input, the excess water will build up at the soil surface (ponding), and excess of this ponded water layer will then move along the soil surface to lower parts and into ditches (runoff). Runoff is unwanted because it may cause erosion and transport sediment and solutes (e.g. nutrients, pesticides, herbicides) from the soil surface into surface waters. Therefore, it is of great importance that soil models describe the infiltration process at the soil surface adequately. One way to test a soil simulation model for this is to check its behaviour either against observations or against known solutions of the flow process. As said before, analytical solutions of the Richards equation are rare. However, for the special case where the water retention and hydraulic conductivity

a)   b)

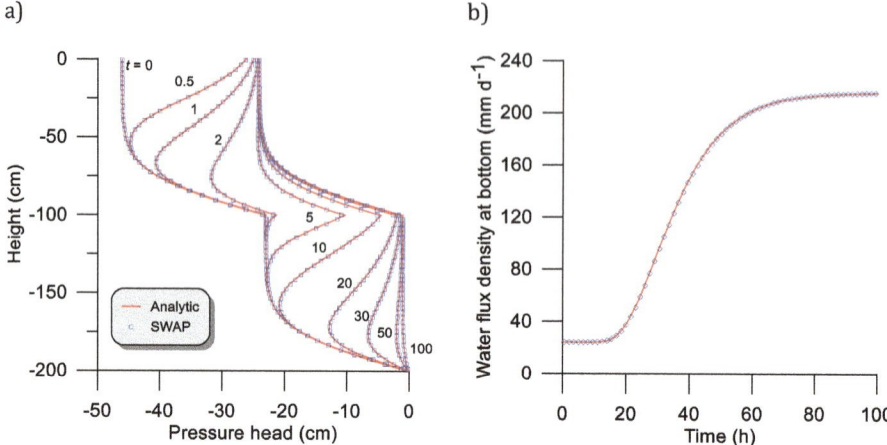

Fig. 3 Comparison between simulated (symbols) and analytical solution for water infiltration in a two-layer soil for case A of Srivastava and Yeh (1991): (**a**) pressure head as a function of depth and time (h) and (**b**) cumulative outflow at the bottom as a function of time

characteristics are described by exponential functions, analytical solutions can be obtained. Such exponential relationships may not describe measured data well, but in essence they show similar behaviour: a sigmoidal-shaped retention curve and a fast decreasing conductivity relationship. Figure 3 shows the strong correspondence between numerically simulated $h(z; t)$ and $q(z = -200$ cm; $t)$ profiles as a function of time in a layered soil and the analytical solution of Srivastava and Yeh (1991). The simulations were performed with SWAP (Kroes et al., 2017) in which the required exponential relationships were included for this purpose. Such a comparison provides confidence in the implementation of the numerical solution. SWAP was also tested against an analytical solution of Basha (1999) which also includes an expression for the time of start of ponding. Here again a very good correspondence was obtained (data not shown; see Heinen et al., 2021).

6   Root Water Uptake

The Richards equation given above includes a sink term S_w (Eq. 3). Although several sinks (or sources) can be considered, here we focus specifically on root water uptake because this forms the link towards coupled soil water–plant production modelling. It is clear that the drier the soil (more negative pressure heads), the more the plant roots will suffer from possible drought stress. Reduced uptake of water means that the plant can no longer fulfil the transpiration demand, which then causes the plant to start closing the stomata. When this occurs, the respiration rate of the crop is negatively affected, which results in reduced photosynthesis and thus reduced growth. Therefore, it is crucial to know whether or not roots can take up water at the required transpiration rate. Root water uptake is a complicated process since it depends on how easily water can move towards the individual roots (depending on soil hydraulic properties), where the roots are located (root length density, spatial distribution, fraction of root wall in contact with soil), how the plant regulates root distribution (more or less root mass production, locating roots in optimal soil regions), what the conductance of the root is (plant property) and how the plant compensates water uptake. Compensation means that if part of the root system cannot fulfil its contribution to the total demand, roots in more favourable conditions might take up extra water to compensate. Because of this complexity, it is not surprising that there are two ways to model root water uptake: (i) simplified or macroscopic models and (ii) more process-oriented or microscopic models. De Willigen et al. (2012) stated: 'Usually, the choice of a one-dimensional model for the sink term is justified by the uniform horizontal distribution assumed for root location and soil moisture. The three-dimensional root architecture is then simplified by means of a root-specific property to distribute the transpiration within the soil–root domain: root length, root surface, and root mass densities have been proposed (Feddes & Raats, 2004). Simple models include one-dimensional vertical profiles of root properties, but two- or three-dimensional models also exist in which the full root spatial distribution has to be defined (e.g. Vrugt et al., 2001). A specific type of three-dimensional sink term also exists in which the full three-dimensional architecture is explicitly accounted for (Javaux et al., 2011)'. It goes beyond the scope of this chapter to provide an in-depth summary of all possible root water uptake models that are available in scientific literature. Here we provide two examples, one for a simplified model and one for a microscopic model.

A well-known macroscopic model is that named after Feddes (e.g. Feddes et al., 1978). The root system is considered to be distributed in a specific way with depth; this can be either uniformly, linearly decreasing or exponentially decreasing. According to this distribution, the total potential transpiration rate is relatively distributed in the same way as the relative root distribution. For convenience here, we consider the one-dimensional situation, but this can be easily applied in two or three dimensions. The required uptake from each numerical soil layer is then given by

$$S_{pot,i} = \frac{L_{rv,i}}{\int_0^{z_r} L_{rv,i}\,dz}\,T_p, \tag{11}$$

where $L_{rv,i}$ is the root length density for layer i (L L^{-3}), z_r is the rooting depth (L) and T_p is the potential transpiration rate (LT^{-1}). Actual root water uptake is computed depending on the magnitude of the soil pressure head according to

$$S_{act,i} = S_{pot,i}\alpha(h_i), \tag{12}$$

Fig. 4 The root water
uptake reduction function
α(h) as a function of soil
water pressure head

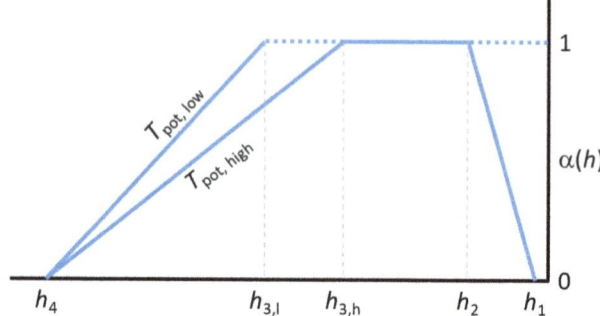

where $S_{act,i}$ is the actual root water uptake from layer i ($L^3 L^{-3} T^{-1}$), $S_{pot,i}$ is the demand for root water uptake from layer i ($L^3 L^{-3} T^{-1}$) and α is the Feddes reduction function shown in Fig. 4. The sum of all $S_{act,i}$ equals the actual transpiration rate T_{act} ($L\ T^{-1}$).

It is clear that this approach has no intrinsic compensation included. Several approaches have been published in the literature (see, e.g., Heinen, 2014a, b) to introduce compensation either at the start of this calculation or to apply it at the end. For example, Bouten et al. (1992) introduced water content scaling in Eq. (11), so that relatively more water is taken up in wet layers and less in more dry layers. An example of redistribution afterwards is the compensation model of Jarvis (1989) which states that if the total reduction in water uptake is less than some critical value, full compensation occurs; otherwise this compensation decreases linearly. Note that in this latter approach, all layers contribute in this compensation mechanism, even layers that were already suffering from drought.

Microscopic root water uptake models are often based on an analysis of flow of water towards the root. This means that gradients in pressure head around the root are taken into account. Because this flow is radially contracting towards the root, the gradients perpendicular to the root may become large. De Willigen and van Noordwijk (1987) analysed this process and showed that the water status around a root can be described well as an analytical expression in terms of the matric flux potential. The matric flux potential (Raats, 1970) is the integral of the hydraulic conductivity. Furthermore, in this concept, the flow of water from the bulk soil around the root towards the soil–root interface is matched with the flux of water across the root wall into the root xylem by introducing the root water potential in the root system (Fig. 5).

Two fluxes are considered: the first water flux is water movement from the bulk soil towards the root–soil interface (the rhizosphere, denoted by rs) and is given by

$$q_1 = \Delta z L_{rv} \pi f(\rho) \left(\overline{\phi}_{bulk} - \phi_{rs} \right), \qquad (13)$$

and the second water flux is the water movement from the rhizosphere into the root xylem and is given by

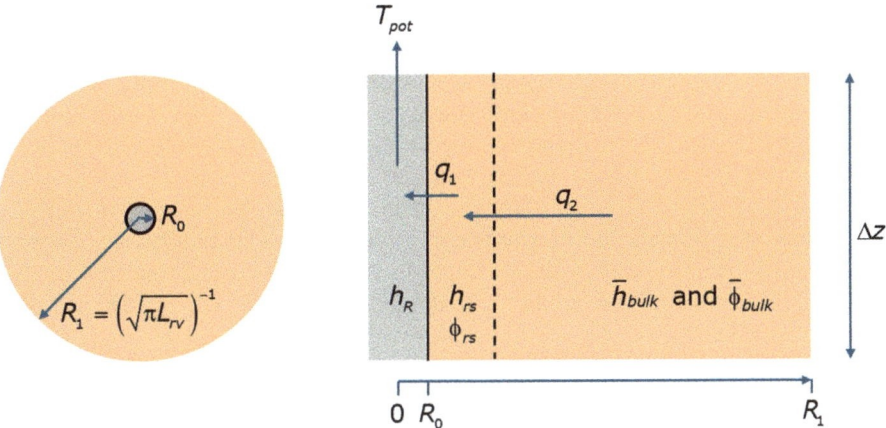

Fig. 5 Conceptualization of the microscopic root water uptake model; see text for further explanation

$$q_2 = \Delta z L_{rv} K_R (h_R - h_{rs}), \qquad (14)$$

where Δz is the length section of the root considered, for example the thickness of the soil layer in a numerical simulation model (L); L_{rv} is the root length density (L L^{-3}); K_R is the root conductance (L T^{-1}); h_R is the root water pressure head (considered uniform inside the whole root system) (L); h_{rs} is the pressure head at the soil–root interface; ϕ_{rs} is the matric flux potential at the soil–root interface (related to h_{rs}) (L^2 T^{-1}); $\overline{\phi}_{bulk}$ is the matric flux potential of the bulk soil surrounding the root, i.e. determined from the average pressure head in the soil (L^2 T^{-1}), and $f(\rho)$ is a function accounting for the geometric conditions and is given by

$$f(\rho = R_1/R_0) = \left(0.5 \left(\frac{1 - 3\rho^2}{4(\rho^2 - 1)} + \frac{4 \ln [\rho]}{(\rho^2 - 1)^2} \right) \right)^{-1}, \qquad (15)$$

where $\rho = R_1/R_0$ is the scaled radial distance (dimensionless), R_0 is the root radius (L) and R_1 is the radial distance to where the soil surrounding each root is present, with R_1 determined by L_{rv} according to $R_1^2 = (\pi L_{rv})^{-1}$.

When h_R is decreasing, it is likely that at lower values the actual root water uptake will be less than the required root water uptake. For example, the following transpiration reduction function could be applied (Campbell, 1991):

$$T_{act} = T_{pot} \left(1 + \left(\frac{h_R}{h_{R,1/2}} \right)^a \right)^{-1}, \qquad (16)$$

where $h_{R,1/2}$ is the root water pressure head when $T_{act} = 0.5T_{pot}$ (L) and a is a dimensionless shape parameter. The solution for root water uptake is found by equating $q_1 = q_2 = T_{act}$. If multiple layers of roots are present, one can set up these equations for each layer, resulting in a set of equations which, because of the non-linear character of the equations, has to be solved iteratively. A slightly alternative approach was used by De Jong van Lier et al. (2008).

De Willigen et al. (2012) showed how the macroscopic Feddes approach and the two microscopic models differ in predictions of root water uptake. They also showed that these differences vary with soil type. In the above microscopic model, it is assumed that roots are in full contact with the soil. De Willigen et al. (2018) provided an analysis when roots are in partial contact with soil.

7   Additional Phenomena: Processes

Although the Richards equation (Eq. 3) is regarded as the state-of-the-art way of describing water movement in soils, it still has some restrictions. Other assumptions were also made, for example, regarding the uniqueness of the closed-form, constitutive hydraulic properties, and we considered only root water uptake as the single sink or source term in the governing flow equation. Here we give a very brief overview of aspects not regarded in the above.

Hysteresis
The closed-form, constitutive relationships for water retention and hydraulic conductivity characteristics are mostly assumed to be unique. However, it appears that this relationship differs from conditions when the porous medium is drying out or wetting up. This phenomenon is called hysteresis and is caused by the varying pore diameters (the so-called ink-bottle effect), differences in advancing and receding water-air meniscus, entrapped air, thermal gradients and swelling and shrinking (Hillel, 1980; Feddes et al., 1988). The relationships $\theta(h)$ and $K(h)$ are known to be hysteretic (Miller & Miller, 1956). According to Topp (1969), the relationship $K(\theta)$ shows negligible hysteresis; therefore, it is preferred to $K(h)$. For describing hysteresis in the $\theta(h)$ relationship, various models have been proposed in the literature (e.g. Mualem, 1984; Jaynes, 1984; Viaene et al., 1994). In general, these models require knowledge about the so-called main drying and main wetting envelop curves. These models can then calculate any intermediate so-called scanning curves if the medium is subject to alternate drying and wetting conditions. The majority of studies determine only the main drying curve and disregard hysteresis. According to Kool and Parker (1987), the main wetting curve can be described well by the same relationship as the MvG model described above. They stated that both curves share the same parameter values, except for α, which should be greater for the main wetting curve than for the main drying curve. As a first guess, they proposed a ratio of $\alpha_{wet}/\alpha_{dry} = 2$. In soil that is not too coarse, hysteresis will be small. It is

important in coarser soils, with extreme hysteresis occurring in coarse horticultural substrates (e.g. Heinen & Raats, 1999; Otten et al., 1999).

Swelling–Shrinking
As stated, the Richards equation is valid only for rigid soils. Therefore, it cannot be applied for soils that are subject to substantial swelling and shrinking, especially when macropores or cracks result from shrinking; these need special attention (see next issue).

Macropores and Cracks
Macropores and cracks may form in soils under shrinking conditions. Macrofauna such as deep-burrowing earthworms (e.g. *Lumbricus terrestris* or *Aporrectodea longa*) may also form biopores. These gaps differ from the porous structure of the soil matrix; therefore, water movement will differ from the theory underlying the Richards equation. Beven and Germann (1982, 2013) reviewed approaches in macroporous soils. They concluded that, thus far, no consistent and coherent theory exists to consider water movement in soils with macropores. They stated that four approaches are used most often in the literature: (1) adapt K near saturation; (2) the soil consists of two continuums: one immobile phase and the other mobile phase and is described according to the Richards–Darcy model with exchange between the two continuums; (3) the soil comprises two conductivity domains each described by their own Richards–Darcy equation (dual-permeability) and (4) the soil comprises two porosity domains (dual-porosity) in which preferential flow in the macropores is described either by simple filling or by a kinematic wave equation, also with exchange between the soil matrix and macropores.

Modifications in Soil Hydraulic Property Description
Several authors have proposed modifications of existing descriptions of the hydraulic properties. These include adding an air-entry value in $\theta(h)$ which also affects $K(\theta)$ (e.g. Vogel et al., 2000; Schaap & Van Genuchten, 2006; Ippisch et al., 2006) or simply changing K near saturation. The MvG model, like many others, is based on the underlying assumption that the pore-size distribution function is uni-modal. Some soils may show, however, multimodal pore-size distributions. Or one could mimic the presence of small macropores by a multimodal approach. Durner (1994) and Priesack and Durner (2006) introduced bimodal expressions for the MvG model.

Drainage
Some soils have drains installed for removing excess water, or sometimes they are used to supply soils with water (adaptive drainage; subirrigation). In multidimensional models drains can be introduced as holes in the domain with boundary conditions applying at their walls (e.g. De Vos et al., 2000). Alternatively, drains can be mimicked with nodes for which the hydraulic conductivity is adapted according to the Vimoke–Taylor concept (Heinen, 2014b). In one-dimensional models, true flow towards drains cannot be considered. However, drains can then be considered as an additional sink term in the Richards equation, and flow towards (or from) drains can then be described as a resistance function. In this way several drainage levels can be considered, for example drainage towards shallow trenches,

drain pipes, edge of field ditches and drainage towards larger open water bodies. See Kroes et al. (2017) for more alternatives as, for example, implemented in the SWAP model.

Other Crop Stress Factors Originating in Soils

Above we described drought stress causing reduction in root water uptake, which may then cause a reduction in crop growth. Two other phenomena can be distinguished that can also cause reduction in root water uptake (functioning of the root system) and thus a reduction in crop growth. Roots need oxygen for their respiration. When soils become too wet for prolonged periods, the air content and thus the oxygen content in the air-filled pore space may become limiting. Roots will no longer function optimally, and they will not be able to take up water at the required rate. Above, in the description of root water uptake, we provided an example of the Feddes reduction function. There, reduction under wet conditions can be considered by introducing the so-called h_1 and h_2 parameters (Fig. 4). Bartholomeus et al. (2008) provided a more mechanistic approach to consider oxygen stress.

The second major phenomenon is salinity stress. When the soil solution becomes more saline, roots can no longer take up water easily due to osmotic effects. In fact, the root wall can be regarded as a membrane, and thus we have to consider the osmotic potential in the soil and root, next to the soil, and root water pressure heads. A microscopic approach will define some salinity reduction function, for example, as proposed by Maas and Hoffman (1977), and add this reduction in Eq. (12) (multiplicative). Alternatively, the osmotic potential can be added in the microscopic description of root water uptake (e.g. Dalton et al., 1975; Heinen, 2001).

Water-Influenced Processes

So far, we have considered the soil water balance per se. We have shown the link to crop models through the exchange of information between crop water demand and water availability in the soil. There are, however, more links between soil water balance terms and other processes. It goes beyond the scope of this chapter to describe these in detail. In brief, these are:

- Solute transport: Solute transport in soils is often described by the classical advection (or convection) – dispersion/diffusion equation. Here information is needed on the volumetric water content θ and the water flux density q. For more detail, the reader is referred to Heinen et al. (2020a), for example. When modelling solute transport, there are several sources and sinks for solutes to be considered. These are often sub-models by themselves. Examples are: denitrification, nitrification and mineralization. In these sub-process descriptions, the process is governed by environmental conditions such as the volumetric water content, soil temperature and soil pH. Examples of soil water reduction functions for θ for denitrification can be found in Heinen (2006). Similar expressions hold for the other processes.
- Soil temperature: Soil temperature is determined by the heat conductivity and heat capacity of the soil. These are determined by heat properties of the components present in soil: soil particles, organic matter, soil water and soil air.

Therefore, θ is needed when soil temperature is considered. In the above we stated that the Richards equation assumes isothermal transport, so that soil temperature is assumed not to influence water movement. Theories exist for non-isothermal water movement or temperature-dependent water vapour transport, but these are not considered here (see, e.g., Ten Berge, 1987; Saito et al., 2006; Jansson & Karlberg, 2011).

8   Soil–Water Simulation Models

In soil science research, many simulation models are being used that include soil water movement; we will not mention them all. In 2014, the International Soil Modelling Consortium was formed with the aim to integrate and advance soil systems modelling, data gathering and observational capabilities (https://soil-modeling.org/). This site gives a list of models that relates to, amongst others, soil physics and hydrology. Some examples are (in alphabetical order) APSIM-SWIMv2.1 (Verburg et al., 1996), APSIM-WEIRDO (Brown et al., 2018), COUP (Jansson & Karlberg, 2011), DAISY (Hansen et al., 1990), FUSSIM-2D (Heinen & De Willigen, 1998, 2001; Heinen, 2001), HYDRUS-1D (Šimůnek et al., 2009), HYDRUS-2D/3D (Šimůnek et al., 2018), MACRO (Larsbo & Jarvis, 2003), RZWQM (Malone et al., 2004) and SWAP (Kroes et al., 2017; Van Dam et al., 2008).

9   Example 2: Water Content in the Root Zone Predicted from Ensemble Weather Forecasts

Precision agriculture, or smart farming, means that plants (or animals) get precisely the treatment they need. For example, with respect to fertilization, much attention is given to the 4R concept – right amount, right source, right place and right time – and technology is used to realize such. This should also be taken into account for water with agriculture being the largest consumer of water worldwide. Irrigation is common practice in (semi-)arid regions, and combined with technology this can be called precision irrigation. Simulation models can help in understanding and investigating the relationship between crop water requirement (demand) and soil water availability. Here we present an example in which the soil water content in the rootzone is simulated based on weather forecasts. This provides insight into the forthcoming changes in water availability and can be used in decision-making by farmers when the next irrigation event for a certain field is required.

For this purpose we used the soil–water–plant–atmosphere model SWAP–WOFOST (Kroes et al., 2017). Based on atmospheric conditions, the Penman–Monteith evapotranspiration is calculated including the required crop information (such as LAI) that comes from the crop model WOFOST. The soil water balance

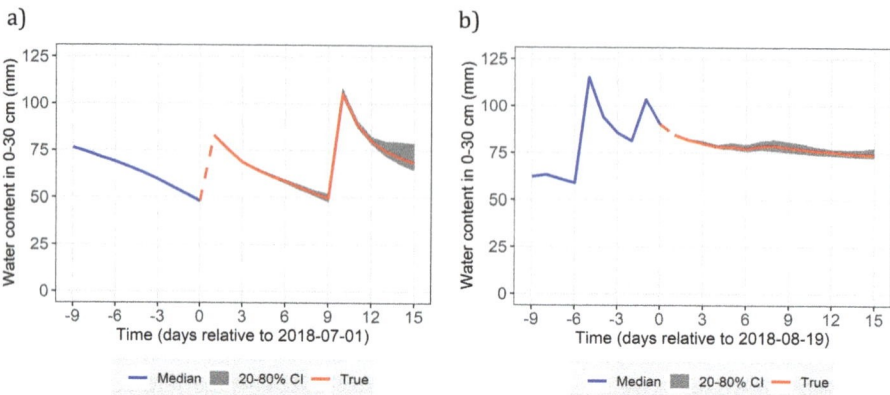

Fig. 6 Simulated time course of water content in the root zone (0–30 cm below soil surface) for two periods during the summer of 2018. The blue line refers to simulations based on actual historic weather data, and the red line is the median of all 51 predictions based on the ensemble weather forecasts and is surrounded by the 20–80% confidence interval based on all 51 predictions

simulation can determine what the actual water uptake will be given the circumstances. Consequently, the crop model will determine to what extent actual crop growth will be less than potential crop growth if root water uptake is limiting. The atmospheric conditions are not restricted to historic data, but forecasts of rainfall, radiation, temperature, wind speed and humidity can also be used. Each day, the European Centre for Medium-Range Weather Forecasts (ECMWF) provides ensemble weather forecasts, i.e. a series of 51 realizations of weather forecasts for the coming 15 days. These can be used in forecast simulations to calculate the time-course of water content in the root zone. By running all realizations of the ensemble forecast provides insight into the spread of the predicted water contents in the root zone. Figure 6 shows two examples of the simulated time-course of water content in the root zone (0–30 cm below soil surface) including the 20–80% confidence interval, referring to two periods in the dry summer of 2018 in the south of the Netherlands.[1] The situation refers to a potato field. Local soil information was used to parameterize the model; historic weather data were obtained from the nearby (approximately 20 km) Royal Dutch Meteorological Institute (KNMI), with rainfall replaced by locally measured rainfall in an automatic rain gauge. Ensemble weather

[1]This example refers to part of the work done the project OPERA as part of the joint Water-JPI/FACCE-JPI work programme WaterWorks2015. We thank the EU and the Ministry of Economic Affairs (the Netherlands), CDTI (Spain), MINECO (Spain), ANR (France), MIUR (Italy), NCBR (Poland) and WRC (South Africa) for funding, in the frame of the collaborative international consortium OPERA financed under the ERA-NET Cofund WaterWorks2015 Call. This ERA-NET is an integral part of the 2016 Joint Activities developed by the Water Challenges for a Changing World Joint Programme Initiative (Water JPI). Details can be found in reports available at http://opendata.waterjpi.eu/dataset/opera-operationalizing-the-increase-of-water-use-efficiency-and-resilience-in-irrigation

forecasts were obtained from ECMWF through a paid-for contract with KNMI.[2] The farmer performed seven irrigations in total. From a hindcast analysis, it appeared that irrigations were performed when the model predicted that water content in the root zone was about 50–60 mm. The increases in water content in Fig. 6a are the result of two irrigations (not of equal amount). For this first graph, the 20–80% confidence interval around the median (red) line is very small until day 12 in the future. From then onwards the confidence interval increases, which is due to the wide spread in predicted rainfall in the 51 ensemble weather forecasts. In Fig. 6b the two increases at days −5 and − 1 are due to irrigations. The predicted time course of water content in the root zone decreases only slowly, and the spread of the ensemble forecasts was small for all 15 days. The slow decrease indicates that during this period the predicted rainfall was slightly lower than the predicted transpiration. Suppose that a proper threshold for irrigation is when the root zone water content decreases below 50–60 mm, then it follows from the analysis performed on August 19 that no irrigation will be needed in the forthcoming 15 days.

This example might not be very useful for farmers in (semi-)arid regions, where the water input from rain, especially in summer, is insufficient by definition, and irrigation is the main source of water. Then irrigation is practised on a daily basis. Farmers in more humid regions are currently facing the first major changes due to climate change. During summer prolonged dry periods might now be experienced, and additional irrigation is required. Often this is necessary every few days, and decisions have to be made as to when and which field needs irrigation. In these circumstances, farmers can benefit from additional information of the predicted time courses of available water in the root zone.

10 Spatial Variation

Soils are heterogeneous at many different scales: from the distribution of soils across the landscape down to microscopic pore networks and the molecular structure of biogeochemical interfaces. Consequently, observed flow rates of water, gases and solutes or the dynamics of state variables, such as volumetric water content, temperature, and biological activity, typically depend on the scale of observation (Vereecken et al., 2016).

One of the important aspects in precision agriculture is the aim to adapt to spatially variable conditions. For example, fertilizer application is reduced in those parts of the field where the native N-content is higher than average. This chapter focuses on water movement in soils, and we have explained the major processes, constitutive relationships, initial and boundary conditions. In particular, the constitutive relationships between volumetric water content, pressure head and hydraulic

---

[2]The ensemble weather forecasts were obtained for research purposes only, and refer to their use in the OPERA project.

conductivity are determined by the soil properties. These latter are generally spatially variable within a field. Clearly, the vertical spatial variation is dominated by the presence of different soil layers (horizons). Such heterogeneity is often taken into account. Soil structure and soil texture and the hydraulic properties may also vary in horizontal direction. Horizontal variation can in principle be taken as input in 2D and 3D simulation models. The user needs to specify the constitutive relationships to use for each grid cell in the model. For example, Mallants et al. (1996) measured the water retention and hydraulic conductivity characteristics along a transect and determined for each location the parameter of the MvG relationships. Based on their data, they concluded that K_s is extremely variable (coefficient of variation, CV, at maximum 897%), θ_r and α were moderately variable (max. CV of 156%), and θ_s and n were weekly variable (maximum CV of 27%). These data were used to determine probability density functions and auto- and cross-variograms to determine spatial dependence and coregionalization, respectively. Such information can then be used to generate multidimensional input for multidimensional models.

Another type of soil heterogeneity is due to structural changes resulting from soil tillage, for example. In a modelling study with three 'virtual' soils, Schlüter et al. (2012) investigated the effect of a loosely aggregated seed bed, a plough pan and local compaction in the wheel tracks, including subscale variation (according to the so-called Miller scaling (Miller and Miller, 1956)). They also investigated the difference between the multidimensional modelling results with a one-dimensional model simulation in which each layer was assigned 'effective' soil hydraulic properties. From that comparison they concluded 'A comparison with one-dimensional, effective representations of these virtual soils demonstrated that upscaling of soil water dynamics becomes inaccurate when lateral fluxes become relevant at the scale of observation'. So, the 1D approach with effective hydraulic properties can be used only in cases where lateral flow can be neglected.

Vereecken et al. (2007) provided a comprehensive review on how to upscale soil water processes from the local to the field scale and how to derive effective hydraulic properties. They considered two main approaches: forward and inverse upscaling. For both approaches, information about the spatial structure of hydraulic properties and the temporal and spatial structure and variability of state variables and boundary conditions is important. Lack of this information and that on the effect of the presence of vegetation are seen as the most limiting factors. Finally, the authors stated that at that time most research was based on stochastic perturbation theories. They suggested, however, other methods should also be explored.

There seems to be no single, universal approach to handle spatial variation in modelling soil water movement. When one-dimensional models are used, possible approaches may include the following: (i) perform multiple simulations with different input variables that represent the spatial variation. The results may then be averaged and the spread in outcome (uncertainty) can be quantified. This could also be extended by using several different simulation models (ensemble) and using Bayesian model averaging to obtain an average output and a measure for uncertainty. The parameters can either be taken from locally measured soil data or drawn from some kind of known stochastic probability density function. (ii) Based on 'effective' (upscaled) parameters, the field can be simulated once (or use Bayesian ensemble modelling). (iii) Multidimensional models can explicitly take into account spatially distributed input parameters. The parameters can either be taken from locally measured soil data or drawn from some kind of known stochastic probability density function.

11   Epilogue

This chapter has focused on modelling the dynamics of water in soils. We have described the main theory of water movement in soils, have shown how well the numerical solution compares to an analytical solution for infiltration and provided an example of how weather forecasts can be used to predict future water contents in the root zone. One aspect that plays an important role in precision agriculture is how to deal with spatial variation in soil processes and soil properties. This aspect has gained attention in several studies, such as described in the previous section. However, we believe that research is still needed on how to cope with spatial variation in a parcel in combination with soil–crop modelling in the area of precision agriculture. Can multiple or ensemble simulations using one-dimensional models be used to represent the spatial variation in a field? To what extent can we make use of one-dimensional models, or when is it necessary to switch to multidimensional models; for example when are lateral flows important?

Conflict of Interest   The author declares that there are no conflicts of interest.

References

Bartholomeus, R. P., Witte, J. P. M., van Bodegom, P. M., van Dam, J. C., & Aerts, R. (2008). Critical soil conditions for oxygen stress to plant roots: Substituting the Feddes-function by a process-based model. *Journal of Hydrology, 360*, 147–165.

Basha, H. A. (1999). Multidimensional linearized nonsteady infiltration with prescribed boundary conditions at the soil surface. *Water Resources Research, 35*(1), 75–83.

Beven, K., & Germann, P. (1982). Macropores and water flow in soils. *Water Resources Research, 18*, 1311–1325. https://doi.org/10.1029/WR018i005p01311

Beven, K., & Germann, P. (2013). Macropores and water flow in soils revisited. *Water Resources Research, 49*(6), 3071–3092. https://doi.org/10.1002/wrcr.20156

Boote, K. J., Sau, F., Hoogenboom, G., & Jones, J. W. (2008). Chapter 3: Experience with water balance, evapotranspiration, and predictions of water stress effects in the CROPGRO model. In L. R. Ahuja, V. R. Reddy, S. A. Saseendran, & Q. Yu (Eds.), *Response of crops to limited water: Understanding and Modeling water stress effects on plant growth processes* (Advances in agricultural systems Modeling) (Vol. 1). ASA-CSSA-SSSAJ. https://doi.org/10.2134/advagricsystmodel1.c3

Bouma, J. (2007). Precision agriculture: Introduction to the spatial and temporal variability of environmental quality. In J. V. Lake, G. R. Bock, & J. A. Goode (Eds.), *Precision agriculture: Spatial and temporal variability of environmental quality: Precision agriculture: Spatial and temporal variability of environmental quality: Ciba foundation symposium* (Vol. 210, pp. 5–17). https://doi.org/10.1002/9780470515419.ch2

Bouten, W., Heimovaara, T. J., & Tiktak, A. (1992). Spatial patterns of throughfall and soil water dynamics in a Douglas fir stand. *Water Resources Research, 28*, 3227–3233. https://doi.org/10.1029/92WR01764

Brown, H., Carrick, S., Müller, K., Thomas, S., Sharp, J., Cichota, R., Holzworth, D., & Clothier, B. (2018). Modelling soil-water dynamics in the rootzone of structured and water-repellent soils. *Computers and Geosciences, 113*, 33–42. https://doi.org/10.1016/j.cageo.2018.01.014

Buckingham, E. (1907). *Studies on the movement of soil moisture* (Bulletin No. 38) (pp. 7–61). USDA Bureau of Soils.

Buttler, I. W., & Riha, S. J. (1992). Water fluxes in Oxisols: A comparison of approaches. *Water Resources Research, 28*(1), 221–229.

Campbell, G. S. (1991). Simulation of water uptake by plant roots. In J. Hanks & J. T. Ritchie (Eds.), *Modeling plant and soil systems* (Agronomy monograph 31) (pp. 273–285). American Society of Agronomy.

Celia, M. A., Bouloutas, E. T., & Zarba, R. L. (1990). A general mass-conservative numerical solution for the unsaturated flow equation. *Water Resources Research, 26*, 1483–1496.

Dalton, F. N., Raats, P. A. C., & Gardner, W. R. (1975). Simultaneous uptake of water and solutes by plant roots. *Agronomy Journal, 67*, 334–339.

Dane, J. H., & Topp, G. C. (Eds.). (2002). *Methods of soil analysis. Part 4. Physical methods*. Soil Science Society Of America.

Darcy, H. (1856). *Les fontaines publique de la ville de Dijon*. Dalmont Paris.

Jong, D., van Lier, Q., van Dam, J. C., Metselaar, K., de Jong, R., & Duijnisveld, W. H. M. (2008). Macroscopic root water uptake distribution using a matric flux potential approach. *Vadose Zone Journal, 7*, 1065–1078. https://doi.org/10.2136/vzj2007.0083

De Vos, J. A., Hesterberg, D., & Raats, P. A. C. (2000). Nitrate leaching in a tile-drained silt loam soil. *Soil Science Society of America Journal, 64*, 517–527.

De Willigen, P., & van Noordwijk, M. (1987). *Roots, plant production and nutrient use efficiency*. Ph.D. thesis, Agricultural University, Wageningen, The Netherlands, 282 p.

De Willigen, P., Heinen, M., & van Noordwijk, M. (2018). Roots partially in contact with soil: Analytical solutions and approximation in models of nutrient and water uptake. *Vadose Zone Journal*. https://doi.org/10.2136/vzj2017.03.0060

De Willigen, P., van Dam, J. C., Javaux, M., & Heinen, M. (2012). Root water uptake as simulated by three soil water flow models. *Vadose Zone Journal*. https://doi.org/10.2136/vzj2012.0018

Durner, W. (1994). Hydraulic conductivity estimation for soils with heterogeneous pore structure. *Water Resources Research, 30*, 211–223.

Feddes, R. A., & Raats, P. (2004). Parameterizing the soil–water–plant root system. In R. A. Feddes et al. (Eds.), *Unsaturated-zone modeling: Progress, challenges and applications* (pp. 95–141). Kluwer Acad. Publ.

Feddes, R. A., Kowalik, P. J., & Zaradny, H. (1978). *Simulation of field water use and crop yield*. Simulation Monograph, PUDOC.

Feddes, R. A., Kabat, P., van Bakel, P. J. T., Bronswijk, J. J. B., & Halbertsma, J. (1988). Modelling soil water dynamics in the unsaturated zone - state of the art. *Journal of Hydrology, 100*, 69–111.

Hansen, S., Jensen, H. E., Nielsen, N. E., & Svendsen, H. (1990). *DAISY – Soil plant atmosphere system model. Report A10*. The Royal Veterinary and Agricultural University. Available at: https://daisy.ku.dk/publications/

Heinen, M. (2001). FUSSIM2: Brief description of the simulation model and application to fertigation scenarios. *Agronomie, 21*, 285–296. https://doi.org/10.1051/agro:2001124

Heinen, M. (2006). Simplified denitrification models: Overview and properties. *Geoderma, 133*(3–4), 444–463. https://doi.org/10.1016/j.geoderma.2005.06.010

Heinen, M. (2014a). Compensation in root water uptake models combined with three-dimensional root length density distribution. *Vadose Zone Journal*. https://doi.org/10.2136/vzj2013.08.0149

Heinen, M. (2014b). Correction of the Vimoke-Taylor concept representing drains in a numerical simulation model. *Vadose Zone Journal.* https://doi.org/10.2136/vzj2014.06.0066

Heinen, M., & Bakker, G. (2016). Implications and application of the Raats superclass of soils equations. *Vadose Zone Journal, 15*(8). https://doi.org/10.2136/vzj2016.02.0012

Heinen, M., & de Willigen, P. (1998). *FUSSIM2 A two-dimensional simulation model for water flow, solute transport and root uptake of water and nutrients in partly unsaturated porous media* (Quantitative approaches in systems analysis no. 20) (p. 140). DLO Research Institute for Agrobiology and Soil Fertility and the C.T. de Wit Graduate School for Production Ecology. https://edepot.wur.nl/4408

Heinen, M., & de Willigen, P. (Eds.). (2001). *FUSSIM2 version 5. New features and updated user's guide* (Alterra rapport 363). Alterra. https://edepot.wur.nl/41784

Heinen, M., & Raats, P. A. C. (1999). Hysteretic hydraulic properties of a coarse sand horticultural substrate. In M. T. van Genuchten, F. J. Leij, & L. Wu (Eds.), *Characterization and measurement of the hydraulic properties of unsaturated porous media, part 1: Proceedings of an international workshop organized by the U.S. Salinity Laboratory* (pp. 467–476). USDA-ARS and the Department of Soil & Environmental Sciences of the University of California.

Heinen, M., Assinck, F., Groenendijk, P., & Schoumans, O. (2020a). Soil dynamic models: Predicting the behavior of fertilizers in the soil. Chapter 8.2. In E. Meers, G. Velthof, E. Michels, R. Rietra, & C. V. Stevens (Eds.), *Biorefinery of inorganics: Recovering mineral nutrients from biomass and organic waste.* Wiley.

Heinen, M., Bakker, G., & Wösten, J. H. M. (2020b). *Waterretentie- en doorlatendheidskarakteristieken van boven- en ondergronden in Nederland: de Staringreeks. Update 2018* (Report 2987) (Vol. 10, p. 18174/512761). Wageningen Environmental Research.

Heinen, M., Dik, P. E., & Cruijsen, J. J. P. (2021). *Aanpassing en toepassing SWAP gericht op bodem en hydrologische maatregelen* (p. 3059). Deelrapport thema Bewuste Bodem in onderzoeksprogramma Lumbricus.

Hillel, D. (1980). *Fundamentals of soil physics.* Academic Press.

Ippisch, O., Vogel, H.-J., & Bastian, P. (2006). Validity limits for the van Genuchten–Mualem model and implications for parameter estimation and numerical simulation. *Advances in Water Resources, 29*, 1780–1789. https://doi.org/10.1016/j.advwatres.2005.12.011

Jansson, P.-E., & Karlberg, L. (2011). *Coupled heat and mass transfer model for soil-plant-atmosphere systems.* COUP manual. at: http://www.coupmodel.com/documentation

Jarvis, N. J. (1989). A simple empirical model of root water uptake. *Journal of Hydrology, 107*(1–4), 57–72. https://doi.org/10.1016/0022-1694(89)90050-4

Javaux, M., Draye, X., & Cl. Doussan, J. Vanderborght, and H. Vereecken. (2011). Root water uptake: Toward 3-D functional approaches. In J. Glinski et al. (Eds.), *Encyclopaedia of agrophysics* (pp. 717–721). Springer.

Jaynes, D. B. (1984). Comparison of soil-water hysteresis models. *Journal of Hydrology, 75*, 287–299.

Kool, J. B., & Parker, J. C. (1987). Development and evaluation of closed-form expressions for hysteretic soil hydraulic properties. *Water Resources Research, 23*, 105–114.

Kroes, J. G., van Dam, J. C., Bartholomeus, R. P., Groenendijk, P., Heinen, M., Hendriks, R. F. A., Mulder, H. M., Supit, I., & van Walsum, P. E. V. (2017). *SWAP version 4. Theory description and user manual* (Report 2780). Wageningen Environmental Research. Available at: https://edepot.wur.nl/416321; see also: swap.wur.nl

Larsbo, M., & Jarvis, N. (2003). *MACRO 5.0. A model of water flow and solute transport in macroporous soil. Technical description.* Department of Soil Sciences, Swedish University of Agricultural Sciences. ISBN 91-576-6592-3. Available at: https://www.slu.se/globalassets/ew/org/centrb/ckb/modeller_dokument/macro-5.0-technical-report-2003.pdf

Leij, F. J., Russell, W. B., & Lesch, S. M. (1997). Closed-form expressions for water retention and conductivity data. *Ground Water, 35*, 848–858. https://doi.org/10.1111/j.1745-6584.1997.tb00153.x

Maas, E. V., & Hoffman, G. J. (1977). Crop salt tolerance-current assessment. *Journal of the Irrigation and Drainage Division, American Society of Civil Engineers, 103*, 115–134.

Mallants, D., Mohanty, B. P., Jacques, D., & Feyen, J. (1996). Spatial variability of hydraulic properties in a multi-layered soil profile. *Soil Science, 161*, 167–181.

Malone, R. W., Ahuja, L. R., Ma, L., Wauchope, R. D., Ma, Q., & Rojas, K. W. (2004). Application of the root zone water quality model (RZWQM), to pesticide fate and transport: An overview. *Pest Management Science, 60*, 205–221. https://doi.org/10.1002/ps.789

McCord, J. T. (1991). Application of second-type boundaries in unsaturated flow modeling. *Water Resources Research, 27*, 3257–3260.

Meijerink, J. A., & Van Der Vorst, H. A. (1977). An iterative solution method for linear systems of which the coefficient matrix is a symmetric M-matrix. *Mathematics of Computation, 31*, 148–162.

Miller, E. E., & Miller, R. D. (1956). Physical theory for capillary flow phenomena. *Journal of Applied Physics, 27*, 324–332.

Mualem, Y. (1976). A new model for predicting the hydraulic conductivity of unsaturated porous media. *Water Resources Research, 12*, 513–522.

Mualem, Y. (1984). A modified dependent-domain theory of hysteresis. *Soil Science, 137*, 283–291.

Nemes, A., Schaap, M. G., Leij, F. J., & Wösten, J. H. M. (2001). Description of the unsaturated soil hydraulic database UNSODA version 2.0. *Journal of Hydrology, 251*, 151–162.

Otten, W., Raats, P. A. C., & Kabat, P. (1999). Hydraulic properties of root zone substrates used in greenhouse horticulture. In M. T. van Genuchten, F. J. Leij, & L. Wu (Eds.), *Characterization and measurement of the hydraulic properties of unsaturated porous media, part 1: Proceedings of an international workshop organized by the U.S. Salinity Laboratory* (pp. 477–488). USDA-ARS and the Department of Soil & Environmental Sciences of the University of California.

Patankar, S. V. (1980). *Numerical heat transfer and fluid flow*. Hemisphere Publishing Corporation.

Press, W. H., Teukolsky, S. A., Vetterling, W. T., & Flannery, B. P. (1992). *Numerical recipes in Fortran 77. The art of scientific computing, second edition*. Cambridge University Press.

Priesack, E., & Durner, W. (2006). Closed-form expression for the multi-modal unsaturated conductivity function. *Vadose Zone Journal, 5*, 121–124.

Raats, P. A. C. (1970). Steady infiltration from line sources and furrows. *Soil Science Society of America Proceedings, 34*, 709–714. https://doi.org/10.2136/sssaj1970.03615995003400050015x

Raats, P. A. C. (1992). A superclass of soils. In M. T. van Genuchten et al. (Eds.), *Indirect methods for estimating the hydraulic properties of unsaturated soils: Proceedings of the International Workshop, Riverside, CA. 11–13 Oct. 1989* (pp. 45–51). US Salinity Lab.

Richards, L. A. (1931). Capillary conduction of liquids through porous mediums. *Physics, 1*, 318–333.

Saito, H., Šimůnek, J., & Mohanty, B. P. (2006). Numerical analysis of coupled water, vapor, and heat transport in the vadose zone. *Vadose Zone Journal, 5*, 784–800.

Scanlon, B. R., Christman, M., Reedy, R. C., Porro, I., Šimůnek, J., & Flerchinger, G. F. (2002). Intercode comparisons for simulating water balance of surficial sediments in semiarid regions. *Water Resources Research, 38(12) 1323*, 59.1-59.16. https://doi.org/10.1029/2001WR001233

Schaap, M. G., Leij, F. J., & van Genuchten, M. T. (2001). ROSETTA: A computer program for estimating soil hydraulic parameters with hierarchical pedotransfer functions. *Journal of Hydrology, 251*, 163–176.

Schaap, M. G., & van Genuchten, M. T. (2006). A modified Mualem–van Genuchten formulation for improved description of the hydraulic conductivity near saturation. *Vadose Zone Journal, 5*, 27–34. https://doi.org/10.2136/vzj2005.0005

Schlüter, S., Vogel, H.-J., Ippisch, O., Bastian, P., Roth, K., Schelle, H., Durner, W., Kasteel, R., & Vanderborght, J. (2012). Virtual soils: Assessment of the effects of soil structure on the hydraulic behavior of cultivated soils. *Vadose Zone Journal, 11*, doi:10.2136/vzj2011.0174.

Shelia, V., Šimůnek, J., Boote, K., & Hoogeboom, G. (2017). Coupling DSSAT and HYDRUS-1D for simulations of soil water dynamics in the soil-plant-atmosphere system. *Journal of Hydrology and Hydromechanics, 66*(2), 232–245. https://doi.org/10.1515/johh-2017-0055

Šimůnek, J., Šejna, M., Saito, H., Sakai, M., & van Genuchten, M. T. (2009). *The HYDRUS-1D Software package for simulating the one-dimensional movement of water, heat, and multiple solutes in variably-saturated media. Version 4.08.* Department of Environmental Sciences University of California Riverside Riverside. Available at: https://www.pc-progress.com/Downloads/Pgm_hydrus1D/HYDRUS1D-4.08.pdf

Šimůnek, J., van Genuchten, M. T., & Šejna, M. (2018). *The HYDRUS software package for simulating the two- and three-dimensional movement of water, heat, and multiple solutes in variably-saturated porous media* (Technical Manual Version 3). PC-Progress. Available at: https://www.pc-progress.com/downloads/Pgm_Hydrus3D3/HYDRUS3D_Technical_Manual_v3.pdf

Srivastava, R., & Yeh, T.-C. J. (1991). Analytical solutions for one-dimensional transient infiltration toward the water table in homogeneous and layered soils. *Water Resources Research, 27*(5), 753–762.

Ten Berge, H. F. M. (1987). *Heat and water transfer at the bare soil surface. Aspects affecting thermal imagery.* PhD thesis, Landbouwhogeschool, Wageningen. Available at: https://edepot.wur.nl/166817

Topp, G. C. (1969). Soil-water hysteresis measured in a sandy loam and compared with the hysteretic domain model. *Soil Science Society of America Proceedings, 33*, 645–651.

Van Dam, J. C., de Rooij, G. H., Heinen, M., & Stagniti, F. (2004). Concepts and dimensionality in modeling unsaturated water flow and solute transport. In R. A. Feddes, G. H. de Rooij, & J. C. van Dam (Eds.), *Unsaturated zone modeling. Progress, challenges and applications* (Wageningen UR Frontis Series) (Vol. 6, pp. 1–36). Kluwer Academic Publishers.

Van Dam, J. C., Groenendijk, P., Hendriks, R. F. A., & Kroes, J. G. (2008). Advances of modeling water flow in variably saturated soils with SWAP. *Vadose Zone Journal, 7*, 640–653. https://doi.org/10.2136/vzj2007.0060

Van Genuchten, M. T., Leij, F. J., & Yates, S. R. (1991). *The RETC code for quantifying the hydraulic functions of unsaturated soils* (Report No. EPA/600/2-91/065) (p. 83). Robert S. Kerr Environmental Research Laboratory, Office of Research and Development, US Environmental Protection Agency.

Van Genuchten, M. T. (1980). A closed-form equation for predicting the hydraulic conductivity of unsaturated soils. *Soil Science Society of America Journal, 44*, 892–898.

Verburg, K. (Ed.). (1996). *Methodology in soil-water-solute balance modelling: An evaluation of the APSIM-SoilWat and SWIMv2 models.* Report of an APSRU / CSIRO Division of Soils workshop held in Brisbane, Australia from 16–18 May 1995, Divisional Report No. 131, Division of Soils, CSIRO, Australia. Available at: https://doi.org/10.4225/08/586d398cbe4e2.

Verburg, K., Ross, P. J., & Bristow, K. L. (1996). *SWIMv2.1 user manual.* CSIRO Division of Soil. Available at: https://www.apsim.info/Portals/0/Documentation/SWIMv21UserManual.pdf

Vereecken, H., Kasteel, R., Vanderborght, J., & Harter, T. (2007). Upscaling hydraulic properties and soil water flow processes in heterogeneous soils: A review. *Vadose Zone Journal, 6*, 1–28. https://doi.org/10.2136/vzj2006.0055

Vereecken, H., Schnepf, A., Hopmans, J. W., Javaux, M., Or, D., Roose, T., Vanderborght, J., Young, M. H., Amelung, W., Aitkenhead, M., Allison, S. D., Assouline, S., Baveye, P., Berli, M., Brüggemann, N., Finke, P., Flury, M., Gaiser, T., Govers, G., Ghezzehei, T., Hallett, P., Hendricks Franssen, H. J., Heppell, J., Horn, R., Huisman, J. A., Jacques, D., Jonard, F., Kollet, S., Lafolie, F., Lamorski, K., Leitner, D., McBratney, A., Minasny, B., Montzka, C., Nowak, W., Pachepsky, Y., Padarian, J., Romano, N., Roth, K., Rothfuss, Y., Rowe, E. C., Schwen, A., Šimůnek, J., Tiktak, A., Van Dam, J., van der Zee, S. E. A. T. M., Vogel, H. J., Vrugt, J. A., Wöhling, T., & Young, I. M. (2016). Modeling soil processes: Review, key challenges, and new perspectives. *Vadose Zone Journal, 15*(5). https://doi.org/10.2136/vzj2015.09.0131

Viaene, P., Vereecken, H., Diels, J., & Feyen, J. (1994). A statistical analysis of six hysteresis models for the moisture retention characteristic. *Soil Science, 157*, 345–355.

Vogel, T., van Genuchten, M. T., & Cislerova, M. (2000). Effect of the shape of the soil hydraulic functions near saturation on variably-saturated flow predictions. *Advances in Water Resources, 24*, 133–144. https://doi.org/10.1016/S0309-1708(00)00037-3

Vrugt, J. A., van Wijk, M. T., Hopmans, J. W., & Šimůnek, J. (2001). One-, two-, and three-dimensional root water uptake functions for transient modeling. *Water Resources Research, 37*, 2457–2470. https://doi.org/10.1029/2000WR000027

Wang, H. F., & Anderson, M. P. (1982). *Introduction to groundwater modeling. Finite difference and finite element methods* (p. 237). W.H. Freeman and Company.

Wösten, J. H. M., Lilly, A., Nemes, A., & le Bas, C. (1999). Development and use of a database of hydraulic properties of European soils. *Geoderma, 90*, 169–185.

Data Fusion in a Data-Rich Era

Annamaria Castrignanò and Antonella Belmonte

1 Introduction

The field of agricultural sciences is changing from a data-limited to a data-rich field, due to the accumulation of research data from international observatory networks and the integration of multi-source sensor data from space-borne remote sensing systems and proximal sensing arrays. In 2011, the agricultural G20 Agricultural Ministers launched the Global Agricultural Monitoring (GEOGLAM) initiative (http://www.geoglam.org/index.php/en/) to 'strengthen global agricultural monitoring by improving the use of remote sensing tools for crop production prediction and weather forecasting'. The Joint Experiment for Crop Assessment and Monitoring (JECAM) (http://jecam.org), which is the research branch of the GEOGLAM initiative, has been operating over almost 10 years to develop algorithms and collect datasets to improve agricultural monitoring and modelling. Moreover, the advent of the Copernicus program of the European Space Agency (ESA) and the observations collected by its Sentinel1 and Sentinel2 satellites provide unprecedented open and free data relevant for agricultural monitoring at a spatio-temporal scale, which might prove to be of interest even for precision agriculture.

A varied plethora of on-the-go sensors, from radiometric to geophysical ones in contact with the ground or at close proximity (proximal sensing), make it possible to collect an enormous amount of 3D data in a noninvasive, easy and cost-effective way.

A. Castrignanò (✉)
Gabriele D'Annunzio University of Chieti – Pescara, Chieti Scalo, (CH), Italy
e-mail: annamaria.castrignano@unich.it

A. Belmonte
CNR-IREA National Research Council—Institute for Electromagnetic Sensing of the Environment, Bari, Italy
e-mail: belmonte.a@irea.cnr.it

Despite considerable research progress in agricultural monitoring and modelling over the last decades, gaps remain in observational capabilities, data processing and scientific knowledge. In particular, combining different types of data sources, such as in situ, meteorological, (geo)statistical, multi-remote and proximal sensor data, with crop model predictions is of crucial interest to improve agricultural practices.

There is then an unprecedented demand to convert this huge amount of raw data into meaningful information, which can improve our fundamental knowledge and modelling of environmental processes and finally support decision-making in precision agriculture (PA). In a data-rich era, the major research activities will then shift focus from measurements and data collection to data processing, analysis and interpretation for improving the accuracy of model prediction and ultimately action. In the past, limited data could be stored and processed with relatively simple tools, whereas nowadays, more sophisticated techniques are required to analyse big data jointly from multiple sensors of different types (data fusion). An ambitious target of the coming decade is determining the response of vegetation and food production to many co-occurring stressors under different cultivation practices that should be optimised. This will require the development and validation of novel modelling methods, based on the use of big and heterogeneous spatial and temporal data. Observations/monitoring and process-thinking are two basic approaches to scientific investigation. The former records the state of a system and draws information about the underlying processes; the latter identifies these processes and expresses them formally in a model. These are two different ways of approaching the study of a biophysical system and making inference; however, they have to be seen as complementary rather than alternative. Only by synergistically combining the two approaches, the more stochastic approach guided essentially by data and the more deterministic approach based on formal equations will it be possible to make significant progress in the knowledge of the processes. This will entail greater control, the possibility of making informed decisions and ultimately managing the agricultural system more efficiently.

In this chapter, we will propose a way of dealing with data fusion using geostatistics, in the perspective that it can be useful for modelling in precision agriculture.

2 Need for Data Fusion in Precision Agriculture

The most recent definition of precision agriculture (PA) adopted by the International Society for Precision Agriculture (ISPA) in 2019 is the following:

> Precision Agriculture is a management strategy that gathers, processes and analyses temporal, spatial and individual data and combines it with other information to support management decisions according to estimated variability for improved resource use efficiency, productivity, quality, profitability and sustainability of agricultural production. (https://www.ispag.org/)

This definition highlights some fundamental issues, such as that PA is based on the analysis of different types of spatio-temporal data for efficient, cost-effective and sustainable management of the agricultural system. However, one of the greatest obstacles to implementing PA derives from the difficulties to determine temporal variation of agricultural inputs locally (Evans et al., 1996). Soil and crop attributes often vary considerably over space and time so that a large number of samples would be needed to provide information at the fine spatial and temporal resolutions required by PA (Adamchuk et al., 2011; Mulla, 2017). Intensive field sampling and laboratory analysis have been the traditional way of determining soil and plant properties for many years. Unfortunately, these methods are generally expensive, destructive and time-consuming, making it difficult, in some cases, to adopt PA practices. Various remote and proximal sensing technologies have been introduced to overcome these limitations (Castrignanò & Buttafuoco, 2020; Muazen et al., 2020; Castrignanò et al., 2017, 2018).

Sensors of different types are mounted on board satellites, air vehicles or on mobile platforms on the ground, although in the latter case, it is referred to as proximal sensing. The sensors used for remote sensing may differ in their spatial, temporal and spectral resolutions, as well as the intrinsic precision characteristics of the radiometer used. The spatial resolution is related to the pixel size; the temporal resolution is related to the revisit time of the same observation area, while the spectral resolution is indicated by the number of bands captured in the range of the electromagnetic spectrum of the radiometer used (Nowatzki et al., 2004). Hyperspectral sensors contain a large number of correlated bands of narrow width (<20 nm) and separated by very small increments in wavelength (2–3 nm) (Teke et al., 2013; Sishodia et al., 2020). The outputs of these sensors, appropriately processed, make it possible to extract useful information on the status of the crops (Chang et al., 2020; Nagasubramanian et al., 2019). More recently, hyperspectral sensors, which use the process of sunlight-induced chlorophyll fluorescence, have been applied increasingly to estimate photosynthesis, crop nutrient content and biotic and abiotic stresses such as disease and water stress (Chang et al., 2020; Nagasubramanian et al., 2019; Zarco-Tejada et al., 2016; Mohammed et al., 2019; Castrignanò & Buttafuoco, 2020; Muazen et al., 2020).

Two other sensors (thermal infrared and microwave sensors), mounted on board satellites or aircraft, are also used in agriculture mainly to estimate crop water stress and irrigation requirements (Palazzi et al., 2019; Babaeian et al., 2019;). Although satellites with high spatial resolution (<5 m) and daily revisit times (Sishodia et al., 2020) have recently been launched, most of the products supplied by remote sensing cannot be used in PA because of their low spatial and temporal resolutions. Indeed, the resolution required in PA depends on many factors, such as the management objectives, the size of the field, the technological level of the equipment used on the farm and the type of agronomic operation used. For example, for crop yield estimation, higher resolutions (1–3 m) are required, compared with the estimation of fertiliser and irrigation water, which require resolutions of about 5–10 m (Sishodia et al., 2020). Weed control and disease detections, however, require centimetre resolutions (Fernández-Quintanilla et al., 2018; Castaldi et al., 2017), as has been demonstrated recently in the fight against Xylella infection of olive trees in Southern Italy (Castrignanò et al., 2020; Zarco-Tejada et al., 2018).

Unmanned aerial vehicle (UAV)-based platforms, capable of providing finer spatial resolutions than the satellites, have been rapidly adopted in agricultural research in the last two decades. They can provide on-demand relatively cheap information during a crop growing season at the spatial scale required by PA operations (Huang et al., 2018).

An efficient application of PA requires an accurate assessment of the spatial and temporal variation of the soil as well as plants. For this purpose, on-the-go proximal sensors of different types, capable of measuring moisture, micro and macro nutrient contents, texture and other important soil properties at very fine spatial resolution, are already widely used (Castrignanò et al., 2017; De Benedetto et al., 2015; Viscarra Rossel et al., 2011). These sensors employ a wide variety of measurement techniques (electromagnetic induction (EMI), electrical resistivity, ground-penetrating radar (GPR) and gamma sensors, multi- and hyperspectral radiometer and fluorimeter). In particular, geophysical sensors are playing a major role in characterising field-scale soil spatial variation in PA (Corwin & Scudiero, 2016; Castrignanò et al., 2018).

Despite the challenges posed by the use of different sensors, with different spatial and temporal resolutions and uncertainties, combining (fusing) them will result in better spatio-temporal information. For example, the combination of satellite images using data from different types of satellite, such as those with high spatial resolution and those with high temporal resolution, will provide a final product that has all the relevant spatio-temporal features of interest (Zhu et al., 2018; Knipper et al., 2019). Sparse ground truth data could be processed jointly with high spatial and temporal resolution data (e.g. UAV data) and then be used as inputs to decision support systems (DSS) in PA.

Adequate data analysis techniques (data fusion) are needed to integrate multi-source spatial data efficiently, in order to obtain more exhaustive and accurate information. An appropriate data fusion approach must resolve several crucial problems, such as those related to the generally large size of data from remote and proximal sensing (big data) and to differences in measurement volume (support) and uncertainty of the various types of equipment. Presently, there is no single solution to the problem of data fusion from a statistical point of view, and various methods have been proposed, based on regression methods (Best et al., 2000; Young & Gotway, 2007), the Bayesian approach (Castanedo, 2013) and more recently artificial intelligence (AI) (Patrício & Rieder, 2018).

Geostatistics offers a set of optimal linear multivariate estimators (i.e. unbiased and with minimum error variance) to treat spatial data of different types by determining a statistical model of coregionalisation of spatial dependence (Castrignanò et al., 2000; Wackernagel, 2003). Moreover, geostatistics enables the crucial problem of 'change of support' to be solved, where support is used in the sense of geometrical size, shape and spatial orientation of the region associated with the measurements (Olea, 1991). The solution involves defining a statistical approach that jointly analyses two or more variables with different supports to permit valid inference about the new target support.

2.1   Principles of Geostatistical Data Fusion and Change of Support

When dealing with a specific spatial problem, it is common to analyse heterogeneous data jointly that are characterised by different degrees of aggregation and uncertainty. It is critical, therefore, to define a robust statistical approach to transform spatial data from one level of aggregation to another. Topics such as scale and scaling continue to encourage research in spatial analysis and modelling (Atkinson & Tate, 2000; Castrignanò & Buttafuoco, 2020). The two approaches, commonly used to integrate spatial data aggregated at different scales, are downscaling and upscaling. The former makes inferences at a lower level of aggregation than the original data, also called disaggregation, whereas the latter makes inferences at a higher level of aggregation. Data aggregation is also known under different names: the modifiable areal unit problem (MAUP) (Openshaw & Taylor, 1979) or the change of support problem (COSP) in geostatistics (Cressie, 1996; Young & Gotway, 2007). The COSP essentially makes inferences about a spatial variable with a specific spatial support but using variables with different spatial supports on the basis of a statistical model.

The COSP involves two critical inferential issues: the first, the 'aggregation effect', is that the data distribution of a variable averaged over a larger support is more bell-shaped and has smaller variance than that of the original variable. Aggregation then tends to reduce spatial heterogeneity and smooths the spatial data, although spatial autocorrelation (i.e. correlation between nearby observations) works as a mitigating factor (Castrignanò & Buttafuoco, 2020).

The second issue, the 'zoning effect' (Wong, 1996; Castrignanò & Buttafuoco, 2020), is the effect on statistical inference resulting from a different grouping of the same spatial units. The effect of this is to produce groups of different shape and orientation but at the same spatial scale.

A more complex problem arises in data disaggregation, where prediction (inference) is made at a finer spatial support than the one of the original data. It is worth emphasising that correlations at the individual level or from aggregated data are generally different and therefore the inferences obtained at a given scale cannot be transferred to another (Castrignanò & Buttafuoco, 2020). This further emphasises the importance of specifying the spatial scale concerning what the possible processes at play refer to.

There is no universal solution to COSP and the current main solutions include spatial smoothing techniques (Kelsal & Wakefield, 2002), statistical regression models (Grose et al., 2007), Bayesian hierarchical models (Brus et al., 2008) and multi-scale tree models (Huang et al., 2002). The GIS techniques, although widely used, do not offer a proper solution from a statistical point of view because in general they do not provide any measure of uncertainty (Neteler & Mitasova, 2008). Furthermore, they do not define a clear disaggregation or aggregation approach when using covariates.

Several geostatistical solutions have been proposed to solve COSP (Atkinson & Tate, 2000; Chilès & Delfiner, 2012; Castrignanò et al., 2017, 2019). Block cokriging has been widely used in the past for data upscaling only, but in its generalised form (Young & Gotway, 2007), it can be used for both upscaling (aggregation) and downscaling (disaggregation) (Castrignanò & Buttafuoco, 2020). For downscaling, however, other solutions were proposed such as: area-to-point kriging (Goovaerts, 2008; Yoo & Kyrialidis, 2009), cokriging for image sharpening (Pardo-Iguzquiza et al., 2006) and a deconvolution-convolution procedure to estimate the local variogram in super-resolution mapping (Atkinson & Jeganathan, 2010). To fuse multi-resolution remote sensing data, Nguyen et al. (2012) proposed a spatial statistical data fusion model. In addition to block (co) kriging, other more advanced geostatistical techniques, such as nonlinear methods and iso-factorial models, have been used more recently (Chilès & Delfiner, 2012).

The advantage of geostatistical methods to solve COSP is that they are formulated in a probabilistic framework for which a measure of uncertainty can be obtained. However, these approaches have two main disadvantages: (i) they can be computationally intensive and (ii) require the assumption of second-order stationarity. The latter implies that the averages of random variables are constant and known and that the spatial covariance function depends only on the distance between observations (Chilès & Delfiner, 2012). For big spatial datasets covering large areas, the assumption of stationarity is often not satisfied. One way to deal with spatial non-stationarity is through local (co)kriging (Hammerling et al., 2012). In this approach, the observations within local windows are used for both estimation of the covariance parameters and spatial predictions. Other computationally efficient approaches exist to reduce the computational burden, such as data dimension reduction (Wikle, 2010) and fixed rank kriging (FRK) and fixed rank filtering (FRF) developed by Cressie et al. (2010).

Therefore, a data fusion approach that takes advantage of the complementary features of the different types of data and combines them in a statistically robust way is proposed hereafter based on geostatistics. Multivariate geostatistics uses an unbiased optimal (with minimum error variance) estimator (cokriging) for interpolation, which requires the previous specification of a spatial model of multivariate dependence (Wackernagel, 2003). Block cokriging enables the estimate of different variables to be referred to the same support (block), thus providing a solution to the crucial problem of COSP. However, with proximal or remote sensing datasets that are often big in size, the required matrix inversion may be difficult if not impossible. A solution is possible with multicollocated cokriging (MCC) (Rivoirard, 2001), which is a simplified version of full cokriging using the auxiliary (grid) variable (s) only at the target point and at all locations where the primary variable of interest is available. This approach requires that the auxiliary variable(s) is (are) known at all (target) locations where the variable of interest has to be estimated. Moreover, MCC makes it possible to integrate the differences between the various sensor data synergistically and to complement them effectively with the point ground truths of the variable of interest (Castrignanò et al., 2012). The outcome of such an approach is a series of thematic maps of the variables considered to be of interest for PA.

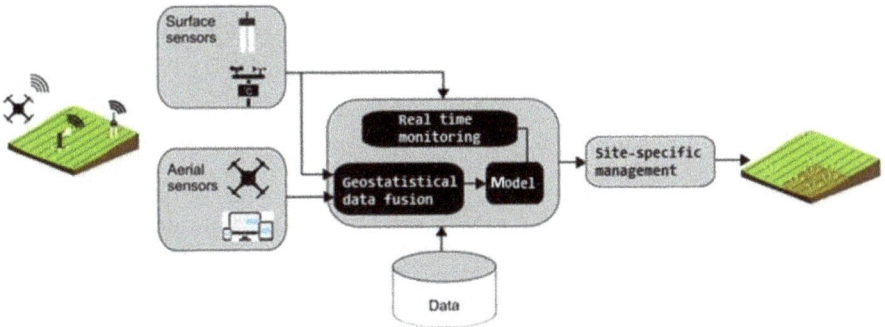

Fig. 1 Integrated framework of geostatistical data fusion and modelling

To delineate homogeneous management zones for site-specific farming, a very effective multivariate geostatistical tool called factorial cokriging can be used. It can synthesise multi-type data that encapsulate different spatial scales, into simple summary scale-dependent indices that can be used effectively for field partitioning. In a fully integrated framework between data fusion and modelling, geostatistics can be considered at the core of all data flows from the various remote and proximal sensors and coupled with a numerical model that in turn uses other data sources and remote sensing data. Geostatistics is then useful for fusing all the different sources of information and for interpreting and modelling their statistical space–time structures, before the transformed data are used as input to a simulation model aimed at site-specific management (Fig. 1).

The following is a brief outline of the proposed approach to data fusion based on geostatistical techniques. The interested reader can learn more about the procedures mentioned in the numerous references provided.

– *Exploratory Data Analysis*

The input data, including both point-based and grid data, generally need some adjustment before undergoing a multivariate analysis. First, the values of the variables on a grid have to be migrated to the nearest sample location to create the coregionalisation dataset. Exploratory data analysis can then be performed on the common dataset by calculating the basic statistics (mean, median, minimum and maximum values, standard deviation, skewness and kurtosis) of all variables in the study.

Skewed data distributions can often be made more suitable for geostatistical modelling by transforming each variable into a normal distribution with mean zero and unit variance using a mathematical function.

However, the transformation function must be corrected if the estimate is to be produced on a larger support (block) by using a support correction coefficient r.

– *Spatial Modelling and a Geostatistical Solution for Support Correction*

Fig. 2 Discretisation of
blocks

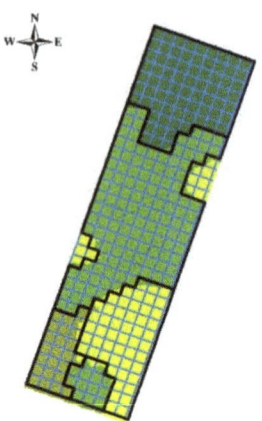

The key tool in geostatistics for expressing spatial dependence is the variogram, both direct and cross- variogram between pairs of variables. All direct and cross-variogram models are expressed as linear combinations of the same basic mathematical functions for each spatial scale under the scope of the linear model of coregionalisation (LMC).

From observations at locations with point support, block cokriging can be used to predict the average values of a multivariate process over a larger support (block), accounting for the specific block attributes (size, shape and orientation). The main difference between point and block (co)kriging is the calculation of the point-to-block covariance:

$$\overline{C}(B, \mathbf{x}_\alpha) = \text{cov}(Z(B), Z(\mathbf{x}_\alpha)) = \int_B \frac{C(\mathbf{u}, \mathbf{x}_\alpha)d\mathbf{u}}{|B|}, \tag{1}$$

where $|B|$ is the volume of the block (spatial support) and $C(\cdot)$ is the point-to-point covariance (spatial covariance is the complement of variogram with respect to the sample variance). The volume of the block B is discretised into points ($\mathbf{u}$) so that the mean covariance ($\overline{C}$) is approximated by a summation through numerical integration. The block variogram is calculated through regularisation, which transforms a variogram model established on points into the corresponding variogram on a given block support. The variogram on a point support is calculated on the smallest support and then regularised. In practice, the regularised LMC is estimated by using a discretisation of the blocks into equal cells, whereby the blocks are replaced by the union of the cell centres (Fig. 2).

After that, a pseudo-experimental variogram is calculated at the fictitious centres of each cell, regularly discretising the new support and then the point variograms are averaged over the whole block. A model of the block regularised variogram is fitted before using it subsequently in cokriging. Using the observations at locations ($\mathbf{x}_\alpha$) with point support, block cokriging predicts the average value of the variable(s) of interest over a larger support (upscaling).

Multi-collocated block cokriging is preferred to standard cokriging to fuse sample variables with gridded (exhaustive) variables for the final prediction on the target support as it reduces the size of the cokriging system to be solved, making it feasible to invert the matrix of spatial covariances.

Block cokriging can be used also in downscaling, when the estimates are on a smaller support than that of observations. However, in this case, a sort of deregulation or deconvolution is needed to transform variograms on areal support into variograms on point support through an iterative fitting process (Goovaerts, 2008; Castrignanò & Buttafuoco, 2020). Unfortunately, there is no unique way to reconstruct statistical properties at the individual level starting from aggregated data (King, 1997), and downscaling is implemented only on few commercial or free software packages.

– *Factorial Block Cokriging*

Factorial cokriging differs from traditional principal component analysis (PCA) (Jackson, 2003) in that it applies to the individual spatial (scale-dependent) components (coregionalisation matrices) instead of applying to the entire variance–covariance matrix, after fitting a linear model of coregionalisation model (LMC) (Castrignanò et al., 2000). As for PCA, it decomposes each coregionalisation matrix into two additional matrices: the eigenvalue matrix and the eigenvector matrix (Wackernagel, 2003). The transformation coefficients (loading coefficients) correspond to the covariances between the variables and the principal component, here called the regionalised factor, for any given spatial scale. They express the influence of each variable on the factor at a given spatial scale and are therefore used to assign a physical meaning to the factor. Mapping of scale-dependent regionalised factors then represents a concise and effective way of showing the spatial relationships among the variables at different spatial scales. It is therefore commonly used in PA to produce a scale-dependent partition of the field into homogeneous areas according to the attributes used. Again, if multiple exhaustive information (i.e. known at each node of the interpolation grid) is available, multicollocated factorial block cokriging is preferred to the full version.

Uncertainty Estimation and Model Evaluation

The final step is to determine prediction uncertainty of the raw variables after back-transformation. It is assessed by calculating the confidence interval (CI) at a given probability level, which reflects the inability to define an unknown value exactly (prediction) (Castrignanò et al., 2019; Manzione & Castrignanò, 2019). The confidence interval (CI) is calculated based on the following equation:

$$CI = UL\text{–}LL, \tag{2}$$

where UL and LL are the upper and lower limits of prediction, respectively. These limits are defined in terms of standard deviation (SD) of a Gaussian variable and for the probability level of 95% are given by

$$UL = \text{predicted value} + 1.96\,SD \qquad (3)$$

and

$$LL = \text{predicted value} - 1.96\,SD. \qquad (4)$$

The CI is then linked to the probability that the unknown value lies within a given interval. The above calculation of CI requires that the data are normally distributed (Castrignanò et al., 2009). Therefore, the block cokriged Gaussian predictions and their corresponding upper and lower limits are back-transformed into the raw values by using the block transformation function, previously estimated.

Figure 3 gives a complete overview of the proposed method.

2.2   Application of Geostatistical Data Fusion in Precision Agriculture

To illustrate the method described previously, an example is given below in which multi-source soil data with different supports are fused to delineate an agricultural field into management zones using regionalised factors as synthetic indicators of multivariate variability (Buttafuoco et al., 2019, 2021). Furthermore, as the processes underlying site-specific management are a function of scale, it is expected that field delineation too will be scale dependent. This particular case study was chosen because in the fusion process the change of support and the critical influence of spatial resolution on the statistical characteristics of the estimates are considered explicitly. The flexibility and effectiveness of the method are also evident in the numerous case studies to which it has been applied and to which the interested reader is advised to refer for further details (Buttafuoco et al., 2019; Castrignanò et al., 2017; Castrignanò & Buttafuoco, 2020; De Benedetto et al., 2013a, b).

The study area (Fig. 4) is in Southern Italy (latitude 40°48′22″N, longitude 14°34′12″E), in a field cropped with *Solanum lycopersicum* L. cv San Marzano. Fifty soil samples were collected at the nodes of a quasi-regular grid with mesh of about 9-m and down to 0.30–m depth to cover the 5500-m^2 field evenly. The locations of soil samples were determined with a GPS system with differential correction (DGPS, HiPer 27 Pro, Topcon, Tokyo, Japan) and sub-metric accuracy in the UTM projection system.

The samples were analysed in the laboratory for total organic carbon content and concentrations of nitrogen, potassium and phosphorous.

In addition, reflectance of each soil sample, in the radiometric range between 350 and 2500 nm, was measured in the laboratory using an ASD FieldSpec IV spectroradiometer and pre-processed according to the procedures described by Conforti et al. (2015). A PCA of spectral data was performed to reduce the dimensionality of the data.

Fig. 3 Flowchart of the
proposed methodology of
geostatistical data fusion

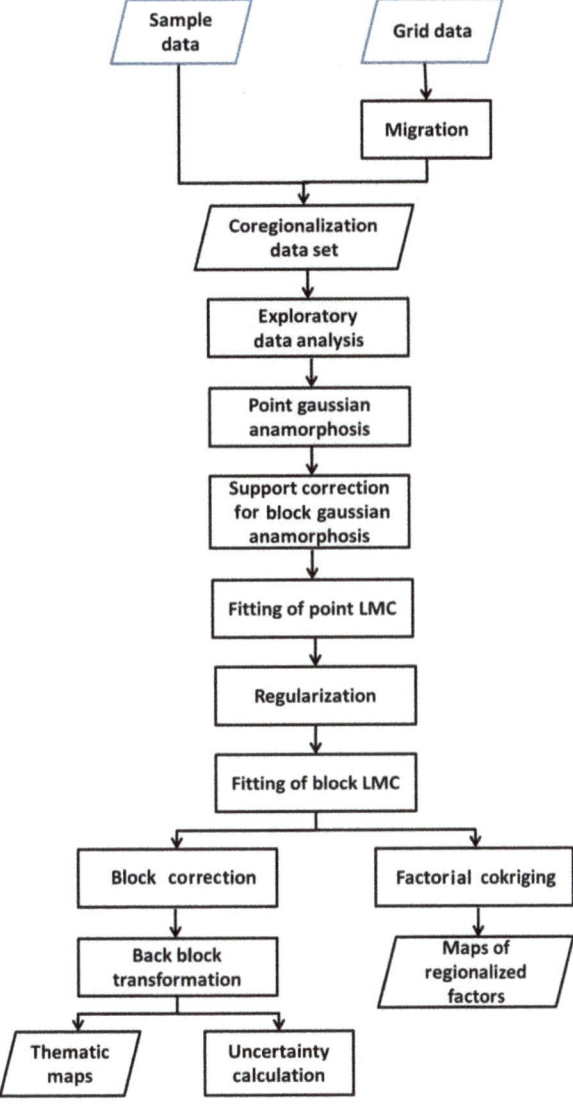

A soil survey was carried out to measure ECa (mS m^{-1}) along 24 parallel transects (Fig. 5) with a multiple frequency electromagnetic profiler GEM300® (Castrignanò et al., 2019). The ECa data, measured with the highest operating frequency only (19,975 Hz), were used in this study to characterise the topsoil. The maximum depth of investigation of this operating frequency was 1.59 m (Huang, 2005), but the effective depth (the depth above which 50% of the signal contribution effect is retained) was about 0.40 m (Sudduth et al., 2005).

Fig. 4 Location of study site (**a**) and the field site (**b**)

Fig. 5 Survey scheme: the
black dotted lines are the
24 transects where the
geophysical surveys were
carried out; the brown points
are the soil sample locations

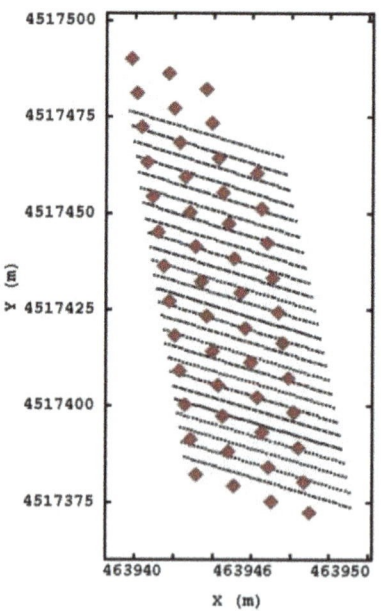

The vertical supports of the three different types of measurements can be considered comparable (0–0.30/0.40-m depth); the surface supports were quite different: point (a circle of about 0.008-m² area) for soil sample data, including reflectance data, and about 1.2 m (distance between the two transmitter–receiver coils) × 1 m (the mean distance travelled in 1 s) for EMI data.

Results of PCA on soil spectra showed that the first principal component (PC1) accounted for about 92% of the total variance; therefore, it was retained and used as a

Table 1 Summary statistics for apparent electrical conductivity (ECa, mS m^{-1}), first principal component of spectral data (PC1), potassium (K, mg kg^{-1}), nitrogen (N, mg kg^{-1}), phosphorous (P, mg kg^{-1}) and total organic carbon (TOC, %)

Variable	Mean	Median	Minimum	Maximum	Standard deviation	Skewness	Kurtosis
ECa	90.69	88.24	80.25	106.69	6.37	0.75	2.81
PC1	0.00	−0.36	−32.26	38.58	14.12	0.23	3.29
K	692.81	678.32	556.74	863.18	70.06	0.73	3.05
N	1.22	1.23	0.90	1.61	0.18	−0.60	2.19
P	135.33	134.32	113.00	171.09	13.88	0.27	2.19
TOC	16.94	16.55	14.18	21.74	2.22	0.76	2.73

synthetic spectral index to be processed together with the other variables. The loading coefficients of PC1 showed that PC1 was mainly influenced by the bands between 750 and 2250 nm, corresponding to the main absorption wavelengths of most chemical soil components (Riefolo et al., 2019). It can therefore be interpreted as an indicator of soil chemistry in addition to laboratory analytical measurements.

A combined coregionalisation dataset, consisting of six variables, four soil attributes (K, N, P, TOC), the first PC of the spectra (PC1) and the migrated ECa data to the nearest soil sample location to a maximum distance of 9 m, was created to be analysed jointly using multivariate analysis. Exploratory data analysis was then performed by calculating the basic statistics (mean, median, minimum and maximum values, standard deviation, skewness and kurtosis).

Table 1 showed that the data distributions of ECa, K and TOC deviate significantly from normality. Moreover, all histograms of the six soil attributes (Fig. 6), except PC1, exhibited clear shifts from the normal distribution with an evident bimodality for N and P. This pattern indicated that at least two zones in the field should be subjected to different fertilisation since the average N and P contents in the soil were different.

All data were then transformed into standardised Gaussian-shaped variables with mean 0 and variance 1 by using a point Gaussian transformation function to mitigate the effect of outliers on variogram fitting. A point LMC was fitted consisting of the following basic structures: a nugget effect, a cubic model with spatial range of 34.95 m and a *K*-Bessel function (Chilès & Delfiner, 2012) with a scale of 14.01 m and second parameter equal to 10.24. The goodness of fit was tested through cross-validation, reporting mean errors for all variables close to zero and error variances within the tolerance interval: 0.4–1.6 (Chilès & Delfiner, 2012) (data not shown).

In Table 2, the block support correction coefficients together with the point and block variances are reported for each variable, highlighting the reduction of variance due to averaging over the block, especially for PC1, N and P.

The above results are quite crucial for site-specific management and emphasise the need to define in advance the spatial scale at which the processes under investigation take place. If, for example, the absorption of N and P by roots occurs

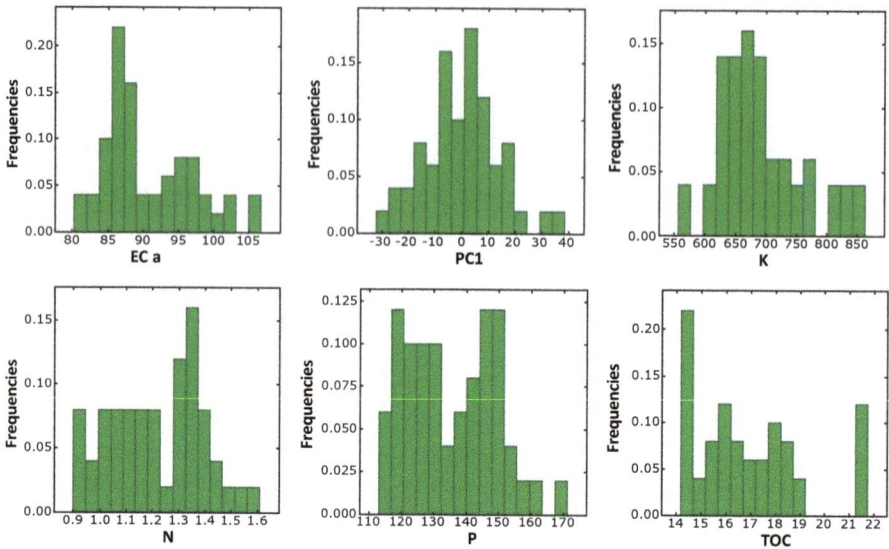

Fig. 6 The histograms of the six soil attributes: ECa, PCI, K, N, P and TOC

Table 2 Block support correction calculations for apparent electrical conductivity (ECa, mS m^{-1}), first principal component of spectral data (PC1), potassium (K, mg kg^{-1}), nitrogen (N, mg kg^{-1}), phosphorous (P, mg kg^{-1}) and total organic carbon (TOC, %)

	Point variance	Block variance	Block support correction
ECa	40.51	39.90	0.99
PC1	198.43	57.31	0.54
K	4892.58	4741.44	0.99
N	0.03	0.01	0.65
P	192.03	74.14	0.63
TOC	4.88	4.39	0.96

on a scale smaller than 1 m × 1 m, producing a composite soil sample over 1 m^2 or more of the field might cause the reduction of more than 40% in variance, which might have a negative effect on the optimisation of local fertilisation because all the information within the scale of 1 m would be lost.

It is important to note that not only is the variance affected in a way that is statistically and easily predictable but also the shape of the whole histogram (Lajaunie & Wackernagel, 2000). The support effect then becomes critical when estimating the conditional probability that the average over a specified block is below a critical value. In PA, such a probability could be used as a criterion to decide when and where nutrients and or water need(s) to be added to the soil.

This highlights the importance of selecting the sensor with a suitable support (footprint) comparable to the scale of the study process, because all predictions can be done only at this scale when using proximal sensing.

Before performing multivariate analysis, an anisotropic nested model was fitted to the point variogram of the Gaussian-transformed ECa data comprising the following three spatial structures: a nugget effect, an isotropic spherical model with range of 20 m and a zonal anisotropic K-Bessel model in the direction $170°$ from north with a range of 40 m and parameter equal to 2. The goodness of fit was tested by cross-validation producing mean error and standardised error variance close to 0 and 1, respectively. The zonal anisotropy, approximatively along the direction of the longer axis of the field, can be explained by its elongated shape, and most mechanical operations were performed in that direction so causing greater spatial continuity (longer range).

The use of point kriging is justified because the surface support of the GEM sensor is about $1.2 \text{ m} \times 1 \text{ m}$; therefore, it is comparable with the one of the interpolation grid cells. Moreover, the preliminary interpolation of ECa on the grid was necessary for the subsequent application of multi-collocated cokriging.

The experimental point variograms of the Gaussian-transformed soil variables were regularised over the $1\text{-m} \times 1\text{-m}$ block, and the fitted block LMC was an isotropic linear model of coregionalisation including the following two basic spatial structures: a spherical model with a range of 18.10 m and a spherical model with range of 35.30 m. As for point LMC, the fitting was reasonable based on mean errors and standardised error variances, quite close to 0 and falling within tolerance interval (0.4–1.6), respectively. The total spatial variance was therefore divided into two spatially correlated components: the former at a shorter range (18.10 m) accounting for 59% of the total variance and the latter at a longer range (35.30 m) accounting for the remaining 41%. Conversely, the spatially uncorrelated component was negligible.

The LMC is a way to describe the mutual spatial relationships between the various variables. The fact that point LMC and block LMC show differences means that the relationships between the variables are scale dependent. Therefore, it is critical, when studying processes occurring in an agricultural field, to decide in advance the scale at which the study is to be conducted.

The two spatial structures of the LMC can be interpreted as the effect of two causes of variation operating at different scales in this agricultural field: one of shorter range, probably of anthropic origin (agricultural operations), and the other of more natural origin related to the physical–chemical characteristics of the soil and subsoil. It is worth noting that the two spatial scales do not differ greatly from those identified in the variogram of ECa. This would suggest that ECa is influenced for the most part by those same processes that determine soil attributes (Fig. 7).

From the proximity of the cross-variogram model to the dotted curve, representing the maximum correlation between the two variables (intrinsic correlation) (Wackernagel, 2003), the significant correlation of ECa with all the other variables except for N can be deduced. More precisely, ECa was positively correlated with K and PC1, therefore with most of the cations present in the soil, but negatively with P and TOC the components most related to mineral and organic soil fertilisation.

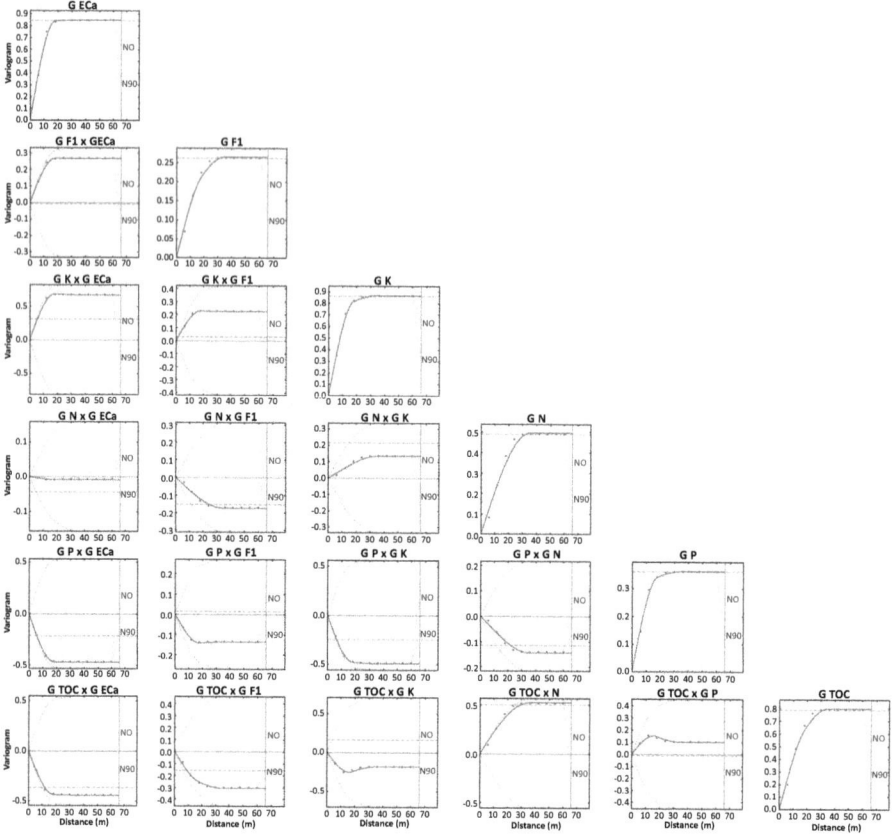

Fig. 7 Regularised LMC of the Gaussian-transformed variables: model (bold line), experimental variogram (dot points) and intrinsic correlation curve (dotted line)

The results confirm the findings of previous studies (Corwin & Scudiero, 2016; De Benedetto et al., 2013a, b) that bulk electrical conductivity measured with an electromagnetic sensor can be used as an indicator of soil variability resulting from the combined action of many physical, chemical and organic properties.

The cokriged maps of PC1, K, N, P and TOC (Fig. 8b–f) appear consistent and reveal greater continuity (uniformity) in the direction 170° from the north. However, for K, N and TOC, it is the western edge of the field that has the largest values, while for PC1 and P, it is the eastern edge (Fig. 8). For P, there are alternate longitudinal strips with different fertility. These characteristics of the spatial variation may be of anthropogenic source because of inhomogeneity in the distribution of fertilisers. A different spatial pattern is shown by ECa (Fig. 8a): on the basis of this variable, the field could be divided roughly into three parts, approximately perpendicular to the longitudinal direction with different dielectric properties of soil. This diversification might be due also to physical and chemical characteristics of the soil, such as texture, porosity and/or moisture, other than those covered by this study. However, from the

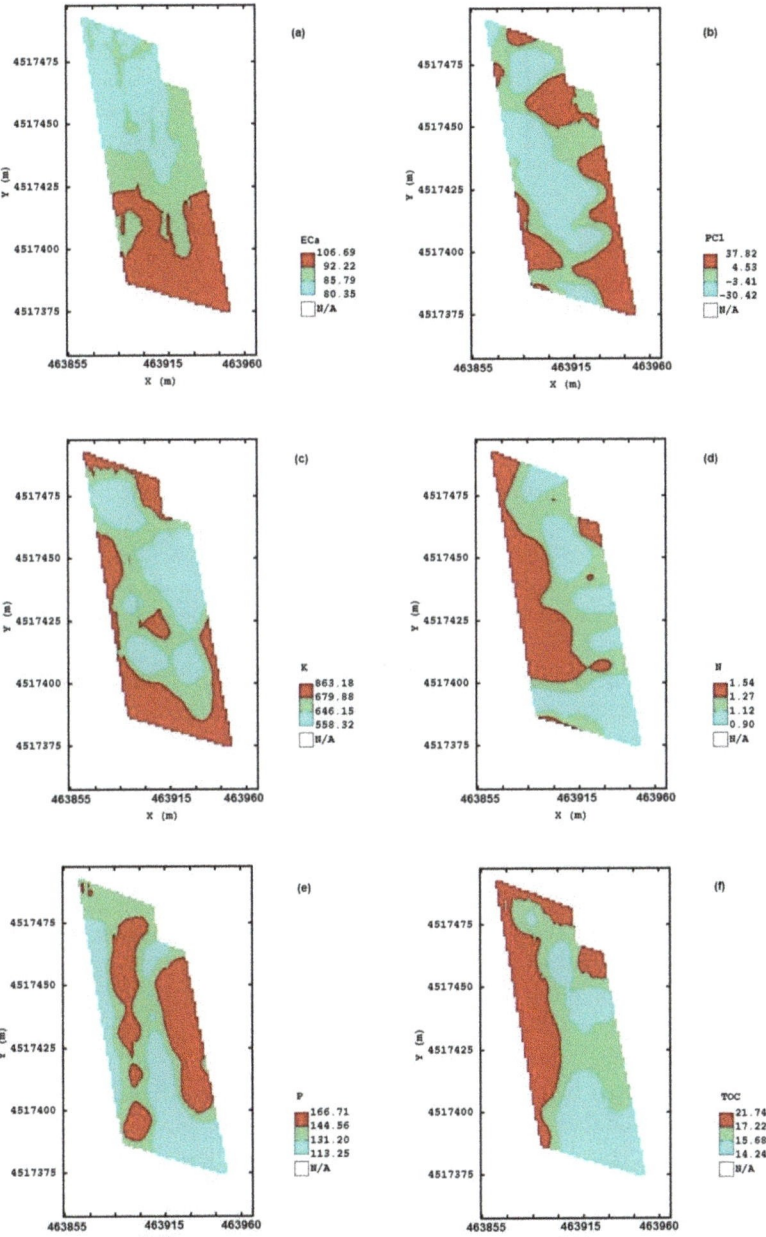

Fig. 8 Maps of the soil properties using three isofrequency classes: (**a**) apparent electrical conductivity (ECa), (**b**) first principal components of spectral data (PC1), (**c**) potassium (K), (**d**) nitrogen (N), (**e**) phosphorous (P) and (**f**) total organic carbon (TOC)

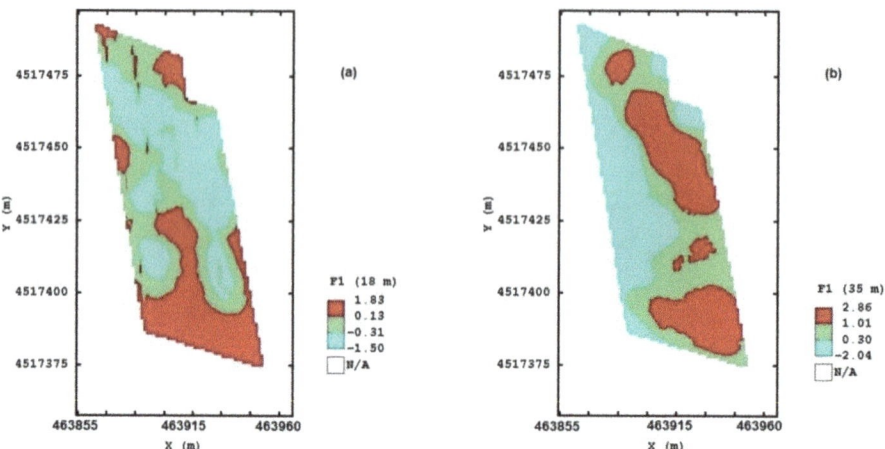

Fig. 9 Maps of the short (**a**) and long (**b**) range regionalised factors using three iso-frequency classes

strong spatial correlation of ECa with K, P and TOC (Fig. 7), the hypothesis may be suggested that these variables also contribute to the partition of the soil in the three macro-areas.

Among the six regionalised factors corresponding to each spatial scale (18.10 and 35.30 m), the first factor only for each scale was retained because it was associated with an eigenvalue greater than 1, therefore explaining a proportion of the variance at each scale greater than the one explained by each variable standardised to unit variance. In particular, the first factor at shorter range explained 97.06% of the variance at that scale and was affected mainly by the Gaussian-transformed variables of ECa and K, which confirms what we stated previously about the strong correlation between these two variables. On the other hand, the first factor at longer range explained 81.75% of the variance, and the Gaussian-transformed variables of TOC and N had a strong negative relation.

The first two regionalised factors have a different pattern (Fig. 9). The map of the first regionalised factor at short range (Fig. 9a) shows a rough partition of the field into three main zones, quite similar to the one produced by ECa although more variable, characterised by different physical properties and different concentrations of K. This map could therefore be used to advantage in site-specific fertilisation of K because the large central zone is characterised by the smallest K values, the southern zone by the largest and the northern zone by the intermediate values.

The first factor at long range (Fig. 9b) shows a similar pattern to that of N and TOC but with an inverse relation as expected (Fig. 8d, f). This map could be used as a support to improve the chemical fertility of the field.

Figure 8 shows that the field delineation may vary as a function of the spatial scale, because most processes occurring in the soil are scale dependent and consequently the variation in soil properties is as well. This means that traditional statistical clustering techniques to delineate homogeneous areas in the field might not be appropriate. Most classical clustering techniques do not consider the proximity of the samples and their spatial correlation, with a few rare exceptions (Oliver & Webster, 1989; Urban, 2004); moreover, they process data jointly regardless of the spatial scale and support.

The results of this case study have shown clearly that not only the variances but also the spatial relationships between the variables vary according to the spatial scale. This has a critical impact on field delineation and then on site-specific management.

In the light of all the above, we would like to warn PA practitioners against using spatial techniques that do not take into account the actual spatial correlation between observations. We would also urge the scientific community to consider the problem of scaling and change of support when analysing heterogeneous data jointly. The case study illustrates that these are not purely academic discussions but crucial topics that can have a significant impact on effectiveness of site-specific management in PA.

2.3   Applications of Data Fusion in Remote Sensing

In remote sensing, data fusion is one of the most commonly used techniques. The current satellites can interpret the spectral, spatial and temporal information of a phenomenon on Earth in different ways. The outcome is a wide variety of data, some repetitive and some integral, proposing various portrayals of a similar physical event. Data fusion aims to integrate the information acquired with different spatial and spectral resolutions from satellite sensors. In general, it is used to merge spatial information of a high-resolution panchromatic (PAN[1]) image with the spectral information of a low-resolution multispectral (MS[2]) or hyperspectral (HS[3]) image of the same area. Taking advantage of having multiple free image datasets (e.g. the Sentinel series ESA's Copernicus program), data fusion has focused increasingly on improving algorithms to merge, in particular, the optical and SAR types of remote sensing data. This section is based on the SAR–optical data fusion aspect. The complementarity of these two types of data can provide enhanced information on Earth observation, particularly in the agricultural field. In the following, an overview of data fusion techniques will be shown, highlighting their uniqueness.

SAR–Optical Data Fusion
Each satellite sensor has different features used for specific tasks. In the Earth observation field, the sensors comprise resolutions from the centimetre to the

---

[1] PAN image: a single greyscale image.

[2] MS image: a colour image with typical 4–12 bands.

[3] HS image: a colour image with typical >100–1000 bands.

kilometre scales. They include both active and passive sensing technologies with a range throughout most of the electromagnetic spectra.

Visible-infrared (VIR) optical sensors are passive elements that acquire the electromagnetic energy reflected and emitted from the surface. By analysing the response detected by the optical sensor, it is possible to discriminate physical and chemical information about terrain cover types and to provide valuable data, for example, about leaf pigments, water content and plants' overall health condition (Zarco-Tejada et al., 2018).

Optical images are relatively easy to interpret but they are affected by weather conditions and solar illumination. In addition, VIR sensors cannot capture images during the night, and they do not record targets hidden by clouds, trees and other ground covers.

The SAR sensors are active elements, and mostly they work in the microwave electromagnetic spectrum, emitting a signal and recording the return echo (back-scattering signal). They can operate at night in the presence of clouds and be sensitive to conditions in the upper few centimetres (about 10 cm for P band, depending on soil texture and soil moisture) of the ground. The SAR images provide information on the geometric and dielectric properties of the target, such as the moisture content and existence of vegetation and the geometry and structure of the features (Gherboudj et al., 2011; Souissi & Ouarzeddine, 2016). However, these sensors suffer from geometric and radiometric decorrelation effects. The former is due to the side-looking acquisition mode of the SAR. This appears as a granular visible effect from the speckle (Goodman, 1976), formed as a result of the coherent radiation. The speckle is a peculiarity of the sensor itself. It makes interpretation of SAR images less intuitive than optical ones.

SAR–Optical Pre-processing

Because of the different geometric and radiometric properties of SAR and optical sensors, it is difficult to match and co-register these two types of images. To improve the performance of SAR–optical data fusion, it is necessary to reduce the sources of geometric and radiometric errors. Therefore, it is necessary to apply appropriate pre-processing algorithms (see Table 3).

Data Fusion Methodologies

There is no universally valid data fusion algorithm because of the variety of image types. The fused image must fulfil three fundamental requirements: (i) maintenance of all relevant information, (ii) removal of irrelevant information and noise and (iii) minimisation of artefacts. An ideal fusion process should preserve the original spectral characteristics while improving the spatial characteristics of the image. Data fusion, depending on the level at which it is performed, is classified into the following types: at pixel, feature and decision levels (see Fig. 10).

- **Pixel Level**

   Pixel level fusion extracts information on a pixel-by-pixel basis to generate a fused image from a set of pixels in original images. This methodology has the

Table 3 Pre-processing data fusion

Optical data	SAR data
Calibration: The procedure allows pixel values to be expressed in units of reflectance. It reduces the number of unwanted artefacts, which are mainly due to atmospheric effects, lighting conditions and soil topography during the acquisition time of images.	*Focalisation*: The SAR image is represented as a matrix, which contains in each row the sampling and the valour of the backscattering signal. During the acquisition, the contribution of a single target is dispersed over several elements of the matrix. The focusing allows us to 'concentrate' the contribution in a single element of the matrix. It assures the correspondence between each matrix element and each ground resolution cell. Thus, the raw data matrix is turned into a real ground image, called a single look complex (SLC). Generally, as SAR data, the product SLC is used.
	De-speckle: This process removes the speckle.
Georeferencing: This algorithm is common to SAR and optical data. It puts all images in the same reference geographical system. This can also be performed before de-speckling for SAR images, although it is preferable afterwards.	
Registration: The process requires that the pixels in the two images (SAR and optical ones) coincide exactly with the same points on the ground.	
Resampling: This step may also be applied. It ensures that the image at low spatial resolution, for example, MS image, has the same spatial resolution as the PAN/SAR image, using various algorithms (e.g. the nearest neighbour or cubic convolution) (Keys, 1982).	

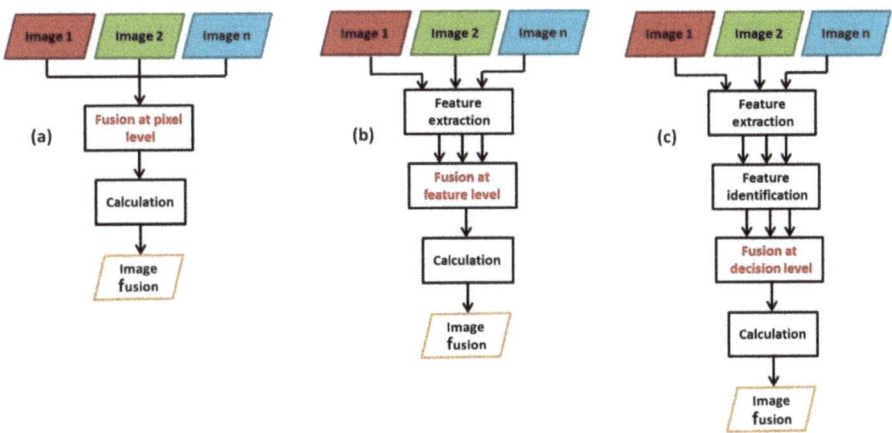

Fig. 10 Data fusion levels: (**a**) pixel level, (**b**) feature level and (**c**) decision level

potential to achieve the highest performance in signal detection, evaluating the spatial and spectral information content present in the recorded images. This data fusion methodology is in turn subdivided into four classes: component substitution (CS), multiresolution analysis (MRA), hybrid methods (a combination of CS and MRA) and model-based algorithms.

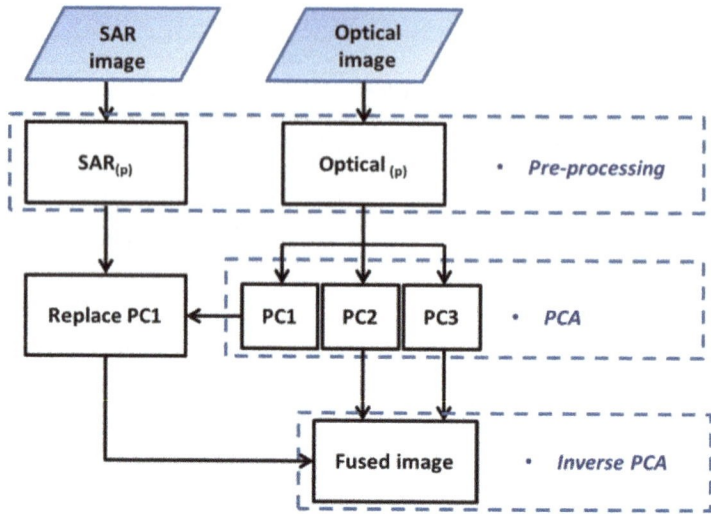

Fig. 11 Flowchart of PCA method

Component Substitution

In CS fusion, one of the transformed components (usually the first component or intensity, IL) is substituted by the high-resolution SAR/PAN image before the inverse transformation is enforced. The most common CS techniques include (i) PCA, principal component analysis; (ii) IHS, intensity-hue-saturation; (iii) GS, Gram–Schmidt; (iv) BT, Brovey transform; (v) HPF, high-pass filtering; (vi) Elhers fusion; and (vii) geostatistical analysis:

– In PCA, the first principal component image, PC1, contains most spatial information of the multispectral channels, which are used as input for PCA, while the remaining components recover the remaining spectral information. The PC1 is replaced by the high spatial resolution image (SAR image) after matching the histogram of the SAR image with one of the first principal components. As the last step, inverse PCA transformation generates the output fused image (see Fig. 11).

– The IHS method aims to enhance the spectral of the final image. It modifies an MS image at low resolution from RGB space into IHS colour space. Thus, the intensity (I) band, which resembles a panchromatic image, is replaced by a high-resolution SAR/PAN image in the fusion. An inverse IHS transform is then performed on the SAR/PAN together with the hue (H) and saturation (S) bands, resulting in an IHS merged image (Fig. 12).

– The GS method can be regarded as a generalisation of PCA. Unlike PCA, the first principal component is chosen arbitrarily while the others are calculated as perpendicular to the first one. A simulated low-resolution SAR band is generated;

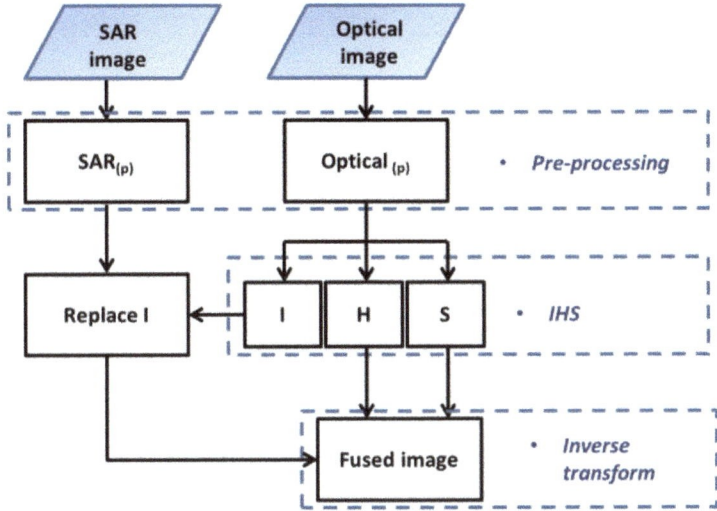

Fig. 12 Flowchart of IHS method

the GS transformation is applied to it and the other multispectral bands; the first GS band is then replaced by the high spatial resolution SAR band; and finally, through inverse GS transformation, a high spatial resolution MS image is produced (Laben & Brower, 2000).

The difference between GS and PCA is based on how the information is distributed: for PCA, the information is mostly in the first component and decreases markedly in the remaining components, and in GS, the transformed components only are orthogonal and the information is not significantly different in each of them. Therefore, to avoid keeping all the information in the first component, this transformation keeps the spatial texture characteristics of the PAN images and the spectral information of the original MS images. However, GS transformation is fairly complex and not applicable for large-scale image fusion.

- In Brovey transformation, the MS bands to be fused are normalised and then multiplied by a high-resolution band as SAR/PAN one. The general equation (see Eq. 5) uses red, green and blue (RGB) and the PAN/SAR bands as inputs to produce the new MS bands. This method is based on spectral modelling and enhances the visual contrast between the low and high ends of the data histogram. It enables the desired spatial information to be obtained but with significant radiometric distortion (Wang et al., 2005).

$$\text{DNB1 new} = [\text{DNB1}/(\text{DNB1} + \text{DNB2} + \text{DNB3})] \times [\text{DN SAR}]$$
$$\text{DNB2 new} = [\text{DNB2}/(\text{DNB1} + \text{DNB2} + \text{DNB3})] \times [\text{DN SAR}] \qquad (5)$$
$$\text{DNB3 new} = [\text{DNB3}/(\text{DNB1} + \text{DNB2} + \text{DNB3})] \times [\text{DN SAR}]$$

Table 4 Advantages and disadvantages of the different CS techniques

Technique	Advantages	Disadvantages
PCA	Very simple, computationally efficient, faster processing time and high spatial quality	Spectral degradation and colour distortion; fusion results may vary depending on the selected datasets
IHS	Very simple, computationally efficient and faster processing and high sharpening ability	It can process only three multispectral bands; colour distortion; the fusion results may vary depending on the selected datasets
GS	Computationally efficient and faster processing	Colour distortion; the fusion results may vary depending on the selected datasets
Brovey	Very simple, computationally efficient and faster processing time. Fused image with higher degree of contrast	Colour distortion
HPF	Very simple and computationally efficient	Sometimes the edges are emphasised too much.
Elhers	Computationally efficient and higher spectral preservation	More or less significant colour shifts

- The HPF method applies a high-pass filter in the frequency domain to the high-resolution image (SAR image) achieving an improvement in spatial resolution of the fusion product (Pohl & Genderen, 1998; Wang et al., 2005). The process is expressed as

$$MS_f = \{W_{high} \cdot HF[SAR]\} \cdot \{W_{low} \cdot LF[MS]\}. \tag{6}$$

where MS_f is the data fusion image and HF and LF correspond to the high- and low-pass filter operators, with W_{high} and W_{low} weights, respectively.

- Ehlers fusion technique combines IHS transformation with fast Fourier transform (FFT) into a frequency domain. The IHS transform is enforced on MS images. Then, low-pass (LP) and high-pass (HP) Fourier filters are applied on the intensity component and SAR image, respectively. At the end a data fusion, image is produced through the inverse FFT^{-1}. This process preserves the spectral characteristics of the original multispectral dataset, without depending on the selection or the order of the bands in IHS transform.
- Geostatistical techniques. This method has been widely debated above: 2.1 (Table 4).

Multi-resolution Analysis (MRA)
Wavelet theory is a popular approach in signal and image processing, with wide-ranging use including remote sensing. It is applied on the multi-resolution analysis (MRA) method. It is based on decomposing the signal into elementary functions called 'wavelets' (Meyer, 1990). In fusing an MS image with a high-resolution SAR/PAN image (see Fig. 13), it is necessary to apply these steps: (i) The SAR/PAN

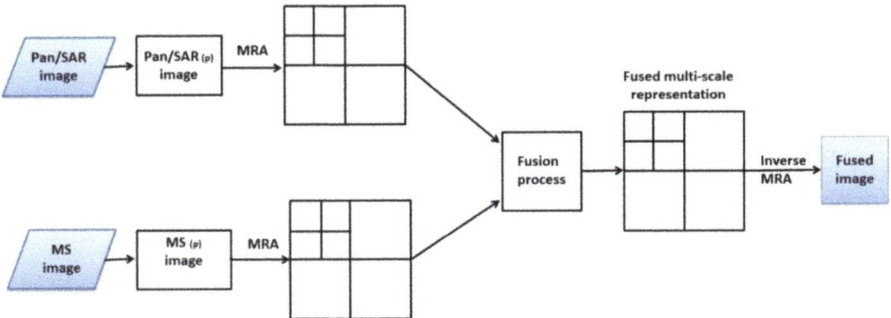

Fig. 13 Flowchart of MRA method

image is first decomposed into a set of low-resolution PAN images, which correspond to wavelet coefficients (spatial details) for each level. (ii) Each band of the MS image then supplants the low-resolution PAN at the resolution of the original MS image. (iii) The SAR/PAN spatial detail is inserted into each MS band by achieving a reverse wavelet transform on each MS band together with the analogous wavelet coefficients.

The MRA methods provide fine resolution in both the spatial and spectral domains, by reducing spatial and spectral distortion, and good signal-to-noise ratio.

Hybrid Methods
Hybrid methods combine the advantages of CS and MRA techniques. Those most frequently used in remote sensing are the high-pass filter (Lu et al., 2011; Zhang, 2010) and the Brovey transformation algorithm (Zhang, 2010). These procedures aim 'to sharpen' low spatial resolution images by linearly combining the pixel values of high and low spatial resolution images.

Model-Based Algorithms (MBA)
Model-based algorithms can be divided into two main classes: (i) 'variational' models (Piella, 2009) and (ii) 'sparse representation-based' (SR) models (Mairal et al., 2008). The first models consider the input images as discrete samples of continuous two-dimensional functions and calculate the unknown values of such two-dimensional functions.

The second models describe signals based on the sparsity and redundancy of their representations. For this method, an image signal is considered as a specific dictionary (i.e. a matrix) that contains k prototype signals, also referred to as atoms (Mairal et al., 2008).

The MBA produce high-resolution merged data as a linear combination of the extracted atoms from an over-complete dictionary constructed by the whole set of the images. A small number of atoms only are essential to regenerate a signal accurately, ensuring a good sparse representation of signals.

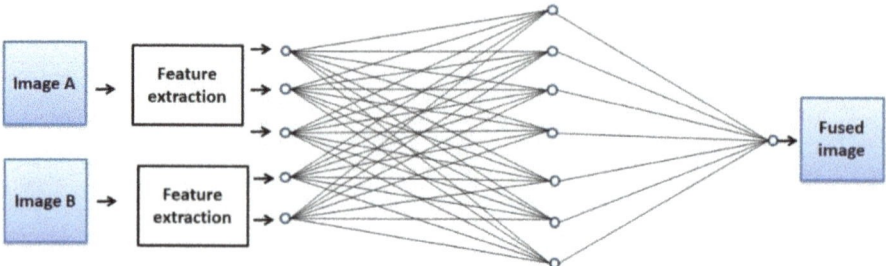

Fig. 14 Flowchart of artificial neural network method

- *Feature Level*

This method works out information on feature level after the recognition and extraction of objects from pixels as intensities, edges or textures in the original data. Contrary to the pixel-level fusion, feature-level fusion provides good results but at the cost of execution time.

- *Decision Level*

Decision-level processes develop information at a higher level of abstraction. In other words, information extraction is achieved using outputs from various algorithms that are then merged using given decision rules. Recently, new procedures based on neural networks have found wide application in remote sensing. Traditional artificial neural networks (ANNs), Fig. 14, are machine learning algorithms that aim to approximate parameterised mathematical functions $y = f(x, \theta)$. These functions map an input x to an output y through some nonlinear transformations: f-$(x) = (f_n \cdot \cdots \cdot f_1)(x)$. Each element f_k represents a simple linear transformation of the previous component's output, tracked by a nonlinear function, $f_k = \sigma_k(\theta_k f_{k-1})$, where σ_k are the nonlinear functions, as Sigmoid, Tanh, ReLU and leaky ReLU (Jones, 2014), and θ_k are matrices of numbers, called the model's weights.

During the training phase, the network makes predictions about the output layer using an intermediate representation of inputs of each network component. The network has to learn how to make use of these representations to create a complex hierarchical scheme of the data to obtain the correct predictions at the end in the output layer.

Nowadays, satellite programmes such as Copernicus allow unlimited access to both SAR and optical images from the Sentinel 1 (S1) and Sentinel 2 (S2) satellites, respectively.

The H2020 SENSAGRI[4]-Sentinels Synergy for Agriculture (2016–2019) project aimed to analyse the capacity of S1 and S2 to develop innovative agricultural monitoring services capable of near real-time operations. In particular, a method

---

[4]https://e2l-coop.eu/en/projects/h2020-sensagri/

was developed to generate an improved time-continuous leaf area index (LAI) product. This fused LAI algorithm is based on a machine learning regression algorithm developed in a Bayesian framework: Gaussian processes regression (GPR).[5]

Quality of Measures

When the assessment of the fused product is left to the visual interpretation of the user, it appears rather subjective, especially when there is no ground truth. To overcome this problem, a 'reference' merged image may be available or created manually. A reference image is produced, downscaling the input MS and PAN/SAR, and then they are merged and related to the original data. Some quality metrics have been defined to evaluate objectively fusion of remote sensing images (Jagalingam & Hegde, 2015). The metrics include two classes: with or without a reference image. In remote sensing, the absence of a reference image is overcome by using Wald's protocol (Wald et al., 1997).

Conclusion

Data fusion is becoming more important because it helps scientists to extract increasingly precise and relevant knowledge from the available information. Fusion of multi-sensor data will enable the maximum use of what is available in the repository of data or of what can be acquired in the shortest possible time. A relevant example for the use of complementary information from the Earth observation sensors is the joint use of SAR and optical data. Fusion of these sensors, with highly complementary characteristics, contributes to a better understanding of the sensed data, and thus, many applications can be improved; but it is not a trivial task.

The future of research in SAR–optical data fusion moves on in several interesting directions: (i) the exploitation of big data; and the ESA Copernicus programme offers access to huge amounts of Earth observation data; (ii) the use of sophisticated processing methods, such as deep learning.

2.4   Summing Up

The above shows clearly how multivariate geostatistics can provide an efficient tool for data fusion in PA. However, modelling of change of support has not yet been studied in detail and verified in the decision support systems (DSS) operated in PA. A possible cause is the lack of awareness of the impact that neglecting spatial and/or temporal autocorrelation and the effect of support may have on agricultural management.

---

[5] https://ipl.uv.es/sensagri/ftp/DELIVERABLES/WP4/SENSAGRI_D4_9_v1.0.pdf

On the other hand, future research is needed to adapt nonlinear geostatistical techniques to PA applications to improve the prediction of plant response, which is affected by so many factors operating on several spatial and temporal scales.

Conflict of Interest The authors declare that there are no conflicts of interest.

References

Adamchuk, V. I., Viscarra Rossel, R. A., Marx, D. B., & Samal, A. K. (2011). Using targeted sampling to process multivariate soil sensing data. *Geoderma, 163*, 63–73.

Atkinson, P. M., & Jeganathan, C. (2010). Estimating the local small support semivariogram for use in superresolution mapping. In P. M. Atkinson & C. D. Lloyd (Eds.), *geoENV VII—geostatistics for environmental applications* (pp. 279–294). Springer.

Atkinson, P. M., & Tate, N. J. (2000). Spatial scale problems and geostatistical solutions: A review. *The Professional Geographer, 52*(4), 607–623.

Babaeian, E., Sidike, P., Newcomb, M. S., Maimaitijiang, M., White, S. A., Demieville, J., Ward, R. W., Sadeghi, M., LeBauer, D. S., Jones, S. B., Sagan, V., & Tuller, M. (2019). A new optical remote sensing technique for high resolution mapping of soil moisture. *Frontiers in Big Data, 2*, 37.

Best, N. G., Ickstadt, K., & Wolpert, R. L. (2000). Spatial Poisson regression for health and exposure data measured at disparate resolutions. *Journal of the American Statistical Association, 95*, 1076–1088. https://doi.org/10.1080/01621459.2000.10474304

Brus, D. J., Bogaert, P., & Heuvelink, G. B. M. (2008). Bayesian maximum entropy prediction of soil categories using a traditional soil map as soft information. *European Journal of Soil Science, 59*(2), 166e177.

Buttafuoco, G., Quarto, R., Quarto, F., Conforti, M., Venezia, A., Vitti, C., & Castrignanò, A. (2019). A geophysical and spectrometric sensor data fusion approach for homogeneous within-field zone delineation. In J. V. Stafford (Ed.), *Precision agriculture '19* (pp. 705–712). Academic.

Buttafuoco, G., Quarto, R., Quarto, F., et al. (2021). Taking into account change of support when merging heterogeneous spatial data for field partition. *Precision Agriculture, 22*, 586–607. https://doi.org/10.1007/s11119-020-09781-9

Castaldi, F. F., Pelosi, F., Pascucci, S., & Casa, R. (2017). Assessing the potential of images from unmanned aerial vehicles (UAV) to support herbicide patch spraying in maize. *Precision Agriculture, 18*, 76–94.

Castanedo, F. (2013). A review of data fusion techniques. *Scientific World Journal.* https://doi.org/10.1155/2013/704504

Castrignanò, A., & Buttafuoco, G. (2020). Chapter 3: Data processing. In A. Castrignanò, G. Buttafuoco, R. Khosla, A. M. Mouazen, D. Moshou, & O. Naud (Eds.), *Agricultural Internet of Things and decision support for precision smart farming* (1st ed., pp. 139–182). Academic. ISBN:978-0-12-818373-1.

Castrignanò, A., Giugliarini, L., Risaliti, R., & Martinelli, N. (2000). Study of spatial relationships among soil physical-chemical properties using Multivariate Geostatistics. *Geoderma, 97*, 39–60.

Castrignanò, A., Costantini, E. A. C., Barbetti, R., & Sollitto, D. (2009). Accounting for extensive topographic and pedologic secondary information to improve soil mapping. *Catena.* https://doi.org/10.1016/j.catena.2008.12.004

Castrignanò, A., Wong, M. T. F., Stelluti, M., De Benedetto, D., & Sollitto, D. (2012). Use of EMI, gamma-ray emission and GPS height as multi-sensor data for soil characterisation. *Geoderma, 175–176*, 78–89.

Castrignanò, A., Buttafuoco, G., Quarto, R., Vitti, C., Langella, G., Terribile, F., & Venezia, A. (2017). A combined approach of sensor data fusion and multivariate geostatistics for delineation of homogeneous zones in an agricultural field. *Sensors (Switzerland)*. https://doi.org/10.3390/s17122794

Castrignanò, A., Buttafuoco, G., Quarto, R., Parisi, D., Viscarra Rossel, R. A., Terribile, F., Langella, G., & Venezia, A. (2018). A geostatistical sensor data fusion approach for delineating homogeneous management zones in Precision Agriculture. *Catena*. https://doi.org/10.1016/j.catena.2018.05.011

Castrignanò, A., Quarto, R., Venezia, A., & Buttafuoco, G. (2019). A comparison between mixed support kriging and block cokriging for modelling and combining spatial data with different support. *Precision Agriculture*. https://doi.org/10.1007/s11119-018-09630-w

Castrignanò, A., Belmonte, A., Antelmi, I., Quarto, R., Quarto, F., Shaddad, S., Sion, V., Muolo, M. R., Ranieri, N. A., Gadaleta, G., Bartoccetti, E., Riefolo, C., Ruggieri, S., & Nigro, F. (2020). A geostatistical fusion approach using UAV data for probabilistic estimation of Xylella fastidiosa subsp. pauca infection in olive trees. *Science of The Total Environment, 752*, ISSN 0048-9697. https://doi.org/10.1016/j.scitotenv.2020.141814

Chang, C. Y., Zhou, R., Kira, O., Marri, S., Skovira, J., Gu, L., & Sun, Y. (2020). An Unmanned Aerial System (UAS) for concurrent measurements of solar induced chlorophyll fluorescence and hyperspectral reflectance toward improving crop monitoring. *Agricultural and Forest Meteorology, 294*, 1–15.

Chilès, J. P., & Delfiner, P. (2012). *Geostatistics: Modeling spatial uncertainty* (2nd ed.). Wiley. https://doi.org/10.1002/9781118136188

Conforti, M., Castrignanò, A., Robustelli, G., Scarciglia, F., Stelluti, M., & Buttafuoco, G. (2015). Laboratory based Vis NIR spectroscopy and partial least square regression with spatially correlated errors for predicting spatial variation of soil organic matter content. *Catena, 124*, 60–67.

Corwin, D. L., & Scudiero, E. (2016). Field-scale apparent soil electrical conductivity. In S. Logsdon (Ed.), *Methods of soil analysis* (pp. 1–29). Soil Science Society of America. https://doi.org/10.2136/methods-soil.2015.0038

Cressie, N. (1996). Change of support and the modifiable areal unit problem. *Geographical Systems, 3*, 159–180.

Cressie, N., Shi, T., & Kang, E. (2010). Fixed rank filtering for spatial-temporal data. *Journal of Computational and Graphical Statistics, 19*(3), 724–745.

De Benedetto, D., Castrignanò, A., Rinaldi, M., Ruggieri, S., Santoro, F., Figorito, B., Gualano, S., Diacono, M., & Tamborrino, R. (2013a). An approach for delineating homogeneous zones by using multi-sensor data. *Geoderma*. https://doi.org/10.1016/j.geoderma.2012.08.028

De Benedetto, D., Castrignano, A., Diacono, M., Rinaldi, M., Ruggieri, S., & Tamborrino, R. (2013b). Field partition by proximal and remote sensing data fusion. *Biosystems Engineering*. https://doi.org/10.1016/j.biosystemseng.2012.12.001

De Benedetto, D., Quarto, R., Castrignanò, A., & Palumbo, D. A. (2015). Impact of data processing and antenna frequency on spatial structure modelling of GPR data. *Sensors (Switzerland), 15*, 16430–16447. https://doi.org/10.3390/s150716430

Evans, R. G., Han, S., Kroeger, M. W., & Schneider, S. M. (1996). Precision center pivot irrigation for efficient use of water and nitrogen. In P. C. Robert, R. H. Rust, & W. E. Larson (Eds.), *Proceedings of the 3rd international conference* (pp. 75–84). ASA/CSSA/SSSA. https://doi.org/10.2134/1996.precisionagproc3.c8

Fernández-Quintanilla, C., Peña, J. M., Andújar, D., Dorado, J., Ribeiro, A., & López-Granados, F. (2018). Is the current state of the art of weed monitoring suitable for site-specific weed management in arable crops? *Weed Research, 58*, 259–272.

Gherboudj, I., Magagi, R., Berg, A. A., & Toth, B. (2011). Soil moisture retrieval over agricultural fields from multi-polarized and multi-angular RADARSAT-2 SAR data. *Remote Sensing of Environment, 115*(1), 33–43. ISSN:0034-4257.

Goodman, J. W. (1976). Some fundamental properties of speckle, JOSA. *Optical Society of America, 66*(11), 1145–1150.

Goovaerts, P. (2008). Kriging and semivariogram deconvolution in the presence of irregular geographical units. *Mathematical Geology, 40*(1), 101–128.

Grose, D. J., Harris, R., Brundson, C., & Kilham, D. (2007). *Grid enabling geographically weighted regression*. In Proceedings of the 3rd international conference on e-Social Science, Ann Arbor.

Hammerling, D. M., Michalak, A. M., O'Dell, C., & Kawa, S. R. (2012). Global CO_2 distributions over land from the greenhouse gases observing satellite (GOSAT). *Geophysical Research Letters, 39*, L08804. https://doi.org/10.1029/2012GL051203

Huang, H. (2005). Depth of investigation for small broadband electromagnetic sensors. *Geophysics, 70*, G135–G142.

Huang, H. C., Cressie, N., & Gabrosek, J. (2002). Fast resolution-consistent spatial prediction of global processes from satellite data. *Journal of Computational and Graphical Statistics, 11*, 1–26.

Huang, W., Lu, J., Ye, H., Kong, W. A., Mortimer, H., & Shi, Y. (2018). Quantitative identification of crop disease and nitrogen-water stress in winter wheat using continuous wavelet analysis. *International Journal of Agricultural and Biological Engineering, 11*, 145–151.

Jackson, J. E. (2003). *User's guide to principal components*. Wiley.

Jagalingam, P., & Hegde, A. V. (2015). A review of quality metrics for fused image. In *International conference on water resources, coastal and ocean engineering* (pp. 133–142).

Jones, N. (2014). Computer science: The learning machines. *Nature, 505*(7482), 146–148, 1.

Kelsall, J., & Wakefield, J. (2002). Modeling spatial variation in disease risk: A geostatistical approach. *Journal of the American Statistical Association, 97*, 692–701. https://doi.org/10.2307/3085705

Keys, R. (1982). Cubic convolution interpolation for digital image processing. IEEE Trans Acoust Speech Signal Process. *IEEE Transactions on Acoustics, Speech, and Signal Processing, 29*, 1153–1160. https://doi.org/10.1109/TASSP.1981.1163711

King, G. (1997). *A solution to the ecological inference problem*. Princeton University Press.

Knipper, K. R., Kustas, W. P., Anderson, M. C., Alfieri, J. G., Prueger, J. H., Hain, C. R., Gao, F., Yang, Y., McKee, L. G., Nieto, H., Hipps, L. E., Mar Alsina, M., & Sanchez, L. (2019). Evapotranspiration estimates derived using thermal-based satellite remote sensing and data fusion for irrigation management in California vineyards. *Irrigation Science, 37*, 431–449.

Laben, C. A., & Brower, B. V. (2000). *Process for enhancing the spatial resolution of multispectral imagery using pan-sharpening*. US Patent 6,011,875.

Lajaunie, C., & Wackernagel, H. (2000). *Geostatistical approaches to change of support problems: Theoretical framework* (IMPACT Project Deliverable Nr 19, Technical Report N–30/01/G). Centre de Géostatistique, Ecole des Mines de Paris.

Lu, D., Li, G., Moran, E., Dutra, L., & Batistella, M. (2011). A comparison of multisensor integration methods for land cover classification in the Brazilian Amazon. *GIScience & Remote Sensing, 48*, 345–370. https://doi.org/10.2747/1548-1603.48.3.345

Mairal, J., Elad, M., & Sapiro, G. (2008). Sparse representation for color image restoration. *IEEE Transactions on Image Processing, 17*(1), 53–69. https://doi.org/10.1109/TIP.2007.911828

Manzione, R. L., & Castrignanò, A. (2019). A geostatistical approach for multi-source data fusion to predict water table depth. *Science of the Total Environment, 696*, 133763. https://doi.org/10.1016/j.scitotenv.2019.133763

Meyer, Y. (1990). *Ondelettes et operateurs I: Ondelettes*. Hermann, 215 pp.

Mohammed, G. H., Colombo, R., Middleton, E. M., Rascher, U., van der Tole, C., Nedbald, L., Goulas, Y., Pérez-Priego, O., Damm, A., Meroni, M., et al. (2019). Remote sensing of solar-induced chlorophyll fluorescence (SIF) in vegetation: 50 years of progress. *Remote Sensing of Environment, 231*, 1–39.

Muazen, A. B., Alexandridis, T., Buddenbaum, H., Cohen, Y., Moshou, D., Mulla, D., Nawar, S., & Sudduth, A. (2020). Chapter 2: Monitoring. In A. Castrignanò, G. Buttafuoco, R. Khosla, A. M. Mouazen, D. Moshou, & O. Naud (Eds.), *Agricultural Internet of Things and decision support for precision smart farming* (1st ed., pp. 35–138). Academic. ISBN:978-0-12-818373-1.

Mulla, D. J. (2017). Spatial variability in precision agriculture. In S. Shashi, H. Xiong, & X. Zhou (Eds.), *Encyclopedia of GIS* (pp. 2118–2125). Springer.

Nagasubramanian, K., Jones, S., Singh, A. K., Sarkar, S., Singh, A., & Ganapathysubramanian, B. (2019). Plant disease identification using explainable 3D deep learning on hyperspectral images. *Plant Methods, 15*, 1–10.

Neteler, M., & Mitasova, H. (2008). *Open source GIS: A GRASS GIS approach* (3rd ed.). Kluwer Academic Publishers/Springer.

Nguyen, H., Cressie, N., & Braverman, A. (2012). Spatial statistical data fusion for remote sensing applications. *Journal of the American Statistical Association, 107*(499), 1004–1018.

Nowatzki, J., Andres, R., & Kyllo, K. (2004). *Agricultural Remote Sensing Basics*. NDSU Extension Service Publication. Available online: www.ag.ndsu.nodak.edu. Accessed 23 Sept 2020

Olea, R. A. (Ed.). (1991). *Geostatistical glossary and multilingual dictionary*. Oxford University Press.

Oliver, M. A., & Webster, R. (1989). A geostatistical bases for spatial weighting in multivariate classification. *Mathematical Geology, 21*(1), 15–35.

Openshaw, S., & Taylor, P. (1979). A million or so correlation coefficients. In N. Wrigley (Ed.), *Statistical methods in the spatial sciences* (pp. 127–144). Pion.

Palazzi, V., Bonafoni, S., Alimenti, F., Mezzanotte, P., & Roselli, L. (2019). Feeding the world with microwaves: How remote and wireless sensing can help precision agriculture. *IEEE Microwave Magazine, 20*(12), 72–86.

Pardo-Iguzquiza, E., Chica-Olmo, M., & Atkinson, P. M. (2006). Downscaling cokriging for image sharpening. *Remote Sensing of Environment, 102*(1–2), 86–98.

Patrício, D., & Rieder, R. (2018). Computer vision and artificial intelligence in precision agriculture for grain crops: A systematic review. *Computers and Electronics in Agriculture, 153*, 69–81. https://doi.org/10.1016/j.compag.2018.08.001

Piella, G. (2009). Image fusion for enhanced visualization: A variational approach. *International Journal of Computer Vision, 83*, 1–11.

Pohl, C., & Genderen, J. L. V. (1998). Multisensor image fusion in remote sensing: Concepts, methods, and applications. *International Journal of Remote Sensing, 19*(5), 823–854.

Riefolo, C., Castrignanò, A., Colombo, C., Conforti, M., Vitti, C., & Buttafuoco, G. (2019). Investigation of soil surface organic and inorganic carbon contents in a low-intensity farming system using laboratory visible and near-infrared spectroscopy. *Archives of Agronomy and Soil Science, 66*(10), 1436–1448. https://doi.org/10.1080/03650340.2019.1674446

Rivoirard, J. (2001). Which models for collocated cokriging? *Mathematical Geology, 332*, 117–131.

Sishodia, R., Ray, R., & Singh, S. (2020). Applications of remote sensing in precision agriculture: A review. *Remote Sensing, 12*, 3136. https://doi.org/10.3390/rs12193136

Souissi, B., & Ouarzeddine, M. (2016). Polarimetric SAR data correction and terrain topography measurement based on the radar target orientation angle. *Journal of the Indian Society of Remote Sensing, 44*, 335–349. https://doi.org/10.1007/s12524-015-0493-x

Sudduth, K. A., Kitchen, N. R., Wiebold, W. J., Batchelor, W. D., Bollero, G. A., Bullock, D. G., Clay, D. G., Palm, H. L., Pierce, F. J., Schuler, R. T., & Thelen, K. D. (2005). Relating apparent electrical conductivity to soil properties across the north-central USA. *Computers and Electronics in Agriculture, 46*, 263e283.

Teke, M., Deveci, H. S., Haliloglu, O., Gürbüz, S. Z., & Sakarya, U. (2013). A short survey of hyperspectral remote sensing applications in agriculture. In *Proceedings of the 2013 6th international conference on Recent Advances in Space Technologies (RAST), Istanbul, Turkey* (pp. 171–176).

Urban, D. L. (2004). *Multivariate analysis: Nonhierarchical agglomeration, course notes, multivariate methods for environmental applications*. Nicholas School of the Environment and Earth Sciences at Duke University. [online]: www.env.duke.edu/landscape/classes/env358/mv_ pooling.pdf. Accessed 21 Jan 2005

Viscarra Rossel, R. A., Adamchuk, V. I., Sudduth, K. A., McKenzie, N. J., & Lobsey, C. (2011). Proximal soil sensing. An effective approach for soil measurements in space and time. *Advances in Agronomy, 113*, 237–282. https://doi.org/10.1016/B978-0-12-386473-4.00010-5

Wackernagel, H. (2003). *Multivariate geostatistics: An introduction with applications*. Springer. ISBN:13:9783540441427.

Wald, L., Ranchin, T., & Mangolini, M. (**1997**). Fusion of satellite images of different spatial resolutions: Assessing the quality of resulting images. *Photogrammetric Engineering and Remote Sensing, 63*, 691–699.

Wang, Z., Ziou, D., Armenakis, C., Li, D., & Li, Q. (2005). A comparative analysis of image fusion methods. *IEEE Transactions on Geoscience and Remote Sensing, 43*(6), 1391–1402.

Wikle, C. K. (2010). Low-rank representation for spatial processes. In A. E. Gelfand, P. Diggle, P. Guttorp, & M. Fuentes (Eds.), *Handbook of spatial statistics* (pp. 107–118). CRC Press.

Wong, D. W. S. (1996). Aggregation effects in geo-referenced data. In D. Griffiths (Ed.), *Advanced spatial statistics* (pp. 83–106). CRC Press.

Yoo, E. H., & Kyriakidis, P. C. (2009). Area-to-point kriging in spatial hedonic pricing models. *Journal of Geographical Systems, 11*(4), 381–406.

Young, L. J., & Gotway, C. A. (2007). Linking spatial data from different sources: The effect of change of support. *Stochastic Environmental Research and Risk Assessment, 21*, 589–600.

Zarco-Tejada, P. J., González-Dugo, M. V., & Fereres, E. (2016). Seasonal stability of chlorophyll fluorescence quantified from airborne hyperspectral imagery as an indicator of net photosynthesis in the context of precision agriculture. *Remote Sensing of Environment, 179*, 89–103.

Zarco-Tejada, P. J., Camino, C., Beck, P. S. A., Calderon, R., Hornero, A., Hernández-Clemente, R., Kattenborn, T., Montes-Borrego, M., Susca, L., Morelli, M., et al. (2018). Previsual symptoms of Xylella fastidiosa infection revealed in spectral plant-trait alterations. *Nature Plants, 4*, 432–439.

Zhang, J. (2010). Multi-source remote sensing data fusion: Status and trends. *International Journal of Image and Data Fusion, 1*, 5–24. https://doi.org/10.1080/19479830903561035

Zhu, X., Cai, F., Tian, J., & Williams, T. K. A. (2018). Spatiotemporal fusion of multisource remote sensing data: Literature survey, taxonomy, principles, applications, and future directions. *Remote Sensing, 10*, 527.

Data Assimilation of Remote Sensing Data into a Crop Growth Model

Keiji Jindo, Osamu Kozan, and Allard de Wit

Abbreviations

APSIM	agricultural production system simulator
CSM	crop simulation model
DA	data assimilation
DSSAT	decision support system for agrotechnology
EVI	enhanced vegetation index
FAO-WRSI	Food and Agriculture Organization-water requirement satisfaction index
FAPAR	fraction of absorbed photosynthetically active radiation
IOT	Internet of Things
LAI	leaf area index
NDVI	normalized difference vegetation index
PAR	photosynthetically active radiation
RS	remote sensing
VI	vegetative index
WDVI	weighted difference vegetation index
WOFOST	world food studies

K. Jindo (✉)
Agrosystems Research, Wageningen University & Research, Wageningen, the Netherlands
e-mail: keiji.jindo@wur.nl

O. Kozan
Center for Southeast Asia Studies, Kyoto University, Kyoto, Japan

A. de Wit
Environmental Research, Wageningen University & Research, Wageningen, the Netherlands

© The Author(s), under exclusive license to Springer Nature Switzerland AG 2023
D. Cammarano et al. (eds.), *Precision Agriculture: Modelling*, Progress in Precision Agriculture, https://doi.org/10.1007/978-3-031-15258-0_8

1 Introduction

Data assimilation (DA) is the overarching term for an ensemble of techniques to combine all possible information (models, observations, a priori data and statistics) to obtain the best possible estimate of the state of a system (Zhang & Moore, 2015). Data assimilation has its origins in meteorology and found its way into operational weather forecasting, oceanography and hydrology, but it is also a valuable technique for estimating variables related to crop growth (soil moisture, LAI (leaf area index), biomass, etc.) by combining models and observations of crop variables.

In precision agriculture, we are interested in "the application of technologies and principles to manage spatial and temporal variability associated with all aspects of agricultural production for the purpose of improving crop performance and environmental quality" (Pierce & Nowak, 1999). Observing the state of crops and soils accurately and precisely is key to precision agriculture. However, many aspects of crops and their environment are difficult to observe. For example, the most accurate way of measuring crop yield is by destructive sampling, which is time-consuming and often not practical at the scale that would be required in commercial farming. Other variables such as crop nitrogen content or the amount of nitrate that has leached beyond the reach of crop roots are even harder to measure.

Moreover, some of the variables are affected by management decisions that have been taken weeks or even months earlier (sowing date, fertilization, etc.). Decisions about agro-management must be based on a good understanding about the interaction between crop management, crop growth and the environmental factors (weather, soil, etc.). Crop simulation models (CSMs) provide a tool for describing these dynamic interactions, and they allow us to make predictions about the impact of agro-management decisions on the crop variables of interest. Many crop simulation models exist, ranging from simple models focusing on crop water use only (e.g. FAO-WRSI) (Frère & Popov, 1986) to models simulating crop growth in a comprehensive manner which include phenology, photosynthesis, growth of plant organs and water/nitrogen use (e.g. DSSAT, decision support system for agrotechnology; WOFOST, world food studies; APSIM, agricultural production system simulator). Plant functional–structural models extend this even further by explicitly describing growth and development at an individual plant level including structural aspects such as branch and leaf geometry and the competition between plants for resources (Evers et al., 2019).

Although crop simulation models can be highly relevant for understanding the interaction between crop growth, weather and management, their application in precision agriculture is often confounded by uncertainties. Such uncertainties reduce the accuracy with which the model can be used to make predictions about the growth and development of the crop, which reduces the benefits of applying a CSM. In precision agriculture, such uncertainties are often related to model processes that are not well simulated due to lack of knowledge about them or intra-field variation of soil properties which cannot be resolved by the available soil maps but which cause differences in crop response to lack of water (drought) or too much water

(water-logging and ponding). Other uncertainties rise from factors not included in the model (e.g. pest and disease) or factors that are difficult to describe accurately (e.g. local frost kill due to varying snow depth as a result of micro-relief).

The application of DA can reduce uncertainties in CSMs by assimilating external observations of crop variables into the model. They can be used to adjust the model for the effect of factors that are not included in the model structure (its biophysics) or for uncertainty about the correct values for certain model inputs. In other words, DA can help to 'bring the model back on track' in cases where discrepancies arise between model predictions and observations in the real world.

The routine collection of observations of crop-related variables at the field scale has become much easier in recent years due to the developments in sensor networks and (remote) sensing techniques. Cheap and accessible sensors to measure soil moisture, light interception and crop spectral properties are now available, and sensor networks using the Internet of Things (IOT) technologies can be used to report in real time. Moreover, continuing advances in imaging technology mean that images of crop fields can be collected routinely by drones or satellites. For example, the Sentinel 1 and Sentinel 2 satellite series, which are part of the European Copernicus programme, are tremendously useful for precision agriculture. The collection of observations through both sensors, which have a high temporal resolution, as well as observations from imaging platforms that can characterize spatial variation provides a valuable opportunity for integrating DA with crop simulation models.

The objective of this chapter is to provide an introduction to the application of DA in precision agriculture. This includes an overview of the different DA techniques available, with their pros and cons, as well as references to relevant papers describing applications of DA in real-world situations. We end the chapter with an outlook and future perspectives of applications of DA for precision agriculture.

2   Data Assimilation Techniques

Overview
Assimilation Methods of Remote Sensing Data into Crop Growth Models

A classification of data assimilation methods can be gleaned from several reviews (Dorigo et al., 2007; Moulin et al., 2010; Jin et al., 2018; Fischer et al., 2009). The classes are forcing, calibrating, updating and re-initiating/re-calibrating (Table 1).

2.1 *Forcing*

Forcing takes place when instead of calculating values for a given state variable in the model, values are used that are observed in the real world. As a concrete

Table 1 The basic features of three methods of assimilating remote sensing data into a crop simulation model

| | Method of assimilating remote sensing data into a crop simulation model | | |
	Forcing	Calibration	Updating
Computation time	Less	More	Less
Flexibility	No	Yes	Yes
Propagation of uncertainty	Possibly	Minimizing errors	Minimizing errors
Number of parameters	Fewer	More	More

example, a value of from LAI obtained from a vegetation index (VI) measured with a satellite can be used instead of the value of LAI simulated by a CSM. It has been established that WDVI (weighted difference vegetation index) has a strong relationship with LAI in several crops (Bouman, 1995; Clevers, 1991; Clevers et al., 2017). The forcing method completely replaces the model logic by an external observation (e.g. fraction of absorbed PAR (photosynthetically active radiation) is prescribed by the external (satellite) data and the model no longer simulates canopy development).

To use the forcing method, Bouman stressed two key criteria: (1) a large number of (remote sensing) observations with a regular frequency (e.g. weekly observations) and (2) the initial point of crop growth in time (sowing or emergence date) must be known in advance. Wagner et al. (2020) mentioned a couple of drawbacks of using a forcing method as the following:

1. Highly demanding of measurements for each simulation step which are occasionally unavailable or require to be interpolated. This is especially so when it comes to integration with observations of optical satellite sensors, which could be limited in the number of available observations due to cloud cover.
2. This method breaks up the simulation loop because it replaces intermediate outcomes with external input.
3. Uncertainties of the observations are not accounted for, implying that possible errors in observations could propagate into the model.

2.2 Recalibrating

This method is applied by adjusting the initial parameters of CSMs to reach consistency with the satellite data. As the first step, recalibration of model inputs and parameters takes place with remote sensing data, and then the optimal scenario is chosen for running the recalibrated model in predictive mode (Baret et al., 2007). Generally, the objective of this method is to minimize the gap between the remote sensing data and the simulated data using an optimization algorithm. Therefore, the result of this method is a new set of parameters which bring a better approximation to the observations. However, two constraints exist in using this method: (1) it requires considerable computation time due to the use of optimization algorithms that usually apply iterative search strategies; (2) calibration settings might represent an unreliable parameter setup (Gumma et al., 2021). To improve the estimated prediction by this approach, it is recommended to (1) collect accurate input data (e.g. nitrogen application rate and timing) and (2) use LAI and other vegetation indices (NDVI or EVI, enhanced vegetation index) rather than use LAI alone (Gumma et al., 2021; Fang et al., 2011). Another approach has also been developed recently on the basis of the four-dimensional variation algorithm (4DVar) considering anisotropy of the background error and time window to overcome the difficulty with certain irregular time intervals of observations and heterogenous effects of observations on the effect of yield simulation (Wu et al., 2021).

2.2.1 Reinitiating

In this method, remote sensing data are used through a runtime calibration of the model. Sometimes, this approach is categorized as a sub-group of calibrating methods (Wagner et al., 2020). In general, two input parameters, at minimum, are used for reinitialization. For example, Maas (1992) recalibrated two parameters of the GRAMI model: sowing date and leaf area index at emergence time. In contrast, Xu et al. (2011) reinitialized emergence date and minimum temperature of plant growth to compare phenological information by using MODIS (moderate resolution imaging spectroradiometer)-LAI before assimilating it into the SWAP model. Both studies achieved a better prediction result compared to a prediction without external observations.

There exist potential uncertainties in the remote sensing data themselves. Therefore, rather than one single variable for the assimilation (e.g. LAI), the approach with multi-objective variables is recommended to obtain a more accurate simulation. Evapotranspiration, soil moisture and fraction of absorbed photosynthetically active radiation (FAPAR) can be those candidates (Gumma et al., 2021) to create a well-calibrated and more robust simulation.

2.2.2 Re-parameterization

Re-parameterizing CSM by time series remote sensing measurement. This is also considered a sub-group of the calibrating methods (Wagner et al., 2020). In this approach, the parameter which connects CSM and remote sensing is the one that most likely determines the quality of the performance of the simulation. Commonly, LAI is known as a parameter of re-parameterization (Bouman, 1995; Maas, 1988). Through the procedure of connecting two sides, canopy reflectance can be simulated with CSM, and then, the model is re-parameterized to adjust simulated canopy reflectance to the measured one (Fig. 1).

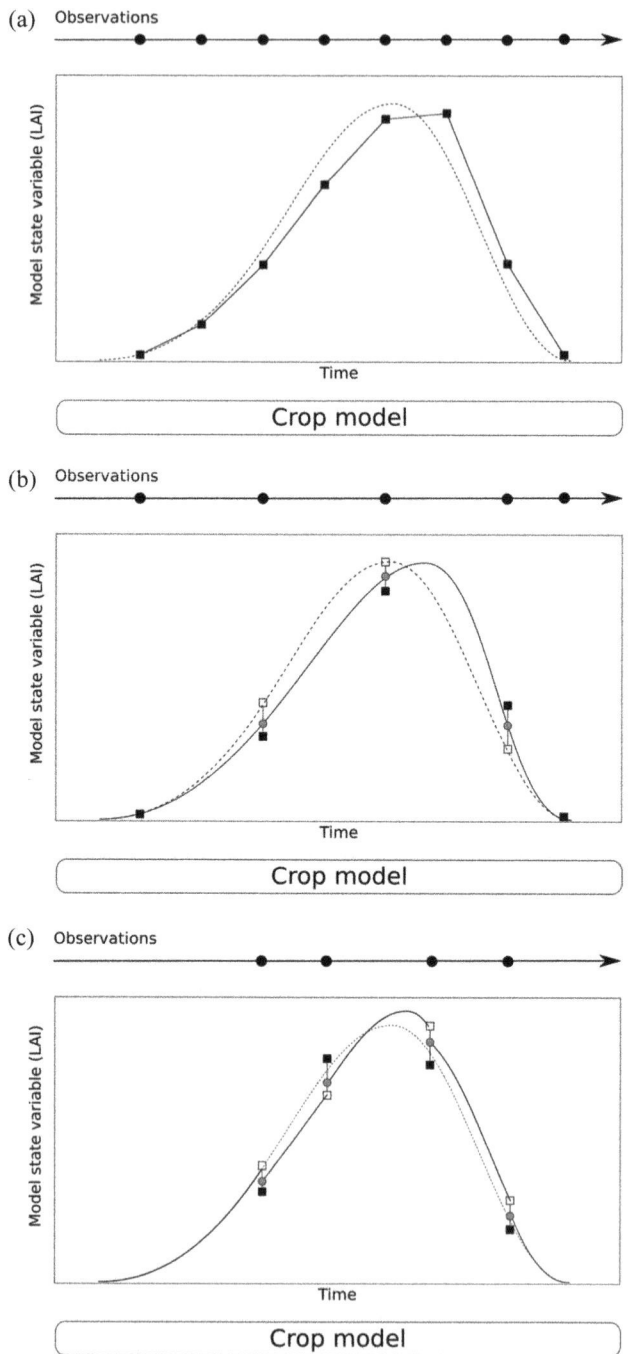

Fig. 1 Schematic representation of three data assimilation methods of remote sensing data into crop simulation models: ((**a**) forcing method, (**b**) recalibrating method, (**c**) updating). Round black symbols represent remote sensing observations, square black symbols represent remote-sensed state variables, open square symbols represent modelled state variables and round grey symbols represent optimized state variables. The dotted line represents the simulated LAI by the model, and the solid line represents LAI by integrating the observations. (Adapted from Jin et al., 2018)

2.3   *Updating*

The updating approach works by directly adjusting model state variables whenever an observation is available. This method is attributed to the assumption that a better estimate of the model state variables at 'day t' by combining a model prediction and an observation will improve accuracy of the simulated variable during the succeeding days. It should be noted that updating techniques can 'break' a model in the sense that mass and energy are not conserved. This implies that special measures often need to be taken to ensure that mass and energy balances close. It is noteworthy to mention that updating two state variables at the same time could provoke imbalances (Jongschaap, 2006).

A key aspect of updating approaches is that uncertainty of both the model prediction and the observation must be estimated. The classical example of an updating method is the Kalman filter which can be applied to linear or mildly nonlinear models. The Kalman filter provides an analytical solution for propagating the uncertainty of the simulated states and computes a new state estimate by weighting the model and observation uncertainties assuming that the errors on both have a Gaussian distribution. This weight factor for updating of the model states is often called the Kalman gain.

For strongly nonlinear models, however, such as crop models, the analytical solution for propagating model uncertainty does not apply. For such models, a numerical version of the Kalman filter has been developed: the Ensemble Kalman filter (EnKF) (Evensen, 2003). The EnKF propagates uncertainty through an ensemble of model simulations, and the model variance is estimated from the ensemble of model states. Assumptions on the Gaussian distribution of the errors for model states and observations still applies. An alternative updating technique is the particle filter which can also be applied in cases when error distributions are strongly non-Gaussian (Mladenova et al., 2020; Xie et al., 2017).

3   A Guideline on Applying DA Techniques

As discussed in the previous section, the application of data assimilation in crop growth models often reduces the error of model predictions compared to the predictions derived solely from the crop growth model. It is worthwhile to address the importance of how users would choose one of the approaches in practice, depending on the objectives and the quality of the dataset as well as remote sensing features. The forcing strategy is merely useful when users have guaranteed observations from a high-quality dataset, and therefore uncertainty in the observations is much less than from the model. In contrast, the recalibration strategy is often used for solving unknown initial conditions at the field scale (e.g. sowing date, initial biomass) or unknown parameter values for the particular cultivar growing at the field (e.g. temperature sum, leaf death).

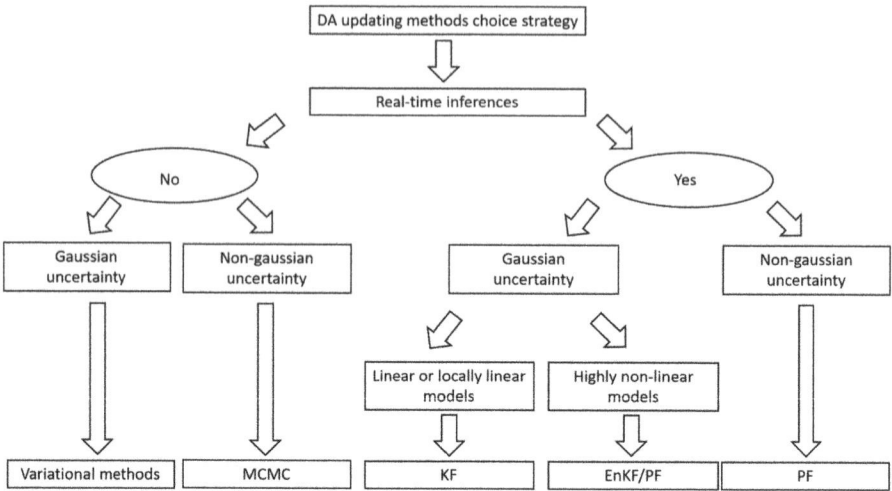

Fig. 2 Decision choice for the updating method of data assimilation (DA) into crop growth models with different approaches including the Markov Chain Monte Carlo (MCMC), Kalman filter (KF), Ensemble Kalman filter (EnKF) and particle filter (PF). (Source: Huang et al., 2019)

The updating strategy is more suitable where some factors are not taken into account at all by the model (e.g. pests and diseases) or are inherently uncertain (e.g. soil moisture due to rainfall variability). The ensemble Kalman filter showed a better performance for yield forecasting using the WOFOST model with MODIS (de Wit & van Diepen, 2007) as well as the CERES model with Landsat (Xie et al., 2017). The schematic flow of the DA method choices, described in Huang et al. (2019), illustrates a useful chart for the selection of the various assimilation methods (Fig. 2).

4 CSM-RS Assimilation in the Framework of Precision Agriculture

Within-field variability in soil fertility, plant growth and yield is the top priority to be resolved in the implementation of precision agriculture. Recently, the number of reports on the use of high-resolution imagery data for monitoring agricultural fields shows that it is becoming increasingly popular. Sentinel 1 and Sentinel 2 are examples of these well-known products, and the range of resolution is from 10 m to m. It is worthwhile to mention that PlanetLab operates with a large constellations of products with 3–5 m resolution. Another emerging technology is the application of proximal sensing in field operations such as data acquisition of reflectance for nitrogen status (e.g. Greenseeker, Yara N-sensor, Cropcircle) from tractor-mounted sensors (Thessler et al., 2011; Van Evert et al., 2017; Jonard et al., 2019) (and vegetation indices measured by an airborne vehicle (e.g. unmanned aerial vehicle, UAV, helicopter and LiDAR)) (Kasampalis et al., 2018). Hyperspectral imaging has the capability of measuring the reflectance in large number of narrow bands and can create a rich dataset to detect phenomena that are not detectable by wide band sensors. Li et al. (2015) present the application of two assimilation variables taken from a hyperspectral sensor in DSSAT-CERES crop model for yield prediction, showing more robust estimation. Yu et al. (2020) incorporated the plant height measurement taken from UAV to the SWAP model to estimate sugarcane yield. Proximal sensing and very high-resolution satellite imagery can contribute to improving the assimilation methods and provide decision support to farmers such as variable rate fertilizer applications.

Another highlight is that there exists great interest in interdisciplinary research between the breeding and CSM + DA, which is highly demanding at present, for example, high-throughput phenotyping (HTP) technologies to develop association mapping at field level as well as predictive genomic selection models in crop improvement (Moreira et al., 2020; Yang et al., 2021).

5 Future Perspective

Pest Management

Development to minimize the difference between observation and simulation is a principal goal for the implementation of DA. In addition, the occurrence of pests and diseases is a major concern for farmers. Several CSMs can simulate the biotic stress related to pests and diseases if they are not properly controlled. In the 1990s, the PNUTGRO and SOYGRO models were created to couple pest damage into the peanut (*Arachis hypogaea*) and soybean (*Glycine max*) growth models, respectively (Batchelor et al., 1993). The DSSAT v.3 is a well-known and long-run crop model with a more holistic approach for decision support taking account of not only pest damage but also other topics (e.g. rotation and crop residues) (Jone et al., 1998). According to Peng et al. (2020)), at least three sub-modules are required for process-based simulation of crop pest and disease management; these include population dynamics, injury and damage and management actions. Combination of the population model with the crop growth model has been developed to create a complex interaction that includes crop and pest or pathogen dynamics. The drawback of this combined approach is the lack of deep understanding of the injuries and damage mechanisms, which requires a wide range of knowledge about plant pathology and physiology to pinpoint the identification of damage mechanisms. Once the population dynamics of pests and diseases are taken into account, simulation of pesticide use can be estimated. Donatelli et al. (2017) proposed a new roadmap that comprised five steps which incorporate the effect of pests and diseases into crop simulation modelling. Evaluation of the dataset as model inputs of the pests and diseases and assessment of the model are the required pre-processing steps before integration to the CSM. Suitable identification of the damage mechanisms is a challenging issue which could partly be tackled by image processing techniques using multispectral or hyperspectral sensors (Liu & Chahl, 2021).

As another example, InfoCrop is a crop simulation model which takes into consideration about the influence of weather, followed by pests and soil properties. And one of the simulation outputs is yield loss due to the pest. Hebbar et al. (2008) applied a satellite data of 24 m resolution into this model to derive the spatial distribution of cotton growth and production. Based on this model, the development of crop loss assessment monitor (CAM) has been released recently on which four different models for yield loss are structured, including RS model, CSM model, weather-based model and hybrid model (Aggarwal et al., 2020).

Metamodel

There is some criticism of the application of CSM and RS due to their complex structures, which are often too complicated for end users such as farmers, policymakers and extension officers. A metamodel is a simpler model derived from a complex model and can be easily operationalized with the aim of decision support to users. Metamodels are often used in hydrology to operationalize complex models for applications such as irrigation management (Galelli et al., 2010). Regarding CSM, crop modelling and PA domains, Florin et al. (2010) applied a metamodel of the APSIM model to estimate soil available water at high resolution by integrating yield data from yield monitors. The use of metamodels could be integrated with DA and would become particularly useful when the metamodels enable the integration of observations throughout the cropping season with a direct relationship between crop managements, such as crop nitrogen content and soil water content. Future study of metamodels in combination with DA needs to be explored with the aim of emulating the outputs of complex models with greater accuracy (Folberth et al., 2019).

6   Conclusion

The combination of CSMs with external observations of crop variables through DA techniques has demonstrated improvement in the monitoring of fields and in making better predictions of crop yield and other variables. This approach makes possible more robust decision-making support service to end users. There are still several issues that need to be addressed, however, such as (1) information provided by farmers regarding field management (e.g. crop variety, sowing date, fertilization time and pest management) which is an essential input to run the models. Only the farmers have full insight into the management activities of their fields; therefore, the CSM + DA framework has to be run in an environment that is operated by the farm or an extension officer, not by scientists and specialists. This puts considerable demand on robustness, ease of use and usability of the CSM + DA framework. No such integrated applications exist today; (2) having existing free remote sensing data at a spatial resolution of 10 m limits the accuracy of estimates of within-field variation in the status of the soil and plants if the farmers have small field sizes. Nevertheless, the use the commercially available imagery from the Planet Labs constellation (3–5 m spatial resolution) combined with other emerging technologies such as proximal sensing, drones and IOT sensors can overcome this challenge.

Conflict of Interest The authors declare that there are no conflicts of interest.

References

Aggarwal, P., Shirsath, P., Vyas, S., Arumugam, P., Goroshi, S., Aravind, S., et al. (2020). Application note: Crop-loss assessment monitor–A multi-model and multi-stage decision support system. *Computers and Electronics in Agriculture, 175*, 105619.

Baret, F., Houles, V., & Guerif, M. (2007). Quantification of plant stress using remote sensing observations and crop models: The case of nitrogen management. *Journal of Experimental Botany, 58*, 869–880. https://doi.org/10.1093/jxb/erl231

Batchelor, W. D., Jones, J. W., Boote, K. J., & Pinnschmidt, H. O. (1993). Extending the use of crop models to study pest damage. *Transactions of the ASABE, 36*, 551–558.

Bouman, B. A. M. (1995). Crop modelling and remote sensing for yield prediction. *Netherlands Journal of Agricultural Science, 43*, 143–161. https://doi.org/10.18174/njas.v43i2.573

Clevers, J., Kooistra, L., & van den Brande, M. (2017). Using Sentinel-2 data for retrieving LAI and leaf and canopy chlorophyll content of a potato crop. *Remote Sensing, 9*(5). https://doi.org/10.3390/rs9050405

Clevers, J. G. P. W. (1991). Application of the WDVI in estimating LAI at the generative stage of barley. *ISPRS Journal of Photogrammetry and Remote Sensing, 46*(1), 37–47. https://doi.org/10.1016/0924-2716(91)90005-G

de Wit, A. J. W., & van Diepen, C. A. (2007). Crop model data assimilation with the ensemble Kalman filter for improving regional crop yield forecasts. *Agricultural and Forest Meteorology, 146*(1–2), 38–56. https://doi.org/10.1016/j.agrformet.2007.05.004

Donatelli, M., Magarey, R. D., Bregaglio, S., Willocquet, L., Whish, J. P. M., & Savary, S. (2017). Modelling the impacts of pests and diseases on agricultural systems. *Agricultural Systems, 155*, 213–224. https://doi.org/10.1016/j.agsy.2017.01.019

Dorigo, W. A., Zurita-Milla, R., de Wit, A. J. W., Brazile, J., Singh, R., & Schaepman, M. E. (2007). A review on reflective remote sensing and data assimilation techniques for enhanced agroecosystem modeling. *International Journal of Applied Earth Observation and Geoinformation, 9*(2), 165–193. https://doi.org/10.1016/j.jag.2006.05.003

Evensen, G. (2003). The ensemble Kalman filter: Theoretical formulation and practical implementation. *Ocean Dynamics, 53*, 343–367. https://doi.org/10.1007/s10236-003-0036-9

Evers, J. B., van der Werf, W., Stomph, T. J., Bastiaans, L., & Anten, N. P. R. (2019). Understanding and optimizing species mixtures using functional-structural plant modelling. *Journal of Experimental Botany, 29*;70(9), 2381–2388. https://doi.org/10.1093/jxb/ery288

Fang, H., Liang, S., & Hoogenboom, G. (2011). Integration of MODIS LAI and vegetation index products with the CSM–CERES–Maize model for corn yield estimation. *International Journal of Remote Sensing, 32*(4), 1039–1065. https://doi.org/10.1080/01431160903505310

Fischer, A., Kergoat, L., & Dedieu, G. (2009). Coupling satellite data with vegetation functional models: Review of different approaches and perspectives suggested by the assimilation strategy. *Remote Sensing Reviews, 15*(1–4), 283–303. https://doi.org/10.1080/02757259709532343

Florin, M. J., McBratney, A. B., Whelan, B. M., & Minasny, B. (2010). Inverse meta-modelling to estimate soil available water capacity at high spatial resolution across a farm. *Precision Agriculture, 12*(3), 421–438. https://doi.org/10.1007/s11119-010-9184-3

Folberth, C., Elliott, J., Muller, C., Balkovic, J., Chryssanthacopoulos, J., Izaurralde, R. C., Jones, C. D., Khabarov, N., Liu, W., Reddy, A., Schmid, E., Skalsky, R., Yang, H., Arneth, A., Ciais, P., Deryng, D., Lawrence, P. J., Olin, S., Pugh, T. A. M., Ruane, A. C., & Wang, X. (2019). Parameterization-induced uncertainties and impacts of crop management harmonization in a global gridded crop model ensemble. *PLoS One, 14*(9), e0221862. https://doi.org/10.1371/journal.pone.0221862

Frère, M., & Popov, G. (1986). *Early agrometeorological crop yield forecasting*. The Food and Agriculture Organization of the United Nations.

Galelli, S., Gandolfi, C., Soncini-Sessa, R., & Agostani, D. (2010). Building a metamodel of an irrigation district distributed-parameter model. *Agricultural Water Management, 97*(2), 187–200. https://doi.org/10.1016/j.agwat.2009.09.007

Gumma, M. K., Kadiyala, M. D. M., Panjala, P., Ray, S. S., Akuraju, V. R., Dubey, S., Smith, A. P., Das, R., & Whitbread, A. M. (2021). Assimilation of remote sensing data into crop growth model for yield estimation: A case study from India. *Journal of the Indian Society of Remote Sensing*. https://doi.org/10.1007/s12524-021-01341-6

Hebbar, K. B., Venugopalan, M. V., Seshasai, M. V. R., Rao, K. V., Patil, B. C., Prakash, A. H., et al. (2008). Predicting cotton production using Infocrop-cotton simulation model, remote sensing and spatial agro-climatic data. *Current Science*, 1570–1579.

Huang, J., Gómez-Dans, J. L., Huang, H., Ma, H., Wu, Q., Lewis, P. E., Liang, S., Chen, Z., Xue, J.-H., Wu, Y., Zhao, F., Wang, J., & Xie, X. (2019). Assimilation of remote sensing into crop growth models: Current status and perspectives. *Agricultural and Forest Meteorology*, 276–277. https://doi.org/10.1016/j.agrformet.2019.06.008

Jin, X., Kumar, L., Li, Z., Feng, H., Xu, X., Yang, G., & Wang, J. (2018). A review of data assimilation of remote sensing and crop models. *European Journal of Agronomy, 92*, 141–152. https://doi.org/10.1016/j.eja.2017.11.002

Jonard, F., Bogena, H., Caterina, D., Garré, S., Klotzsche, A., Monerris, A., Schwank, M., & von Hebel, C. (2019). Ground-based soil moisture determination. In *Observation and measurement of ecohydrological processes* (pp. 29–70). Ecohydrology. https://doi.org/10.1007/978-3-662-48297-1_2

Jones, J. W., et al. (1998). Decision support system for agrotechnology transfer: DSSAT v3. In G. Y. Tsuji, G. Hoogenboom, & P. K. Thornton (Eds.), *Understanding options for agricultural production. Systems approaches for sustainable agricultural development* (Vol. 7). Springer. https://doi.org/10.1007/978-94-017-3624-4_8

Jongschaap, R. E. E. (2006). Run-time calibration of simulation models by integrating remote sensing estimates of leaf area index and canopy nitrogen. *European Journal of Agronomy, 24*(4), 316–324. https://doi.org/10.1016/j.eja.2005.10.009

Kasampalis, D., Alexandridis, T., Deva, C., Challinor, A., Moshou, D., & Zalidis, G. (2018). Contribution of remote sensing on crop models: A review. *Journal of Imaging, 4*(4). https://doi.org/10.3390/jimaging4040052

Li, Z., Jin, X., Zhao, C., Wang, J., Xu, X., Yang, G., Li, C., & Shen, J. (2015). Estimating wheat yield and quality by coupling the DSSAT-CERES model and proximal remote sensing. *European Journal of Agronomy, 71*, 53–62. https://doi.org/10.1016/j.eja.2015.08.006

Liu, H., & Chahl, J. S. (2021). Proximal detecting invertebrate pests on crops using a deep residual convolutional neural network trained by virtual images. *Artificial Intelligence in Agriculture, 5*, 13–23. https://doi.org/10.1016/j.aiia.2021.01.003

Maas, S. J. (1992). *GRAMI: A crop model growth that can use remotely sensed information; USDA-ARS*. ISSN: 1052-5386.

Maas, S. J. (1988). Using satellite data to improve model estimates of crop yield. *Agronomy Journal, 80*(4), 655–662. https://doi.org/10.2134/agronj1988.00021962008000040021x

Mladenova, I. E., Bolten, J. D., Crow, W., Sazib, N., & Reynolds, C. (2020). Agricultural drought monitoring via the assimilation of SMAP soil moisture retrievals into a global soil water balance model. *Frontiers in Big Data, 3*. https://doi.org/10.3389/fdata.2020.00010

Moreira, F. F., Oliveira, H. R., Volenec, J. J., Rainey, K. M., & Brito, L. F. (2020). Integrating high-throughput phenotyping and statistical genomic methods to genetically improve longitudinal traits in crops. *Frontiers in Plant Science, 11*, 681. https://doi.org/10.3389/fpls.2020.00681

Moulin, S., Bondeau, A., & Delecolle, R. (2010). Combining agricultural crop models and satellite observations: From field to regional scales. *International Journal of Remote Sensing, 19*(6), 1021–1036. https://doi.org/10.1080/014311698215586

Peng, B., Guan, K., Tang, J., Ainsworth, E. A., Asseng, S., Bernacchi, C. J., Cooper, M., Delucia, E. H., Elliott, J. W., Ewert, F., Grant, R. F., Gustafson, D. I., Hammer, G. L., Jin, Z., Jones, J. W., Kimm, H., Lawrence, D. M., Li, Y., Lombardozzi, D. L., Marshall-Colon, A., Messina, C. D., Ort, D. R., Schnable, J. C., Vallejos, C. E., Wu, A., Yin, X., & Zhou, W. (2020). Towards a multiscale crop modelling framework for climate change adaptation assessment. *Nature Plants, 6*(4), 338–348. https://doi.org/10.1038/s41477-020-0625-3

Pierce, F. J., & Nowak, P. (1999). Aspects of precision agriculture. In *Advances in agronomy* (Advances in agronomy) (Vol. 67, pp. 1–85). https://doi.org/10.1016/s0065-2113(08)60513-1

Thessler, S., Kooistra, L., Teye, F., Huitu, H., & Bregt, A. K. (2011). Geosensors to support crop production: Current applications and user requirements. *Sensors, 11*(7), 6656–6684. https://doi.org/10.3390/s110706656

Van Evert, F. K., Gaitán-Cremaschi, D., Fountas, S., & Kempenaar, C. (2017). Can precision agriculture increase the profitability and sustainability of the production of potatoes and olives? *Sustainability, 9*(10), 1863. https://doi.org/10.3390/su9101863

Wagner, M. P., Slawig, T., Taravat, A., & Oppelt, N. (2020). Remote sensing data assimilation in dynamic crop models using particle swarm optimization. *ISPRS International Journal of Geo-Information, 9*(2). https://doi.org/10.3390/ijgi9020105

Wu, S., Yang, P., Ren, J., Chen, Z., & Li, H. (2021). Regional winter wheat yield estimation based on the WOFOST model and a novel VW-4DEnSRF assimilation algorithm. *Remote Sensing of Environment, 255*. https://doi.org/10.1016/j.rse.2020.112276

Xie, Y., Wang, P., Bai, X., Khan, J., Zhang, S., Li, L., & Wang, L. (2017). Assimilation of the leaf area index and vegetation temperature condition index for winter wheat yield estimation using Landsat imagery and the CERES-wheat model. *Agricultural and Forest Meteorology, 246*, 194–206. https://doi.org/10.1016/j.agrformet.2017.06.015

Xu, W., Jiang, H., & Huang, J. (2011). Regional crop yield assessment by combination of a crop growth model and phenology information derived from MODIS. *Sensor Letters, 9*(3), 981–989. https://doi.org/10.1166/sl.2011.1388

Yang, S., Zheng, L., He, P., Wu, T., Sun, S., & Wang, M. (2021). High-throughput soybean seeds phenotyping with convolutional neural networks and transfer learning. *Plant Methods, 17*(Maro), 50. https://doi.org/10.1186/s13007-021-00749-y

Yu, D., Zha, Y., Shi, L., Jin, X., Hu, S., Yang, Q., Huang, K., & Zeng, W. (2020). Improvement of sugarcane yield estimation by assimilating UAV-derived plant height observations. *European Journal of Agronomy, 121*. https://doi.org/10.1016/j.eja.2020.126159

Zhang, Z., & Moore, J. C. (2015). Chapter 9: Data assimilation. In Z. Zhang & J. C. Moore (Eds.), *Mathematical and physical fundamentals of climate change* (pp. 291–311). Elsevier. ISBN 9780128000663. https://doi.org/10.1016/B978-0-12-800066-3.00009-7

Part III
Case Studies

Adapt-N® (Yara International)

Harold van Es, Rebecca Marjerison, and Muhammad Barik

1 Introduction

Among essential crop nutrients, nitrogen is the most challenging to manage but also the most likely to benefit from precision management technologies (van Es et al., 2007). It causes major water quality concerns, including nitrate contamination of groundwater in agricultural regions and hypoxia in estuaries. Also, gaseous forms of N are the largest contributor to agricultural greenhouse gas emissions (EPA, 2020) and fine particulate air pollution (Pinder et al. 2007). Anthropogenic N pollution, therefore, has large social costs, with estimated potential ecosystem and health damages of $157 billion annually in the USA alone (Sobota et al. 2015).

The 4R idea promotes field application of nutrients with the right source, at the right place, right time and right rate (TFI, 2018). As a concept, 4R is straightforward, but under the complexity of production environments in farmer fields, the 'right' answer to each R factor, especially N rate, remains an elusive concept. Tools based on dynamic simulation models, like Adapt-N, can integrate these concepts and tailor N management to the localized conditions of the production environment (Sela & van Es, 2018). At a global scale, this is especially relevant for maize (corn, *Zea mays* L.) due to its widespread cultivation, high N needs and sensitivity to weather (Tremblay et al., 2012).

H. van Es (✉)
Cornell University, Ithaca, NY, USA
e-mail: hmv1@cornell.edu

R. Marjerison · M. Barik
Yara International ASA, Oslo, Norway

© The Author(s), under exclusive license to Springer Nature Switzerland AG 2023
D. Cammarano et al. (eds.), *Precision Agriculture: Modelling*, Progress in Precision Agriculture, https://doi.org/10.1007/978-3-031-15258-0_9

2 Technique

Adapt-N is a decision support framework that facilitates the monitoring of soil N, crop N and other agronomic and environmental elements based on daily model simulations (Sela et al., 2016). The tool was developed at Cornell University and is now licensed by Yara International ASA. Adapt-N is enabled by both deterministic and stochastic model components to derive optimum N recommendations at either field, zone or grid (20×20 m) scale. The model to derive N recommendations is based on a mass-balance equation that incorporates the following:

(i)   The N needed to achieve a yield target at the end of the season.
(ii) The amount of N in the crop and soil (from dynamic model simulations).
(iii) A probabilistic estimate of projected future N gains and losses until the end of the growing season, based on historical climate data.
(iv) An integrated economic, risk and efficiency factor that accounts for fertilizer and grain prices, asymmetric profit risks and a yield-efficiency relationship.
(v)   A rotation effect credit.

Components (ii) and (iii) are the most complex and dynamic and are addressed through the application of the precision nitrogen management (PNM) model (Melkonian et al. 2007), a set of process-based computational routines that simulate soil hydrology, chemical transformations and movement of N in the soil, as well as crop growth and yield. The PNM model was developed from two earlier models, the LEACHN soil model (Hutson, 2010) and an unnamed maize crop growth model by Sinclair and Muchow (1995). Transformations of C and N are primarily modelled using functions modulated by soil temperature and soil water content as predicted from high-resolution climate data. The model also reports environmental N losses in the form of gases or leachate.

The evolution of the Adapt-N solution followed a series of developmental steps: (i) PNM model development (Melkonian et al., 2007); (ii) model calibration and validation efforts based on detailed (mostly lysimeter-based) multi-year studies involving N rates, rotations, manure applications and cover crops (Sogbedji et al., 2001a,b,c; Sogbedji et al., 2006; Melkonian et al., 2010; Marjerison et al., 2016); (iii) development of the integrated Adapt-N tool and a user interface (Melkonian et al., 2008); and (iv) independent validation on commercial farms – more than 200 strip trials – in broad geographical regions in the USA (Sela et al., 2016, 2017, 2018a; Osmond et al., 2018). Also, Graham et al. (2010) first applied the model to generate within-field site-specific N recommendations.

3 Usage

Adapt-N was launched in 2008 (Melkonian et al., 2008) and made publicly available through XML web services. In 2013, it was licensed to, and commercialized by, a start-up company which in turn was acquired by Yara International ASA in 2017.

Adapt-N is cloud-based and provides real-time N recommendations for maize-based systems (being expanded to other crops) using daily high-resolution climate data and regionally adapted soil parameters under a diverse range of management scenarios. It accounts for different maize systems (grain, silage, sweet corn), fertilizer formulations, tillage systems and residue levels, rotations, cover crops, manure or compost applications, cultivars and enhanced efficiency products. It also facilitates selected field observations to enhance prediction accuracy (date of crop emergence, in-season soil nitrate test results, irrigation applications). In addition to daily tracking of soil and crop N status, the model also provides estimates of crop stage and soil water status (facilitating irrigation planning). The primary client group comprises fertilizer retailers and crop consultants, with many now accessing Adapt-N through application programming interface (API) technology with third-party software.

4   Advantages to Growers

Adapt-N recommendations were compared with grower practice under a variety of management scenarios through on-farm strip trials in New York and Iowa ($n = 113$; Sela et al., 2016). Marginal profits were on average \$64 ha^{-1} higher and N inputs 45 kg ha^{-1} lower with Adapt-N rates. The monetary benefits were associated with fertilizer savings when Adapt-N rates were lower and achieved similar yields to grower rates (83% of cases; Fig. 1) and prevented yield reductions from N deficits when the tool recommended higher rates (17% of cases, mostly in wet years).

Another 16 on-farm multi-N rate trials in New York and 23 trials in the Midwest (Indiana, Ohio and Wisconsin; Sela et al., 2017) estimated the ex post economic optimum N rate (EONR). In the New York trials Adapt-N achieved RMSE values of 33 kg ha^{-1} and an average bias of -12 kg ha^{-1} relative to the EONR, compared to 85 kg ha^{-1} and 64 kg ha^{-1} for a conventional static N calculator. For the Midwest trials, Adapt-N had an RMSE of 33 kg ha^{-1} and bias of -10 kg ha^{-1} compared to 49 kg ha^{-1} and 39 kg ha^{-1}, respectively, for the regional static N calculator. Profits

Fig. 1 Yield results from 113 Adapt-N vs. grower trials. When Adapt-N recommended lower rates (87%), average yields were the same as grower yields; when Adapt-N recommended higher rates (17%), yields generally increased

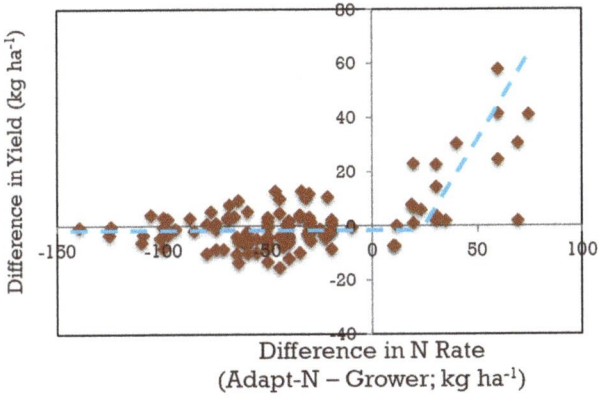

were \$64 ha^{-1} higher for Adapt-N for the New York trials and \$25 ha^{-1} higher for the Midwest trials. Adapt-N on average recommended 13 kg N ha^{-1} less than experienced crop consultants in 38 site years of on-farm strip trials in North Carolina, with identical profits (Osmond et al., 2018).

5 Environmental Benefits

The environmental impacts were assessed from the above trials by simulating leaching (beyond bottom of root zone) and gaseous losses (from soil surface) under the trial conditions. For the Adapt-N vs. grower trials, estimated post-sidedress N losses averaged 36% less leaching and 39% lower gaseous losses for Adapt-N (Sela et al., 2016). In the multi-N rate studies, Adapt-N reduced simulated N leaching losses by 53% and gaseous losses by 54% compared to the conventional N calculator in New York trials and by 3% and 22%, respectively, in Midwest trials (Sela et al., 2017). A separate field lysimeter study in New York corroborated these results (van Es et al., 2020) with mean leached NO_3 concentrations reduced by 37% and 41% for a clay loam and loamy sand soil, respectively, while yields were not negatively affected. The above studies only addressed N rate differences, and larger gains might be achieved when all 4R practices are optimized simultaneously (Sela et al., 2019).

6 Transparency

Adapt-N provides N recommendations at the field, zone and grid scale. Users receive tabular simulation results to gain insight into, and understanding of the N rate advisory (Fig. 2). This includes crop stage and N uptake, N in the soil (root zone and top 30 cm), N needed at harvest, N mineralized from organic sources (inherent and applied), N lost to the environment, predicted future N losses, rainfall received and soil water status and irrigation needs. In addition, more detailed graphical outputs (by date) are provided for N leaching, N gaseous losses, N mineralization, inorganic N in soil, precipitation (cumulative, daily), crop stage development and growing degree days (Fig. 2). The interface has review and editing features for input data, map displays, aggregated field, and export utilities tailored to products (including prescription shapefiles). Users can further adjust the tool by specifying current fertilizer and grain prices, N contents of mature crops and threshold levels for daily alerts by text or email.

7 Adoption

Customer surveys (internal Yara research, 2020) rated the three top-ranked criteria for the use of Adapt-N to be scientific validity, usability, and compatibility with other software (through API). Other stated advantages included: better recommendations and more profitability, ability to fine-tune N management, customization

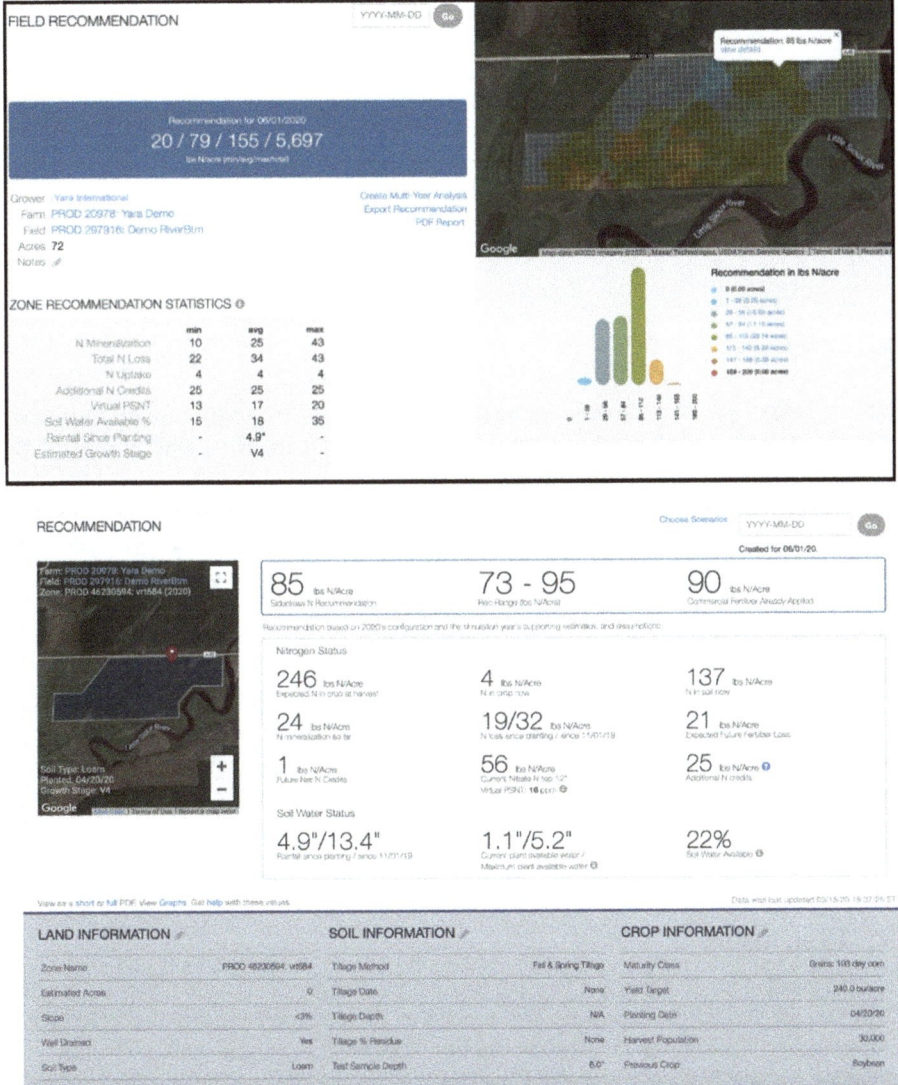

Fig. 2 Selected examples of the Adapt-N user interface with recommendations (here: 20 × 20 m grid based) and simulation results

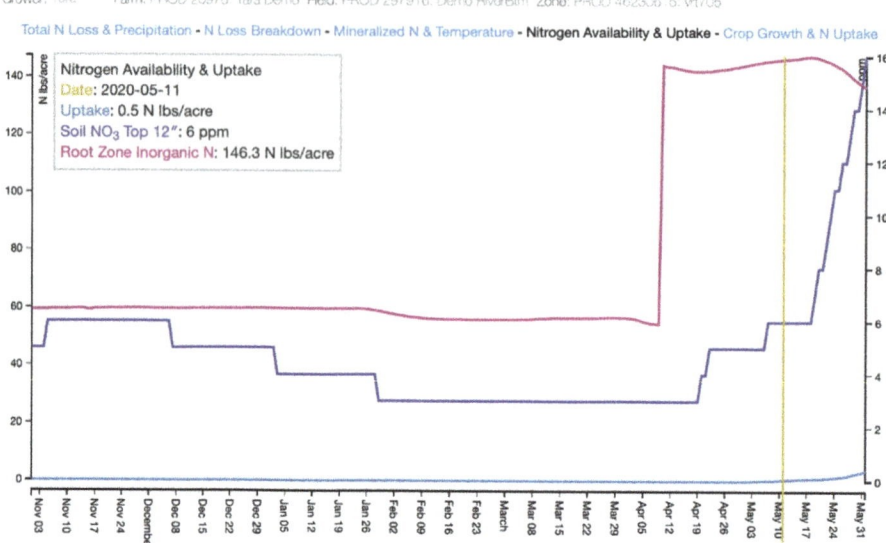

Fig. 2 (continued)

options with GIS, spatial recommendations, full-season N programming, improved
grower relations, ability to show benefits of variable-rate management, improved 4R
practices and remote field monitoring capabilities.

Conversely, the most prominent adoption challenge is limited grower understanding of the opportunities with data-driven models. Another challenge relates to
regulatory standards which remain focused on simple static and generalized calcu-
lators, despite generally inferior performance (Sela et al., 2017, 2018a; van Es et al.,
2020). Adverse results from incorrect usage can be a problem but is addressed
through user training.

The Adapt-N model framework has also been applied for in silico evaluations of
environmental benefits of 4R management scenarios in support of policy development (McLellan et al., 2018; Sela et al., 2018b, 2019). They demonstrated opportunities for the use of model tools to meet environmental standards for N use efficiency
and carbon targets.

8 Conclusion

The use of model-driven tools for precision N management can be effectively
implemented in commercial crop production with clear economic and environmental
benefits.

Conflict of Interest The authors declare that the Adapt-N technology was developed at Cornell University and is commercially marketed by the Yara International ASA. Co-authors Marjerison and Barik are fully employed by the company, while author and lead inventor van Es receives remuneration for advisory services.

References

EPA. (2020). *Inventory of US greenhouse gas emissions and sinks*. United States Environmental Protection Agency. EPA 430-R-19-001.

Graham, C.J., H.M. van Es, J.J. Melkonian, D.A. Laird. 2010. Improved nitrogen and energy use efficiency using NIR estimated soil organic carbon and N simulation modeling. In: D.A. Clay J. Shanahan. GIS applications in agriculture – Nutrient Management for Improved Energy Efficiency. pp. 301–325, Taylor and Francis, LLC.

Hutson, J. L. (2010). *LEACHM: Leaching Estimation and Chemistry Model: A process-based model of water and solute movement, transformations, plant uptake and chemical reactions in the unsaturated zone* (Version 4. Res. Ser. R03–1). Cornell Univ.

Marjerison, Melkonian, R. D. J., Hutson, J. L., van Es, H. M., Sela, S., Geohring, L. D., & Vetsch, J. (2016). Drainage and nitrate leaching from artificially drained maize fields simulated by the Precision Nitrogen Management model. *Journal of Environmental Quality, 45*, 2044–2052.

McLellan, E. L., Cassman, K. G., Woodbury, P. B., Sela, S., Eagle, A. J., Tonitto, C., Marjerison, R. D., & van Es, H. M. (2018). The nitrogen balancing act: Tracking the environmental performance of food supply chains. *Bioscience, 68*(3), 194–203.

Melkonian, J., van Es, H. M., DeGaetano, A. T., Sogbedji, J. M., & Joseph, L. (2007). Application of dynamic simulation modeling for nitrogen Management in Maize. In T. Bruulsema (Ed.), *Managing crop nutrition for weather* (pp. 14–22). Plant Nutrition Institute Publ.

Melkonian, J. J., van Es, H. M., DeGaetano, A. T., & Joseph, L. (2008). ADAPT-N: Adaptive nitrogen management for maize using high-resolution climate data and model simulations. In R. Kosla (Ed.), *Proceedings of the 9th International Conference on Precision Agriculture, July 20–23, 2008, Denver, CO (CD-ROM)*.

Melkonian, Geohring, J. L. D., van Es, H. M., Wright, P. E., Steenhuis, T. S., & Graham, C. (2010). *Subsurface drainage discharges following manure application: Measurements and model analyses*. Proc. XVIIth World Congress of the Intern. Commission of Agric. Engineering, Quebec City, Canada.

Osmond, D. E., Shelton, S., Autin, R., van Es, H., & Sela, S. (2018). Evaluation of adapt-N and realistic yield expectation approaches for maize management in North Carolina. *Soil Science Society of America Journal, 82*, 1449–1458.

Pinder, R. W., Adams, P. J., & Pandis, S. N. (2007). Ammonia emission controls as a cost-effective strategy for reducing atmospheric particulate matter in the eastern United States. *Environmental Science and Technology, 41*, 380–386.

Sela, S., & van Es, H. M. (2018). Dynamic tools unify fragmented 4Rs into an integrative nitrogen management approach. *The Journal of Soil and Water Conservation, 73*, 107A–112A.

Sela, S., van Es, H. M., Moebius-Clune, B. N., Marjerison, S. R., Melkonian, J. J., Moebius-Clune, D., Schindelbeck, R., & Gomes, S. (2016). Adapt-N outperforms grower-selected nitrogen rates in Northeast and Midwest USA strip trials. *Agronomy Journal, 108*(4), 1726–1734.

Sela, S., van Es, H. M., Moebius-Clune, B. N., Marjerison, R., Moebius-Clune, D., Schindelbeck, R., Severson, K., & Young, E. (2017). Dynamic model improves agronomic and environmental outcomes for corn N management over static approaches. *Journal of Environmental Quality, 46*(2), 311–319.

Sela, S., Woodbury, S. P., & van Es, H. M. (2018a). Dynamic model-based N management reduces surplus nitrogen and improves the environmental performance of corn production. *Environmental Research Letters, 13*, 054010. https://doi.org/10.1088/1748-9326/aab908

Sela, S., van Es, H. M., Moebius-Clune, B. N., Marjerison, S. R., & Kneubuhler, G. (2018b). Dynamic model-based recommendations increase the precision and sustainability of N fertilization in midwestern US maize production. *Computers and Electronics in Agriculture, 153,* 256–265.

Sela, S., Woodbury, P. B., Marjerison, R., & van Es, H. M. (2019). Towards applying N balance as a sustainability indicator for the US cornbelt: Realistic achievable targets, spatio-temporal variability and policy implications. *Environmental Research Letters, 14,* 064015.

Sinclair, T. R., & Muchow, R. C. (1995). Effect of nitrogen supply on maize yield.1. Modeling physiological responses. *Agronomy Journal, 87,* 632–641.

Sobota, D. J., Compton, J. E., McCrackin, M. L., & Singh, S. (2015). Cost of reactive nitrogen release from human activities to the environment in the United States. *Environmental Research Letters, 10.*

Sogbedji, J. M., van Es, H. M., Hutson, J. L., & Geohring, L. D. (2001a). N rate and transport under variable cropping history and fertilizer rate on loamy sand and clay loam soils: II. Performance of LEACHMN using different calibration scenarios. *Plant and Soil, 229*(1), 71–82.

Sogbedji, J. M., van Es, H. M., Hutson, J. L., & Geohring, L. D. (2001b). Fate of N fertilizer and green manure in clay loam and loamy sand soils: I Calibration of the LEACHM model. *Plant and Soil, 229*(1), 57–70.

Sogbedji, J. M., van Es, H. M., Klausner, S. D., Bouldin, D. R., & Cox, W. J. (2001c). Spatial and temporal processes affecting nitrogen availability at the landscape scale. *Soil & Tillage Research, 58*(3–4), 233–244.

Sogbedji, J. M., van Es, H. M., Melkonian, J. M., & Schindelbeck, R. R. (2006). Evaluation of the PNM model for simulating drain flow nitrate-N concentrations under manure-fertilized maize. *Plant and Soil, 282,* 343–360.

TFI (The Fertilizer Institute). (2018). *The 4R of nutrient stewardship.* Available at (https://nutrientstewardship.com/4rs/. Accessed 22 Nov 2018.

Tremblay, N., Bouroubi, Y. M., Bélec, C., Mullen, R. W., Kitchen, N. R., Thomason, W. E., Ebelhar, S., Mengel, D. B., Raun, W. R., Francis, D. D., Vories, E. D., & Ortiz-Monasterio, I. (2012). Corn response to nitrogen is influenced by soil texture and weather. *Agronomy Journal, 104,* 1658–1671.

van Es, H. M., Kay, B. D., Melkonian, J. J., & Sogbedji, J. M. (2007). Nitrogen management under maize in humid regions: Case for a dynamic approach. In T. Bruulsema (Ed.), *Managing crop nutrition for weather* (pp. 6–13). Plant Nutrition Institute Publ.

van Es, H. M., Ristow, A., Nunes, M., Schindelbeck, R., Sela, S., & Davis, M. (2020). Nitrate leaching decreases with adaptive-dynamic nitrogen management and reduced tillage. *Soil Science Society of America Journal.* https://doi.org/10.1002/saj2.20031

Granular Agronomy Nitrogen Management

Robert Gunzenhauser

1 Introduction

Granular Agronomy Nitrogen Management is a software tool to provide spatially optimal nitrogen rate recommendations for maize fields in the United States. Utilizing a proprietary dynamic crop growth model (Granular Crop Model), thousands of simulations can be run using weather, soil properties, management and genetic information to estimate nitrogen requirements. Available via the Internet, the software is primarily used by advisors on behalf of farmers.

Granular Agronomy Nitrogen Management was started in 2013 as Encirca Nitrogen Management, part of the suite of the Encirca Services offered through DuPont-Pioneer, a division of DuPont. In 2017, at the same time as DuPont and Dow merged to form first DowDuPont and later the spin-off Corteva Agriscience, Granular, a farm business management software company, was also purchased. Encirca Services were then moved to Granular as Granular Agronomy. Over this time, millions of hectares of growers' maize have been serviced using Nitrogen Management. Nitrogen Management was one of the first commercial softwares on the market to offer dynamic nitrogen recommendations based on soils, weather, management and genetics with a crop model as the core of the offering.

This is a description of the service offering of Nitrogen Management, the important components of the software, and future opportunities to be investigated.

R. Gunzenhauser (✉)
Corteva Agriscience, Johnston, IA, USA
e-mail: robert.gunzenhauser@corteva.com

© The Author(s), under exclusive license to Springer Nature Switzerland AG 2023
D. Cammarano et al. (eds.), *Precision Agriculture: Modelling*, Progress in Precision
Agriculture, https://doi.org/10.1007/978-3-031-15258-0_10

2 Service Offering

Granular Agronomy Nitrogen Management is provided in the United States through certified service agents (CSAs). These agents range from independent agronomists to associate sellers working with Pioneer sales representatives. The CSAs receive yearly agronomic training to keep their skills up to date. As a CSA, they should not sell products and services that compete with Granular, Pioneer or Corteva offerings. Many offer complimentary services, including soil sampling and analysis, crop scouting, irrigation scheduling and custom farming services.

Growers engage with Granular through the CSAs on a yearly basis. Granular Agronomy Nitrogen Management is offered as part of the Granular Agronomy service, which includes soil fertility and seeding management; Nitrogen Management is an additional package. The suggested retail price for Granular Agronomy + Nitrogen Management is \$32.11 per hectare. However, CSAs have the freedom to adjust this price to suit the needs, purchase volume and service-level requirements of their grower-customers.

The value delivered to the farmer depends greatly on the level of service delivered by the CSA. Farmers rate the Granular Agronomy Nitrogen Management service higher when the CSA spends the necessary time to discuss goals and outcomes with the farmer, adjusting the inputs based on the changing conditions in the field (e.g. changes in planting, nitrogen and irrigation dates). The service also has greater value in years when the prices of both maize and nitrogen are higher than usual; Granular Agronomy Nitrogen Management helps to adjust nitrogen requirements based on the farmer's economic situation at the time.

3 Setup

To ensure accuracy, Nitrogen Management requires a certain level of information be gathered and input into the software. A dedicated CSA as part of the initial setup has been found to be a key to the success of the service to the grower.

On initiation of the service, a CSA will visit with the grower and gather some initial information about his or her operation. What kind of management do they use? How much nitrogen do they typically apply to their maize, in what forms and at what times? This helps the CSA to put together the correct management plans.

3.1 Field Boundaries

Field boundaries of the growers' fields must be digitized into the system. In many cases, this can be done from machine data, namely, from on-the-go planting, application, or yield monitors with GPS receivers. If these are not available, field boundaries can be obtained with a data layer called the United States Department of Agriculture (USDA) common land unit (CLU) layer. Digitized in 2005–2006, this layer captured the outline of most agricultural fields in the United States at the time. It was publicly available for a brief time before being removed from public circulation shortly after. If neither machine data nor CLU data is available, boundaries may be manually digitized using a commonly available background layer of recent imagery. Boundaries can be edited from year to year.

3.2 Data Hierarchy

Data are arranged in an Operation–Farm–Field hierarchy in Granular Agronomy. The Operation corresponds to the farmer; the Farm is a collection of fields – some operations have one Farm with all fields linked to it, while others may have a Farm designated for each Field, and others somewhere in between these two extremes. A Field is the area of work.

Once the Fields have been set up with Maize as the primary crop for the current year, the CSA will enroll the fields into Granular Agronomy Nitrogen Management. This initiates a set of processes in the background that enable the usage of Nitrogen Management on the field for that year.

3.3 Decision Zones

Once a crop has been assigned for a field, a layer called 'Decision Zones' is generated automatically for the field. This digital layer is polygon-based and is an intersection of two or more layers:

- Environmental Response Units (digitized soil data)
- Multi-Year Yield Analysis (MYYA) (optional, based on the assumption of access to yield maps)
- Irrigation Zones (optional)
- Other user-supplied polygon-based layers to segment and capture further spatial variation

For the service to generate estimated soil nitrogen amounts and recommendations, this key spatial information is required for Decision Zones:

- Soil properties (water-holding capacity, drainage, organic matter)
- Yield Targets – aspirational yield goals for each Decision Zone
- Irrigation zones – where irrigation takes place or not

3.3.1 Yield Targets

Yield Targets are used in nitrogen to determine the target level of nitrogen that should be in the soil at plant growth stage VT/R1 (change from vegetative to reproductive stage). They are a way for CSAs and growers to indicate the potential productivity of each zone and to what levels of nitrogen each should have available. These Yield Targets should be judicially selected; having extremely low or high Yield Targets may result in disappointment and or nitrogen loss.

Yield Targets are assigned to each Decision Zone. In some fields, there may be dozens of Decision Zones, so two approaches are used to 'spread' a single field yield goal across the Decision Zones:

- National Yield Index
- Grower Yield Index

The National Yield Index uses a data source called the National Crop Commodity Productivity Index or NCCPI. This is created and maintained by the USDA's Natural Resources Conservation Service (NRCS); this productivity index rates each soil type on its potential productivity for a number of commodity crops, including maize, soyabeans, wheat and cotton. The maize productivity index value is used from the NCCPI data set for this purpose. The NCCPI has a range from 0 to 1, with 1 being the most productive for that crop. It is developed using each soil type's physical, chemical and biological measurements and crop modelling in the environment where the soil type is found.

The Grower Yield Index is based upon Multi-Year Yield Analysis, or MYYA. The MYYA requires at least 1 year of machine yield data collected from on-the-go yield monitors with GPS receivers (multiple years are preferred). To create an MYYA layer, each year's yield data is normalized using a raster approach. Then, each of the layers is merged and averaged together. This allows multiple crops (maize, soybeans, wheat) to be considered together. Through the normalization and averaging, spatial productivity trends can be identified. Polygon zones of low, average and high are generated from the mean, normalized data with these ranges: low, < 90%; average, 90–110%; and high, > 110%.

Through either approach, a relative productivity value is generated for each Decision Zone. Based on the area of each Decision Zone, using the Field Yield Goal, Yield Targets will be generated by Decision Zone such that their area-weighted average will equal the Field Yield Goal. Once these Yield Targets have been generated, a user of the software can edit individual Decision Zone Yield Targets to their satisfaction.

3.3.2 Soils

Soil data are a key factor in crop model simulations. In many small-plot simulations, actual soil measurements are used for the site. For a set of large-scale simulations across millions of hectares and thousands of growers, a broader set of soil data are utilized.

The USDA has generated the Soil Survey Geographic Database or SSURGO and made it publicly available. This database contains spatial and tabular information describing the soils of the United States at varying levels of detail. Individual area-level soil surveys (typically at a state–county level) have been performed multiple times over the last 40–50 years by individual soil scientists. The SSURGO harmonizes these area-level soil surveys into one source that can be used by many consumers.

As a way to improve further upon the SSURGO data, team members developed a process to re-position SSURGO map units based spatially on high-resolution digital elevation models (DEMs), typically from Light Detection and Ranging (LiDAR) or other sources. By comparing the SSURGO map data with topographic derivatives from the DEMs and determining, by sampling, where SSURGO map units were typically found based on the topographical derivatives (e.g. slope, curvature, wetness index and more), a re-mapping of the map units resulted in a new product called Environmental Response Units or ERUs (Fig. 1). This is a Corteva-proprietary process to improve the spatial accuracy of the SSURGO map units and is used in the Nitrogen Management offering. The SSURGO map units may also be used as an alternative if so desired.

3.4 Management Plans

A key feature of Nitrogen Management is the ability to set up several nitrogen application plans. These may have different application timings, methods and products from the grower's typical plans. Plans may be assigned to fields and allow the user to explore how changes in management may affect their outcomes.

Nitrogen applications may be flat-rate or variable-rate and utilize commercially available fertilizers or manure. The software user may designate the placement

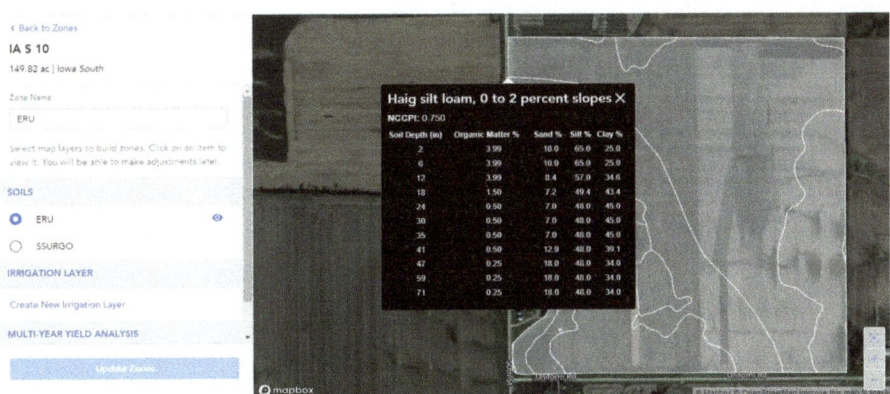

Fig. 1 Example ERU (Environmental Response Units) soil layer with soil property details of a selected soil map unit

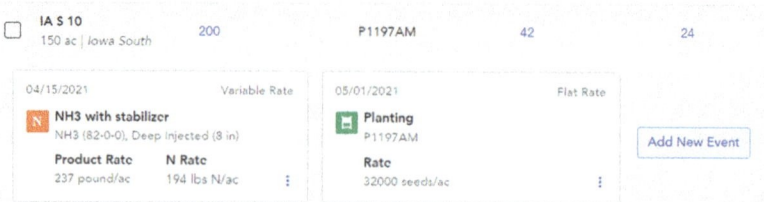

Fig. 2 Management 'cards' displaying the various management activities performed on a field

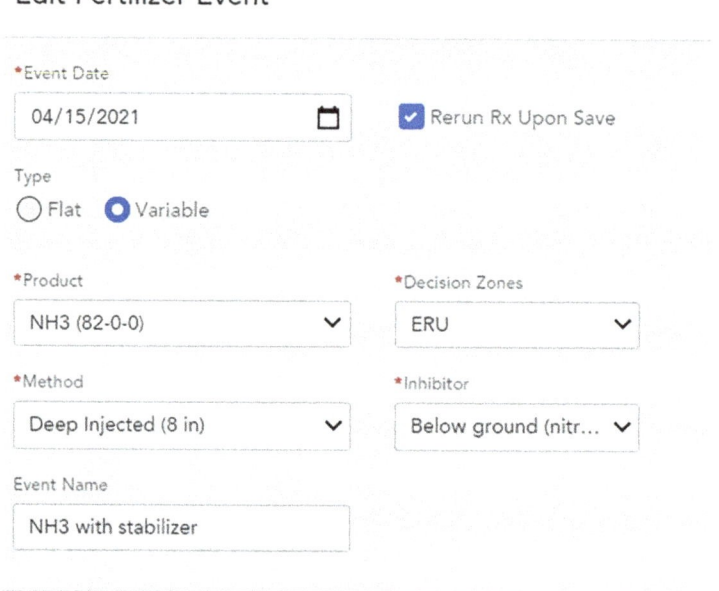

Fig. 3 Fertilizer application information user interface

method of each application (broadcast not incorporated, broadcast incorporated or surface banded, shallow injected and deep injected), if a nitrogen stabilizer is being used and what type of stabilizer and if using a flat rate, the rate applied by the farmer (Figs. 2 and 3). Once this plan is created, it can be applied to any of the operations and fields the user has access to.

3.5 Initial Nitrogen

The user can specify the amount of nitrogen that is present in a 0–60 cm (0–24″) depth from soil tests as a way to initialize the simulation. Based on the soil test, the amount of ammonium (NH4-N) and nitrate (NO3-N) can be entered in parts per million. This provides the model with values from which to start simulation.

3.6   Soil Monitoring Depth

While the soil data used are modelled to approximately 180 cm (about 6 ft), the user can specify a particular depth at which to monitor soil nitrogen amounts, particularly for indicating if the target soil nitrogen is available at VT/R1. The soil depth can be set in at 30, 45, 60, 90 or 120 cm (about 1, 1.5, 2, 3 or 4 ft).

3.7   Planting Information

The seeding date, seeding rate and relative maturity of the maize seed are critical information. Initially, this may be considered as the planned planting information, but as the crop is planted, actual values may be adjusted into the process. As an extra feature, if a Pioneer brand hybrid is selected, hybrid-specific genetic parameters are used in the simulation; if a non-Pioneer brand hybrid is selected, average values based on relative maturity are used instead.

3.8   Irrigation

For fields that are irrigated, the user can enter a schedule of irrigation events prior to the cropping season and later can edit them as the irrigation events are applied. A range of dates for irrigation events, the frequency and amount per irrigation (inches) can be set so that the user does not need to enter each event one by one (sometimes 12–18 events are applied each season).

In addition, nitrogen applied through irrigation can be accounted for in this feature. Liquid nitrogen fertilizer products, such as urea-ammonium nitrate (UAN), can be included in irrigation events as 'fertigation'; quantities can be applied in mass per area of nitrogen or liquid volume per area of product. Also, nitrates existing in the irrigation water may be accounted for; the user may enter a water test with nitrate-N, and the system will include nitrogen additions for each irrigation event based on the quantity of water and the nitrate concentration.

4   Operation

Once a field has been enrolled in Granular Agronomy Nitrogen Management and the appropriate setup information has been entered, the system will run on a daily basis, utilizing daily weather information to re-simulate the results each day (Fig. 4). Over the winter period, this daily simulation is reduced or turned off to save computational cycles, but the user can initiate a simulation on demand.

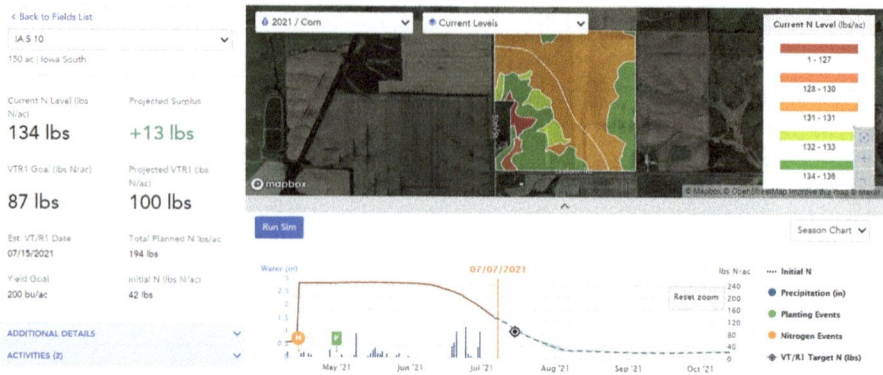

Fig. 4 Example use interface of current soil nitrogen levels and season chart (bottom)

4.1 Granular Crop Model

The Granular Crop Model (GCM) is a mechanistic, daily time-step crop growth model that simulates both below-ground processes (soil temperature, soil moisture, nitrogen transformation and movement and root development) and above-ground processes (plant development and growth, water and N uptake, light interception and more) and powers the Granular Nitrogen Management software. The GCM combines below-ground functionality originally found in DSSAT (Hoogenboom et al., 2019; Jones et al., 2003) with proprietary above-ground functionality developed by Corteva and the University of Queensland. Additional functionality and improvements have been made by Granular and Corteva scientists to the below-ground functionality since the original development of the Granular Crop Model. GCM can execute yearly crop growth simulations within a fraction of a second.

4.2 Season Chart

A dynamic season chart is provided for each enrolled field. This chart shows the estimated and projected soil nitrogen amounts throughout the season, together with key additional information, including nitrogen application events, planting date, precipitation and irrigation events and soil nitrogen target at VT/R1 (Fig. 4).

To indicate the uncertainty of soil nitrogen amounts in the future (past the current date plus 7 days of forecasted weather), confidence bands are indicated around the median projected soil nitrogen amounts. Moving the mouse over the season chart can show the estimated soil nitrogen levels and confidence values for each day.

4.3   Weather

To simulate crop growth properly, daily weather data are a necessary input into the crop growth model. Air temperature minimum and maximum, precipitation and solar radiation are key inputs into the model.

Three weather data sets are used in combination:

- Current year
- Forecast
- Historical

The combination of these sources allows Nitrogen Management to calculate potential outcomes of soil nitrogen amounts based on historical ranges.

4.3.1   Current Year

The crop simulation in Nitrogen Managements starts on October 1 of the previous year and ends on December 31 of the current year. As each day progresses after October 1, the available weather for the field is used and added to the simulation. Weather data are provided through Corteva's own network of weather stations and publicly available weather stations.

4.3.2   Forecast

For 8 days from the current day onwards, forecast weather is used in the simulation. This is useful to indicate effects to the soil nitrogen levels if a large rain event is forecast in the next few days. The same sources for current year are also used for forecast.

4.3.3   Historical

To provide a range of potential outcomes, historical weather found for the field centroid is used. Twenty years of historical weather, maintained on a 30 km grid and obtained from a service developed by Corteva, is utilized. In effect, for each Decision Zone, there will be 20 model runs each day, each with a different set of historical weather (See Fig. 5).

4.4   Variable-Rate Prescriptions

A key feature of the service is the ability to create variable-rate nitrogen prescriptions for export to most application controllers in order to apply nitrogen at different rates

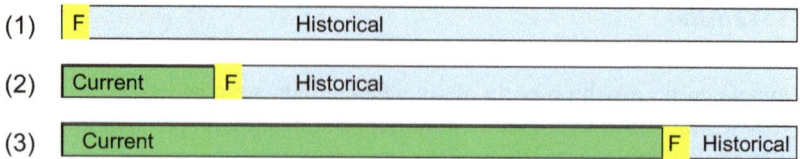

Fig. 5 Schematic diagram of the use of weather data in simulations. (1) Beginning of the crop season. (2) Around January 1 of crop season. (3) Late in the crop season

across the field. The nitrogen rates are applied at the resolution of each Decision Zone polygon.

To generate a nitrogen recommendation, all pertinent information must be entered for the field (genetics, planting, irrigation, fixed-rate nitrogen applications). The variable-rate application is added as part of the Management Plan, with the date, product, application method and stabilizer use entered.

For each Decision Zone and one of 20 previous weather years, a series of simulations is run with varying amounts of nitrogen applied in the simulation for the variable-rate application. The first run is with zero additional units of nitrogen. The estimated soil nitrogen at VT/R1 is compared to the target soil nitrogen level for the Decision Zones (based on Yield Target); if the estimated soil nitrogen amount is above or below within a set range (4.5 kg of N per hectare or 5 pounds per acre) of the soil N target, the simulation stops, posts the proposed nitrogen rate and moves to the next Decision Zone x Yield. If the difference is greater than the set range, the difference is determined and added to the next attempted nitrogen rate. This combined rate is run again through the process, and again the same logic is applied. This iterative process is allowed to run up to 15 times, but generally, 1–5 cycles are necessary to determine a converged value (Fig. 6).

Once all 20 years of simulations for a given Decision Zone are run, the median of the yearly proposed N rates is determined and posted as the recommended nitrogen rate for the Decision Zone; this process is run for all Decision Zones in the field.

In the user interface (Fig. 7), suggested nitrogen recommendations may be edited by Decision Zone. Information about the Decision Zone (soil type, yield goal target soil N and recommended N rate) is displayed. The rate may be adjusted by setting the value or by adjusting the rate up or down by a percentage. In addition, multiple Decision Zones may be selected, and the same process can be applied. This allows a user to adjust the rates uniformly if necessary.

Variable rate prescriptions may be exported from the software to various common data formats used in application controllers. One or many fields' prescriptions may be selected at export time. The rates may be exported as either units of nitrogen per area or in units of the nitrogen product per area that corresponds to the nitrogen rate. This includes dry products such as urea and ammonium nitrate or liquid products such as urea-ammonium nitrate (UAN). Conversions to product units per area are based on the density of each product.

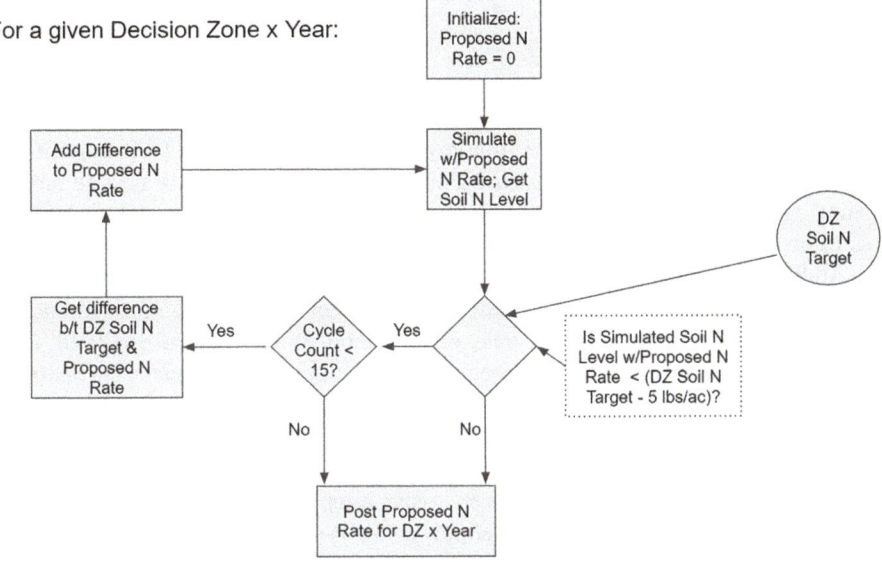

Fig. 6 Flowchart showing the process to calculate N-rate recommendation for a given Decision Zone in a given year

Fig. 7 Example user interface of variable-rate nitrogen recommendation

4.5 Daily Reports

In season, the CSA will receive daily emails from the system, providing them with a list of fields they have enrolled, their current nitrogen amounts and the projected soil nitrogen levels at growth stage VT/R1. This provides a quick report for the CSA to scan through and to note any fields that may be trending towards insufficient amounts of nitrogen.

5 Future Development

5.1 Crops

As described thus far, Granular Agronomy Nitrogen Management focuses primarily on maize farmers in the Midwestern United States. This is where a large area of maize is grown, together with a relatively large nitrogen use and technology adoption.

Adding additional crops to Nitrogen Management is being considered. Wheat, barley, canola and sunflowers are also consumers of commercial nitrogen and are planted extensively across large areas of the globe. While maximizing yield is an important aspect of nitrogen management, these crops are different from maize in that there are yield penalties for over-application of nitrogen. Many of these crops will have lodging issues if too much N is applied. Sunflower and canola attract more insects and diseases and have maturity issues at harvest. Wheat and barley have two issues with excessive nitrogen; one is excessive lodging and difficulty of harvest, the other is a change in grain protein that may make it unsuitable for certain uses, like malting.

5.2 Environmental Concerns

When over-applied, nitrogen can become a pollutant, both to groundwater and to the atmosphere. Modelling the losses of nitrogen prior to application is a way to provide growers with information about the impact they may have on the environment. Providing this information would allow them to try alternative approaches that can be modelled to reduce losses.

Currently, the Granular Crop Model does simulate leaching, denitrification, nitrification and volatilization. These values may be exposed in the user interface to help show the user how nitrogen may be lost (Fig. 8). In addition, as more research becomes available, monetary values may be assigned to these losses and used in the decision support process, balancing potential yield and gross income versus nitrogen fertilizer and environmental costs.

5.3 Variable-Rate Nitrogen Approaches

Presently, the soil nitrogen target approach through Decision Zone Yield Targets is the approach taken to apply nitrogen variably across the field. However, as the Granular Crop Model is further tuned and improved through data collection and analysis, alternative approaches may be better suited.

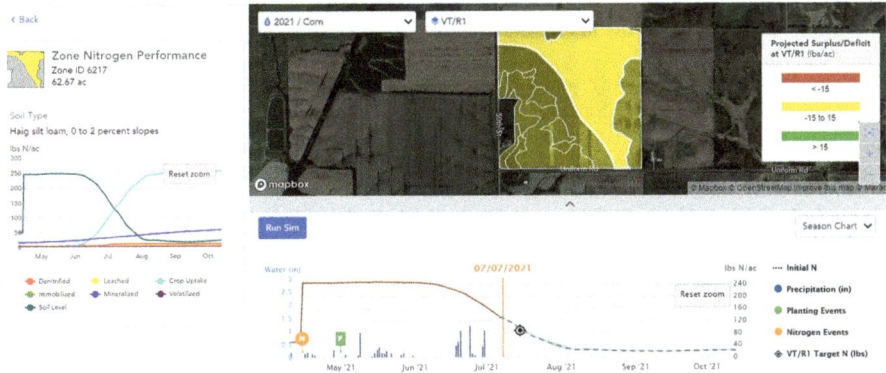

Fig. 8 User interface showing an individual zone's estimated nitrogen mineralization, leaching, denitrification, soil N levels and plant N uptake (left)

The Economic Optimum Nitrogen Recommendation (EONR) approach uses simulated yield responses to nitrogen application, together with the market price of grain and cost of nitrogen fertilizer, to determine the optimal amount of nitrogen to apply that provides the best return for the investments. This approach is easier to understand by growers (best nitrogen value) and does not require Yield Targets to be assigned at each Decision Zone (an often time-consuming process performed by the CSAs). This approach is undergoing further study with field tests to validate it.

6   Conclusion

Granular Agronomy Nitrogen Management is a comprehensive nitrogen management tool, relying on a proprietary crop growth model as the engine, wrapped in features to enable growers to address their particular needs. This approach takes the crop model out of the laboratory and into a production setting, accessible by many growers. While there are still areas of improvement that need to be addressed, Nitrogen Management covers a wide range of scenarios for many growers. Ultimately, this tool can be used to not only improve the grower's profitability but also to start to address environmental concerns, providing guidance towards improved management.

Conflict of Interest   The author declares that he is employed by Corteva Agriscience, which owns Granular Nitrogen Management and the Granular Nitrogen Model.

References

Hoogenboom, G., Porter, C. H., Boote, K. J., Shelia, V., Wilkens, P. W., Singh, U., White, J. W.,
Asseng, S., Lizaso, J. I., Moreno, L. P., Pavan, W., Ogoshi, R., Hunt, L. A., Tsuji, G. Y., &
Jones, J. W. (2019). The DSSAT crop modeling ecosystem. In K. J. Boote (Ed.), *Advances in
crop modeling for a sustainable agriculture* (pp. 173–216). Burleigh Dodds Science Publishing.
https://doi.org/10.19103/AS.2019.0061.10

Jones, J. W., Hoogenboom, G., Porter, C. H., Boote, K. J., Batchelor, W. D., Hunt, L. A., Wilkens,
P. W., Singh, U., Gijsman, A. J., & Ritchie, J. T. (2003). DSSAT cropping system model.
European Journal of Agronomy, 18, 235–265.

xarvio® Digital Farming Solutions

Manuel Nolte, Andreas Tewes, and Holger Hoffmann

1 Introduction

In this chapter, we illustrate the practical relevance of crop model systems for agricultural businesses by presenting a case study. The framework described (Tewes et al., 2020a, b) is part of the tools used by the digital farming platform xarvio® FIELD MANAGER offered by the xarvio® Digital Farming Solutions. We introduce the case study with some conceptual remarks on how to integrate model outputs into agricultural decision-making. This is followed by a summary of potential precision farming uses enabled by crop models. We conclude by providing insights into the current state of crop modelling-based products at xarvio® Digital Farming.

Modelling crop development plays a key role in precision agriculture digital systems. In such systems, models are used to calculate, interpolate, extrapolate and translate between relevant variables that indicate crop status and/or crop health. Additionally, data modelling is used to identify patterns in the data (see Table 1 for a list of example use cases).

With these capabilities and by using a range of environmental inputs (soil, weather, remote sensing, etc.), management data (sowing date, crop, variety, etc.), metadata (field boundary, variety details) and other crop models connect such data for a digital farming system. As a result, they can be used to supplement or deliver relevant data layers for digital farming recommendations and decision-making as described in the following:

M. Nolte (✉) · A. Tewes · H. Hoffmann
xarvio, BASF Digital Farming GmbH, Köln, Germany
e-mail: manuel.nolte@xarvio.com

© The Author(s), under exclusive license to Springer Nature Switzerland AG 2023
D. Cammarano et al. (eds.), *Precision Agriculture: Modelling*, Progress in Precision
Agriculture, https://doi.org/10.1007/978-3-031-15258-0_11

Table 1 Example uses of crop models

#	Action	Example use case
1	Calculate	Calculate crop status, e.g. growth stage, or complex indices such as water stress
2	Interpolate	Interpolate biomass in between satellite images
3	Extrapolate	Extrapolate LAI beyond the saturation in satellite images
4	Translate	Translate measurable variables into other variables (e.g. NDVI to LAI)
5	Identify data patterns	Identify and weigh relevant variables in sensor fusion and identify spatial patterns

1. Deliver data for smarter decisions (all information presented). One example of extremely relevant information as a basis for making smarter decisions are crop growth stages. These are used for most management decisions during the crop season.
2. Deliver data for recommendations (defined option space). A defined set of options is derived based on model output.
3. Deliver data for decision-making (single option). A unique solution is proposed.

The examples above describe how information can be processed straightforwardly from a few inputs to output and directly to application. However, crop management decisions are often taken under complex situations with several interdependencies with other decisions. This is illustrated by the following case study.

2 Possible Use Cases

Crop yield is the most relevant contributor to a farmer's income. Having access to an early, field specific and accurate prediction of yield is of strategic importance in managing agricultural businesses. Therefore, in digital farming applications, it is the most prominent target variable derived from crop models.

Yield estimates are useful for a variety of cases. An early yield forecast helps a farmer to evaluate whether management decisions are economically justified. This is especially valuable when a crop model service operates at sub-field level. Understanding in-field variation of yield allows the farmer to identify low- and high-yielding zones. The farmer is then able to optimize inputs such as fertilizers, seeding rates or irrigation by accounting for these variations.

Yield forecasting can also be used to derive an early estimate of the farmer's revenue and thus enables better informed investment decisions. This is done by multiplying the forecasted yield of a given crop by its expected market price. It may include not only an average estimate but in addition uncertainty measures such as upper and lower confidence intervals of the overall revenue. This allows the farmer to have a more transparent view of his options with regard to future investments and business strategy.

When including historical data, a crop model service can help to assess the general yield potential of a field. This helps when a farmer wants to invest in new farmland without having information on historic yields – allowing him to evaluate better if the price for a specific area is justified. Likewise, yield potential assessment makes insurance management more transparent. Knowing the actual value of his fields helps the farmer to decide whether a yield protection insurance programme is necessary or not.

Finally, crop modelling also may improve farm management in the context of sustainability. Measures such as implementing flowering strips in a field help improve the biodiversity of agro-ecological systems. However, there is substantial uncertainty regarding the associated costs due to yield loss, which may prevent many farmers from introducing such measures. Crop models can help to reduce the conflict between improving the ecological footprint of a farm and yield loss. By assessing the variation of yield potential in each field, low-yielding areas can be identified and selected as potential areas to use for ecological purposes.

Besides yield, additional crop model outputs may be useful in the decision-making process. Depending on the complexity of the model framework at hand, these can range from nutrient uptake and the estimation of water balance, for example, to estimating the magnitude of influences of abiotic and biotic stresses.

If the model framework contains, for example, a routine describing plant nutrient uptake, this can be used to predict optimum nutrient requirements of a crop at given times and field locations. By combining optimal nutrient uptake with information about actual nutrient uptake (e.g. from remote sensing data), one can then derive highly specific fertilizer recommendations.

3   Crop Modelling at xarvio® Digital Farming Solutions

xarvio® (https://www.xarvio.com/) is the digital farming brand of the BASF Digital Farming and a leader in the digital transformation of the agricultural industry. xarvio® is represented by several digital products, addressing different problems, customer profiles and farming systems. They include xarvio® FIELD MANAGER, SCOUTING, HEALTHY FIELDS, CONNECT and the SMART SPRAYING joint venture with Bosch.

xarvio® FIELD MANAGER is a digital farm management platform aimed at improving and automating farming decisions. By combining manifold data sources and modelling approaches, a deeper understanding of field and subfield processes is gained and used to optimize farming inputs while reducing their ecological footprint. xarvio® FIELD MANAGER covers all key areas of farming – including seeding, crop nutrition, crop protection and weed management.

In such a digital farming system, crop models can play an important role by integrating available data sources and simulating various aspects of crop growth including the forecasting of harvested yields. As presented in the beginning, multiple

Fig. 1 Visualization of crop model feature as displayed to the customer in xarvio® FIELD MANAGER (2021 season)

levels of digital farming support can be thought of ranging from 'decision support' to 'decision-making'. The latter undoubtedly requires higher confidence in the quality of input data and an appropriate representation of reality by the model.

The crop model solution which is currently developed within xarvio® FIELD MANAGER represents a 'decision support tool'. The crop model solution provides yield forecasts of the field average from the beginning of the season until harvest. To account for uncertainty in the weather data, localized weather scenarios are calculated from historical weather data and used to display a plausible yield range within the software. The model is currently validated for winter wheat and winter barley and extended to additional crops. Figure 1 shows a visualization of the yield forecast feature implemented in xarvio® FIELD MANAGER (beta), season 2021. The display enables interpretation of the ultimate yield prediction (in t ha^{-1}) as well as the investigation of temporal yield development under different weather scenarios.

Further developments of the solution may include the following features:

- Yield forecast at subfield level: This feature would allow the farmer to identify low- versus high- and stable versus unstable yielding zones and adapt farming inputs accordingly. Zone-based management decisions (e.g. varying nitrogen rate applications) could be derived directly from the crop model output.
- Reference simulated yield of locations surrounding the field or locations of similar environmental conditions. This would then provide context (benchmark) to the given yield forecast and allow for interpretation if attained yields are above or below what can be achieved under given environmental conditions.

4 Crop Model Framework Used by xarvio® Digital Farming

When selecting a crop model for precision farming applications, several aspects need to be considered. The model needs to address the complexity of growing conditions in the field at a sufficient level to make accurate predictions for the crop variable of interest. This depends on the target customer and the farming environment simulated. In highly mechanized, industrialized agricultural systems, one can assume that, in most cases, crops are not grown under nitrogen deficiency (in fact rather the opposite). Therefore, the model will probably not need to consider nitrogen-limited conditions. Hence, leaving out this aspect will not have a strong negative effect on the accuracy of prediction. The chosen complexity of the crop model is limited by the range of environmental data layers that are available in an operational context. For example, if data concerning pest and disease pressure cannot be provided, then the model will not attempt to account for these effects. Furthermore, computational scalability and costs need to be considered in the choice of model.

At xarvio® FIELD MANAGER, models like the LINTUL-5 crop model are explored, which includes components to account for water-, nitrogen-, phosphorus- and potassium-limited growing conditions (Wolf, 2012). In its current implementation for testing, water limitation is the only growth-limiting factor considered. The model is implemented by using SIMPLACE, a crop model framework that provides a modular architecture, allowing for flexible design of crop model solutions (www. simplace.net, accessed September 24, 2021). Crop growth is simulated in daily time steps, that is, biomass dynamics are calculated from emergence to maturation.

Five different data sources are fed into the model framework:

1. Management data such as crop, crop variety and planting date are provided by the user.
2. Growth stage data is provided by an in-house developed, separate phenology model.
3. Weather data input for the phenology model and the crop model itself are provided by a commercial weather data vendor (precipitation, solar radiation, temperature).
4. Soil data required by the soil water module are provided by a public soil database.
5. Leaf area index (LAI) data derived from Sentinel-2 optical satellite data, for example, provide the possibility for downscaling yield forecast on a subfield level.

If data sources 1–4 only were used, one could only predict the average yield per field because of the coarse resolution of these layers. However, if in-field variation of yield is required, satellite imagery data can bridge the gap to downscale yield estimates to subfield level. The LAI is a state variable of LINTUL-5 and is updated at every timestep of the simulation. This allows assimilation of LAI data into the crop model every time a new cloud-free satellite image becomes available. At each pixel, the crop model is shifted to the local LAI level considering both the error of the

Fig. 2 Comparison of
simulated (**a**) and observed
(**b**) spatial yield pattern for a
winter wheat field (2018) in
North-Rhine Westphalia,
Germany. Red dots indicate
field sampling locations

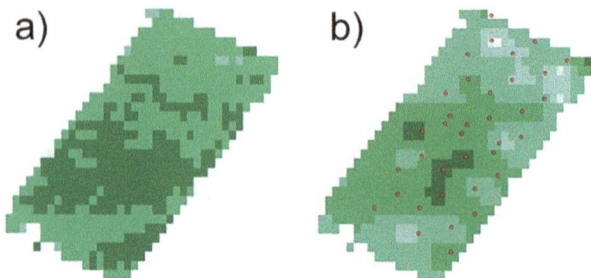

satellite product and the crop model. The pixel-wise assimilation ultimately results in individual yield estimates for each pixel. While there are multiple assimilation methods and strategies that have been tested for crop modelling, in this case, a weighted ensemble mean assimilation is implemented (Tewes et al. 2020b).

Figure 2 shows an example for a German winter wheat field in North-Rhine Westphalia, where yield was predicted in a hindcast simulation using the crop model setup described above. The reference map represents 36 equally distributed ground measured grain yield data points from which we extrapolated over the whole area to illustrate spatial yield variation. As shown by the graphics, the crop model simulations correctly capture the spatial patterns in yield distributions.

Conflict of Interest The authors declare to be employed by the BASF Digital Farming GmbH.

References

Tewes, A., Hoffmann, H., Nolte, M., Krauss, G., Schäfer, F., Kerkhoff, C., & Gaiser, G. (2020a). How do methods assimilating Sentinel-2-derived LAI combined with two different sources of soil input data affect the crop model-based estimation of wheat biomass at sub-field level? *Remote Sensing, 12*, 925.

Tewes, A., Hoffmann, H., Krauss, G., Schäfer, F., Kerkhoff, C., & Gaiser, G. (2020b). New approaches for the assimilation of LAI measurements into a crop model ensemble to improve wheat biomass estimations. *Agronomy, 10*(3), 446.

Wolf, J. (2012). *User guide for LINTUL5: Simple generic model for simulation of crop growth under potential, water limited and nitrogen, phosphorus and potassium limited conditions.* Wageningen UR.

WatchITgrow

Isabelle Piccard, Anne Gobin, Sven Gilliams, Laurent Tits, and Jürgen Decloedt

Abstract WatchITgrow is a platform for crop monitoring that provides farmers, contractors, advisors, researchers, traders, processing companies, suppliers, etc. with weather and soil data and information on crop growth and development derived from satellite images. It provides tools for improved soil sampling and variable-rate planting, fertilization, irrigation, and crop protection based on the intra-field variations visible from the satellite images. WatchITgrow also includes a field registration tool and can be used to store and exchange information on field management practices, field observations, plant or soil analyses, etc. in a digital way. The final objective of WatchITgrow is to combine the various data sources (satellite, weather, soil, crop, yield data) using new technologies such as big data analytics and machine learning to provide farmers with personalized advice to optimize field practices to increase their yields sustainably.

1 Introduction

WatchITgrow is a Belgian platform launched to support farmers to monitor the status and evolution of arable crops and vegetables, in view of increasing yields, both qualitatively and quantitatively. It integrates various types of data such as satellite, weather, soil, and field management data. The platform is freely accessible after registration at www.watchitgrow.be (Fig. 1).

I. Piccard (✉) · A. Gobin · S. Gilliams · L. Tits · J. Decloedt
VITO Remote Sensing, Mol, Belgium
e-mail: isabelle.piccard@vito.be

© The Author(s), under exclusive license to Springer Nature Switzerland AG 2023
D. Cammarano et al. (eds.), *Precision Agriculture: Modelling*, Progress in Precision
Agriculture, https://doi.org/10.1007/978-3-031-15258-0_12

Fig. 1 WatchITgrow
platform for crop
monitoring

The WatchITgrow platform was developed in the framework of the research project, iPot, a collaboration between researchers (VITO, CRA-W, and ULg-Arlon) together with the Belgian potato trade and processing industry association Belgapom. After the research phase, the platform became operational in 2017 and was rebranded as WatchITgrow. WatchITgrow is an independent platform, hosted by VITO, guaranteeing data privacy, security, and access. Belgapom members and the Flemish farmers' organization Boerenbond are actively using and promoting the platform among their farmers in the context of the "digitization of agriculture" in Belgium. Several Flemish and European H2020 projects ensure the further development of the platform. In collaboration with various partners such as the Belgian Soil Service, machinery producer AVR, agricultural research organization and advisor Inagro, the Flemish Research Institute for Agriculture, Fisheries and Food (ILVO), etc., new data layers, improved products, and tools have been added, integrating the results of joint research projects and using the latest big data and cloud processing technologies. Although the platform was initially oriented towards potato monitoring, WatchITgrow can now be used for monitoring a broad range of arable crops, vegetables, and grasslands. Since 2017, more than 1600 users have registered on WatchITgrow, including farmers, traders, processing companies, suppliers, contractors, advisors, researchers, etc. In 2020, the platform was extended to enable monitoring of fields across the border in France, the Netherlands, Germany, and Italy. In 2021, about 2400 fields, approximately 14,000 ha, were monitored with WatchITgrow.

2 Content and Functionalities

2.1 Monitoring Crop Productivity and Health with Satellite Data

WatchITgrow provides information on crop growth and development as observed from Sentinel-1 and Sentinel-2 satellite images. Light reflectances measured by the Sentinel-2 satellite sensor are converted into fAPAR (fraction of absorbed photosynthetically active radiation; Weiss & Baret, 2016). In WatchITgrow, fAPAR is referred to as greenness. The fAPAR is a measure of the crop's productivity and health and is often used as an indicator of the state and evolution of crop cover. Low greenness values indicate that there is no crop growing on the field (bare soil, fAPAR $= 0$). When the crop emerges, the greenness will increase until the crop has reached maturity (fAPAR $= 0.95$–1.00). Thereafter, the index will decrease again until harvest. WatchITgrow produces new greenness maps every time the satellite passes. Archived images are available from July 2015.

The Sentinel satellites have a revisit frequency of 5 days, at higher latitudes even every 2–3 days. The images have pixel sizes of 10 m × 10 m. Under cloud-free conditions, the availability of such frequent high-resolution satellite images allows a close monitoring of the fields and an early detection of possible anomalies.

The satellite shows a picture "from above," and this may reveal variability in crop growth within the field that is not always visible when visiting the field from the road. The reasons for this variability can be diverse and can include (natural) soil heterogeneity, climate-induced problems such as drought or waterlogging and local damage due to pests or diseases, and emergence problems, among others.

In addition to the greenness maps, greenness graphs can be retrieved showing fAPAR evolution throughout the season at field level. By combining Sentinel-2 fAPAR and Sentinel-1 backscatter data in a unique VITO algorithm called CropSAR, we can provide cloud-free fAPAR time series. CropSAR enables us to combine the best of both worlds: understanding what happens in a plant using an optical image and the cloud-penetrating capacity of radar. From these fAPAR or greenness curves, key stages in crop development such as plant emergence, canopy closure, and harvest can be detected even in cloudy periods, which would otherwise be impossible (Fig. 2).

2.2 Improved Sampling

Satellite-based information on variation within the field allows farmers to manage the fields according to different field zones and take representative soil or yield samples. Farmers can define the desired location for soil sampling by combining information from satellite images, aerial pictures, soil scans, and elevation maps and send a request for a soil analysis to the Belgian Soil Service or to Inagro via WatchITgrow.

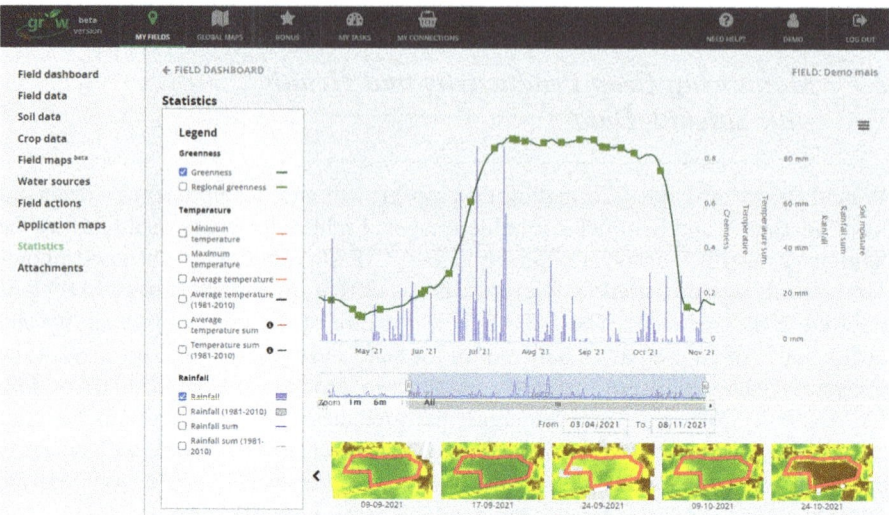

Fig. 2 Crop monitoring at field level with greenness maps and graphs

2.3 Application Maps

Based on the satellite greenness, application maps can be generated in WatchITgrow for variable-rate fertilization, irrigation, and haulm killing. In practice, the farmer enters the product dose (fertilizer, water, herbicide, etc.) that he or she would normally apply on the field. The dose is then redistributed according to the differences in greenness detected from the satellite image. For N-fertilization, the farmer can choose between two strategies, either applying a higher dose of a fertilizer to zones with a higher greenness index or applying a lower dose to these zones. For farmers, variable-rate applications are crucial in order to make cost savings in inputs, to increase yields, and to use their fields in a more sustainable way. Similarly, application maps can be created for variable-rate planting, by accounting for the shadow zones within the field.

The current approach whereby variable-rate application maps are based only on satellite greenness has some limitations. Satellite images may reveal variation within the field, but to make sound decisions on how to respond, it is important to find out what is causing the variation. Therefore, it is recommended that satellite data are combined with other sources of information (also available via WatchITgrow) such as soil maps or soil scans, elevation maps, weather data, or field observations.

Most satellite-based decision support tools use only the most recent images to generate application maps for variable-rate irrigation and fertilization, but historical satellite images collected in dry periods such as the summer of 2018 may also contain valuable information, for example, on the occurrence of dry spots in the field. Moreover, from previous research (Janssens et al., 2020), it was found that the

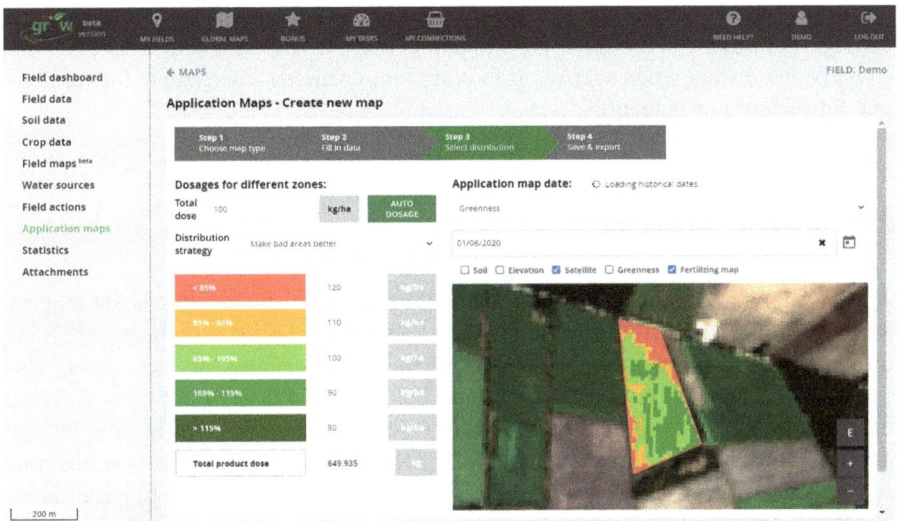

Fig. 3 Application map for variable-rate fertilization

correlation between the spectral indices and in situ observations was often only significant towards the end of the growing season, and hence not at the time of fertilization, which hinders the use of actual satellite images. For this reason, historical greenness maps were added to WatchITgrow. On the basis of these maps, it can be decided to apply less fertilizer on dry spots or to increase the irrigation dose at these spots (Fig. 3).

2.4 Yield Potential Maps

In addition to the greenness (fAPAR) maps, WatchITgrow also derives other maps from the cloud-free images. Variability maps show the difference in crop performance within the field at the time the image was taken. They are derived by comparing the greenness index value for each pixel within the field with the median index value of the field and classifying the differences into a number of categories according to their relevance.

Spatial variation in crop performance can also be examined over an entire growing season. In that case, we determine the average deviation of a pixel from the median over the season. When generating variability maps for several seasons, stable patterns may be detected. Recurrent good or bad zones may indicate the yield potential of the field. Therefore, we also call these multiyear variability maps "yield potential maps."

Where multiyear recurrent variation can be caused by differences in soil, surrounding vegetation, and hydrology, more abrupt and unexpected field variation could be caused by extreme weather events such as drought, hail, storm, or flood damage. A farmer can use these yield potential maps to review a field's history – this could be interesting when a farmer is looking for new fields – or as input for variable-rate fertilization or irrigation.

2.5   Remote Planning

As the satellite images cover large areas, they are ideal to monitor fields that are located further away from the farm. They can also be used to compare and rank different fields within a region, from low to high greenness or vice versa. This information can then be used to organize field visits in a more efficient way, for instance, by prioritizing fields that show greenness anomalies. During the harvest period, information on the greenness of the fields can be useful for harvest planning and logistics. For instance, in Northern Italy, WatchITgrow is used by seed multi-plication companies to estimate the optimal harvest time for the seeds based on the greenness evolution of the field. This is saving them time and fuel.

2.6   Weather and Soil Data

Crop production is largely influenced by the weather. By monitoring temperature and rainfall, production risk or quality losses can be assessed. WatchITgrow provides maps showing temperature and rainfall deviations from the long-term average. In addition, graphs can be retrieved showing the evolution of these meteorological variables at field level.

In case of drought and if there is a need for irrigation, farmers can find information on WatchITgrow about alternative water sources such as treated wastewater from domestic use and food companies with the link from the WaterRadar platform. Depending on the field size and crop, an indication of irrigation need is provided, and potential water suppliers are listed in a search radius around the field.

The WatchITgrow platform also provides access to basic information layers such as soil type, erosion risk, and elevation maps. These may help when trying to find the cause of variation in crop growth within a field. With WatchITgrow, farmers can also obtain information about soil fertility in Flanders at municipality level. This includes information on pH, organic carbon, phosphorus, potassium, magnesium, and calcium content of the soil derived from soil analyses from the Soil Service of Belgium.

Further, information on the field history is provided, showing the main crops that have been grown on the fields. This information is retrieved from the farmers' declarations in the framework of the European Integrated Administration and Control System (IACS).

2.7 Yield Forecasting

WatchITgrow also provides yield forecasts for the two main potato varieties grown in Belgium, Bintje and Fontane, by combining weather, plant-soil, and satellite data in a crop growth model. To generate yield estimates, two crop growth models are used: the AquaCrop plug-in (Raes et al., 2012) and the WOFOST model (van Diepen et al., 1989) implemented in the Python Crop Simulation Environment (PCSE) (http://pcse.readthedocs.io). Both models make use of meteorological-, soil-, and crop-specific data. AquaCrop also integrates satellite-based FCover estimates. To calibrate and validate the models for Belgian conditions, yield data were collected in 2015 and 2016 for about 150 fields distributed across the potato-growing areas in Belgium. Biweekly field visits were organized to collect information on phenological development, crop growth, and field management. From mid-July until harvest yield, samples were weighed, and tuber distribution was recorded at two weekly intervals. The conversion from model-based yield estimates to yield forecasts is done using three possible weather scenarios reflecting bad, average, and good weather conditions to the end of the season. As the crop growth models estimate dry matter yield, a variety-specific empirical function is applied to convert dry matter to fresh weight yields. The performance of yield models was assessed by comparing simulated total tuber dry weight with dry yield estimates from field sampling in 2015 and 2016 (Piccard et al., 2017) and by comparison of fresh yield estimates with yields recorded by the farmers from 2017 to 2019.

Farmers don't always appreciate that yield forecasts are made available to the general public (and hence also to the buyers of their products), even if it is at an aggregated level. For this reason, yield forecasts in WatchITgrow are only made available at field level and for the registered farmer only.

2.8 Field Registration

WatchITgrow also includes a field registration tool. The platform can be used to store and exchange field data in a digital way. This may include basic information such as the variety, planting date, date of haulm killing, harvest date or more specific information on the application of fertilizers, pesticides, irrigation, etc. or on damage assessed in the field. Yield data collected with WatchITgrow are used to train yield forecasting models.

It is also possible to upload pictures or maps of EC, pH, or organic matter derived from soil scanning or to add soil, manure, or plant analysis reports to a field in WatchITgrow. Also drone imagery and analytics maps can be added as a separate map layer. In this way, all the information related to a field can be stored digitally in one place.

Fig. 4 Combination of map layers available for a field (soil type, electrical conductivity measured by soil scanning, yield potential map, greenness map)

It is important to note that the farmer remains the owner of the data at all times. He or she decides whether to share these data or not with advisors, contractors, buyers, etc (Fig. 4).

2.9 Automatic Data Exchange

Farmers generally do not want to spend a lot of time registering their field activities. Data exchange should be automated and standardized as much as possible. Therefore, we are setting up connections with machinery producers such as AVR and John Deere. For example, after connecting WatchITgrow to AVR Connect, yield maps from AVR's Puma 4.0 potato harvesters are automatically added to WatchITgrow. This makes it easy for a farmer to compare the yield maps with soil and satellite data that are available on the platform. Conversely, field boundaries entered in WatchITgrow or application maps for variable-rate planting are automatically transferred to AVR Connect.

To facilitate data sharing and to avoid farmers having to enter their data several times for different platforms, WatchITgrow also connects with data hubs such as DjustConnect, managed by ILVO. Using the DjustConnect dashboard, Flemish farmers can give permission to share their data safely with WatchITgrow.

2.10   Connecting Different Actors in the Chain

WatchITgrow is intended as a collaboration platform where data can be shared among farmers and advisors or researchers, to follow up on what is happening in the field, to provide advice or to monitor field trials. Through close collaborations with research organizations, we ensure valorization of their research results by integrating them in WatchITgrow and making them available for farmers.

With WatchITgrow, farmers can also exchange field information with traders or processors. Sharing fields among farmers and contractors may facilitate the organization and registration of fieldwork activities such as planting, fertilization, crop protection, or harvesting.

Through WatchITgrow, a connection can be made with Mapeo, VITO's drone-based phenotyping solution for research and breeding. This allows researchers from seed companies to compare drone-based information collected on trial fields with weather, soil, and satellite data in WatchITgrow. But it also offers opportunities to follow up on how new varieties are doing later during seed multiplication or on the farmer's fields, how they behave under certain conditions, or how they compare to other varieties and to start interacting on this with the farmer through WatchITgrow.

3   Outlook: Towards Data-Driven Advice

The final objective of WatchITgrow is to combine the various data sources (satellite, weather, soil, field, machinery data, etc.) using new technologies such as big data analytics and machine learning to provide farmers with personalized advice to increase their yields sustainably. This implies a shift from traditional crop growth simulation models to big data models. To train these models, large amounts of field data are needed. Today, many farmers are still reluctant to share data. Uncertainty about what will happen with their data and the fear that they will be used against them are some of the reasons why farmers do not share their data. In other cases, farmers do register data via a field registration tool and do not want to repeat it when using a crop monitoring tool, for example. Data sharing among different platforms and tools, through direct exchange or via data hubs, may be a way to enable and promote the exploitation of these agricultural data and to use them to the benefit of the farmer.

Conflict of Interest The authors declare that they are employees of the company (VITO) that developed WatchITgrow.

References

Janssens P., Reynaert, S., Piccard, I., Pauly, K., Garré, S., Dumont, G., von Hebel, C., Van der Kruk, J., Neumann, A. M., Manevski, K., Peng, J., Korup, K., Kamp, J., & Booij, J. (2020). *Variable rate irrigation and nitrogen fertilization in potato; engage the spatial variation (POTENTIAL).* Available online: https://www.bdb.be/files/rap202001.pdf. Accessed on 19 Apr 2022.

Piccard I., Gobin A., Wellens J., Tychon B, Goffart J.-P., Curnel Y., Planchon V., Leclef A., Cools R., & Cattoor N. (2017). *Potato monitoring in Belgium with "WatchITgrow".* Proceedings of the 9th international workshop on the analysis of multitemporal remote sensing images (MultiTemp), Bruges, 27–29 June 2017. https://doi.org/10.1109/Multi-Temp.2017.8035229. IEEE catalogue number: CFP17608-ART, ISBN: 978-1-5386-3327-4.

Raes, D., Steduto, P., Hsiao, T. C., & Fereres, E. (2012). *Reference manual: AquaCrop plug-in program* (version 4.0). FAO. http://www.fao.org/nr/water/aquacrop.html

van Diepen, C. A., Wolf, J., Keulen, H., & van. (1989). WOFOST: A simulation model of crop production. *Soil Use and Management, 5,* 16–24.

Weiss, M., & Baret, F. (2016). *S2ToolBox level 2 products: LAI, FAPAR, FCOVER.* Version 1.1. Date issued: 03.05.2016. Available online: https://step.esa.int/docs/extra/ATBD_S2ToolBox_L2B_V1.1.pdf. Accessed on 11 May 2020.

Precision Agriculture in Rice Farming

Satoshi Iida

Abstract Japanese agriculture has reached a significant turning point due to factors such as a considerable decline in the number of farmers because of aging. In such circumstances, it is necessary to change Japanese agriculture into an attractive and profitable business for the upcoming generations of professional farmers and to ensure the local development of agriculture. To reach these goals, Kubota has been making efforts to develop "smart agriculture," namely [1], precision farming based on the use of data and [2] ultra-labor-saving approach based on automation, using Information and Communication Technology (ICT) and the Internet of Things (IoT) and its diffusion as the next generation's agriculture. To this point, Kubota has launched the KSAS (Kubota Smart Agri System) with regard to precision farming and autonomous/unmanned agricultural machinery such as a rice transplanter with automatic steering, a combine harvester with automatic operating assistance function, and a fully autonomous tractor.

Furthermore, we give an outlook on smart agriculture considering several approaches of precision farming and autonomous solutions to answer today's challenges toward the realization and fusion of these approaches.

Keywords Precision farming · Rice farming · Smart agriculture · KSAS · Autonomous agricultural machinery

1 Introduction

Agriculture around the world faces many common issues, such as reduced availability of arable land due to climate change and soil degradation and reductions in the number of farm workers, especially skilled workers. In addition to improving productivity, there is also a need for sustainable food production combining high food quality with a low environmental impact. To address these issues, precision

S. Iida (✉)
Kubota Corporation, Osaka, Japan
e-mail: satoshi.iida@kubota.com

farming (PF) has been studied in the West since the 1990s. For example, it is becoming essential for producers to improve their production methods by making use of technologies such as auto-steering devices that use global navigation satellite systems (GNSS), yield maps produced by combine harvesters with yield monitoring functions, and satellite imaging for soil analysis and crop yield prediction. This enables farmers to succeed in today's global markets.

Farming in Japan faces similar issues, and Kubota is approaching these issues by developing and utilizing the next generation of smart farming technologies. The smart agriculture technologies are based on ICT and IoT (autonomous and unmanned farm machinery) for data-driven Japanese-style precision farming and considerable labor saving. In this way, the aim is not only to sell and service agricultural machinery but also to provide an overall solution that benefits the entire value chain.

This paper describes the status quo and the future vision of smart agriculture at Kubota and discusses the opportunities of using big data analysis and artificial intelligence (AI) to achieve further advances.

2   Kubota's Role in Japanese Smart Agriculture

2.1   Current Situation and Issues of Japanese Agriculture

Japanese agriculture currently faces many challenges and has reached a major turning point. For example, between 2000 and 2015, the number of commercial farms in Japan decreased by almost 50%, falling from 2.3 million to 1.3 million. The average Japanese farmer is today over 67 years of age, and it is predicted that the number of active farmers will decrease by another 50% within the next decade.

On the other hand, there are a growing number of farming cooperatives and professionals whose main business is agriculture, and their farms have increased in size by assimilating smaller fields from people who have left farming. The government has taken various steps to increase the proportion of farmland owned by professional farmers from its current level of 56–80% by 2023. Since 2018, the long-standing policy of reducing acreage for rice cultivation has also been abolished, forcing Japanese farmers to finally become financially independent.

In this situation, several challenges need to be faced to ensure the sustainable development of Japanese agriculture:

(a) Japanese agriculture should become an attractive and profitable business.
(b) The agricultural working environment should be reformed to release farmers from heavy work and encourage young people to enter the business.
(c) Rural areas, including mountainous regions, should be revitalized and developed to maintain multifunctional roles of agriculture.

2.2 The Challenges Faced by Professional Farmers and Kubota's Position

The expansion of farming operations presents many challenges to professional farmers and farming corporations who support Japanese agriculture, as outlined below:

[Challenges Faced by Professional Farmers]

(a) Problems in managing multiple farm plots:

  (i) Decrease in yield and quality
  (ii) Management of an increasing number of workers

(b) Labor savings and workload reductions and reduction of production costs
(c) Offering products with higher added value (branding)
(d) Human resource development (transferring know-how to successors)
(e) Development and expansion of sales channels

   To make Japanese agriculture more attractive and profitable, it will also be necessary to visualize the entire agricultural system and create mechanisms within the food value chain whereby farms can produce products precisely that are required in time and amount (i.e., minimizing waste). To achieve this, it is essential to develop and deploy smart agriculture systems based on ICT and IoT technology.

3 Using Data for Precision Farming

3.1 Farming Support System: Kubota Smart Agri System (KSAS)

The KSAS is a support system for farm management and services, launched in 2014 by Kubota. It allows farmers to implement a profitable PDCA (Plan, Do, Check, Act) cycle of management by using agricultural machinery and ICT to gather and utilize information about work activity and crop data (yield and quality(taste)). Fig. 1 shows the overall structure of this system which comprises "KSAS agricultural machinery" equipped with wireless LAN hardware and direct communication units, "KSAS mobile equipment" that workers can use to record their work and relay other information, and "KSAS cloud server system" that stores and analyzes information.
   A farming support system and machinery service system operates on top of these components with the aim of providing additional value in the following ways:

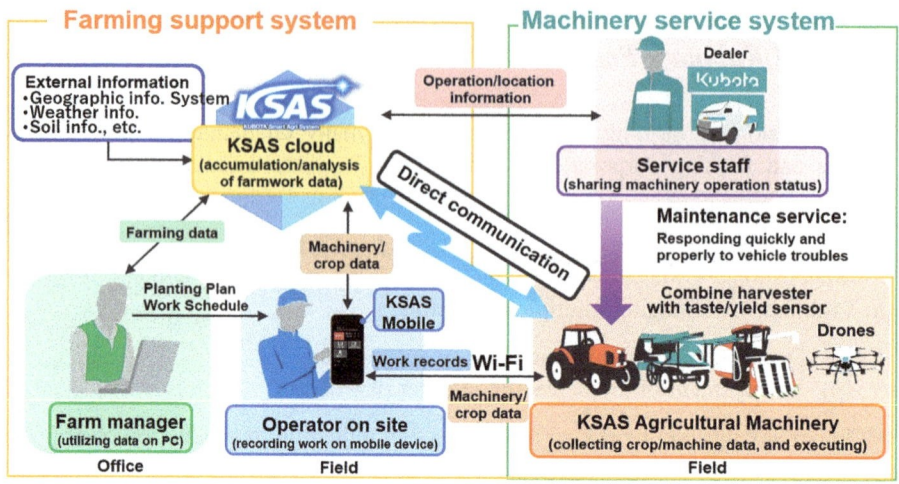

Fig. 1 The overall picture of the current KSAS

[Farming Support System]

(a) Producing tasty rice with high yields
(b) Producing crops in a safe and secure manner (ensuring traceability)
(c) Allowing farmers to work more efficiently and transfer their cultivation know-how
(d) Providing stronger foundations for farm management (cost analysis and reduction)

[Machinery Service System]

• Reducing downtime during the busy season by providing quick and appropriate services based on the location operation and error information provided by agricultural machinery

[PDCA Agriculture Based on Data]

The combine harvester, which plays one of the most important roles in the current KSAS system, is equipped with load cells and near-infrared spectroscopic analysis sensors (NIR) that perform real-time measurements of not only the weight of rice in the grain tank but also its protein and moisture content, which are the main substitutional characteristics for the taste of rice (Fig. 2). Every time a rice field is harvested, these measurement data are transmitted to a cloud server together with the combine harvester's operating data. So far, this transfer was done via a KSAS mobile device, but since 2019, it has been automated by direct communication.

With this system, the farmer can use a PC in the office to see at a glance the data stored on the cloud server, including work records and the dispersion of yield/quality (taste) in each individual plot (left side of Fig. 3). By combining this information with the results of soil analysis, for example, it is possible to implement soil

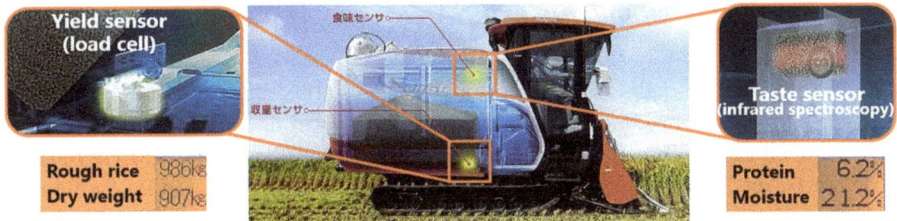

Fig. 2 Combine harvester equipped with yield and quality (taste) sensors

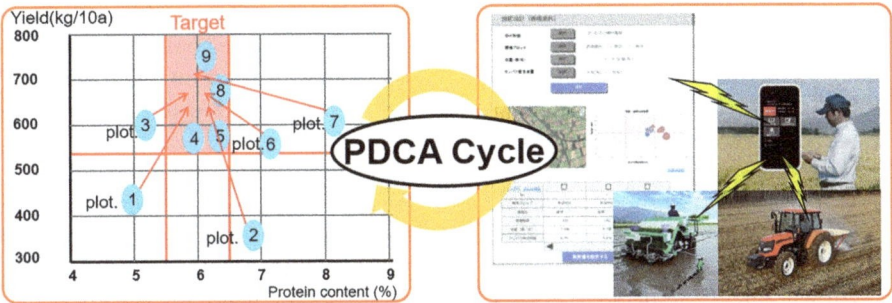

Fig. 3 The PDCA cycle by KSAS

improvement measures tailored to the characteristics of each plot and to create fertilizing plans for the following year. These planned fertilizer application data can be transmitted to KSAS rice transplanters and tractors via their operators' mobile phones. On receiving this information, KSAS agricultural machinery can automatically meter the amount of fertilizer, so that even an unexperienced operator can easily fertilize over a hundred rice fields.

In this way, by repeatedly iterating through a cycle where data gathering leads to work planning, followed by cultivation/harvesting, and then by more data gathering, it is possible to make continuous improvements to both the yield and taste of crops by optimizing the use of fertilizer and the allocation of human effort. This corresponds to the PDCA approach for agriculture based on the use of data that has not been applied to Japanese agriculture before (Fig. 3).

During a three-year trial in Niigata prefecture and other parts of Japan, this approach enabled an improvement of yields by 15% and delivery of a product of better and more stable quality. Furthermore, farmers were able to sell premium quality rice at a higher price due to its superior taste and were able to stabilize the quality of their product and sort it according to its moisture content to reduce the drying costs.

The KSAS system is highly rated by users, who have reported achieving greater field management efficiency and better rice quality and yield. Over the roughly 7-year period since the service began in June 2014, there were over 15,000 subscribers (including service systems) whose average area is around 48 ha with an average number of 225 plots per farm.

3.2   Future Developments in KSAS

Step 1 (Fig. 4) is to realize PDCA agriculture by establishing data links with every item of agricultural machinery in an integrated rice farming mechanization system.

Farmland consolidation has been promoted by the government, resulting in plots that are larger in size; therefore, it is becoming increasingly important to manage inhomogeneity within an individual plot. Therefore, Step 2 focuses on enabling more precise cultivation by developing agricultural machinery systems that enable sensing of soil and growing environments, growth conditions, and yield dispersion in the field and enable more precise cultivation as listed below (Fig. 5).

1. Combine harvesters with precise taste/yield sensors became commercially available with the launch of Kubota's WRH1200 general-purpose combine harvester in April 2018, and from January 2019, the DR series of head-feeding combine harvesters were available.
2. Remote sensing drones are in field trials throughout Japan.
3. WATARAS (Water for Agriculture Remote Actuated System), which is a remote monitoring system and can automatically control the water level of a paddy field by sensing the water level/temperature in the field, was released in 2018.
4. Growth models to determine adequate content of work and timing are in development and a plan to link with the Agricultural Data Collaboration Platform WAGRI.

As part of these efforts, it might be possible to use AI in the design of fertilizer application, which is currently performed based on diverse sensing data and farmer's experience. For example, AI could suggest how much fertilizer to use based on taste/yield data obtained by a taste/yield-sensing combine harvester or judge the crop's stage of growth based on remote sensing images or something similar and adjust the amount and timing of topdressing accordingly. By adding other information such as

Japanese DATA-DRIVEN Precision Farming with the fully mechanized rice farming
①**Farming support system cooperated with a field map and various farming data**
②**PDCA-based agriculture linked with various sensing data from Agri-machinery**
③Expanding from rice to various crops (wheat, soybeans, etc.)

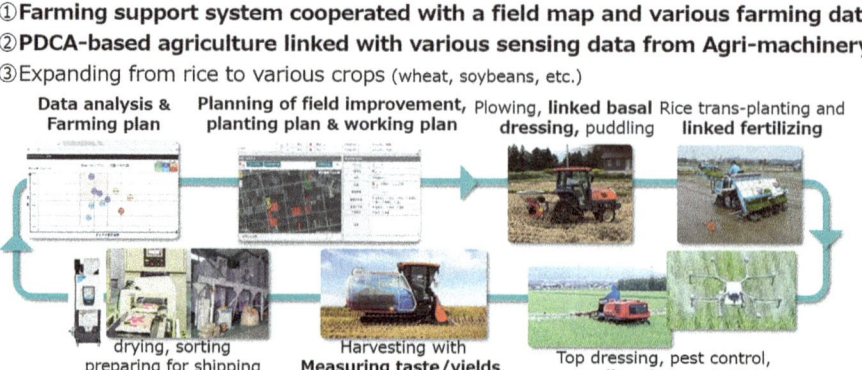

Fig. 4 The KSAS Step 1 (in service)

DATA

- **Taste/Yield data from harvester**(Mesh map)
- **Dispersion of growing** (remote sensing by drones)
- **Field environments**(water level, soil fertility, etc.)
- **Meteorological info, Growth model** (WAGRI), etc.

Precise Farming

- **Precision/variable fertilizing,**
- **Chemical application,**
- **Water control/management**
- **Prediction of harvesting time/yield**

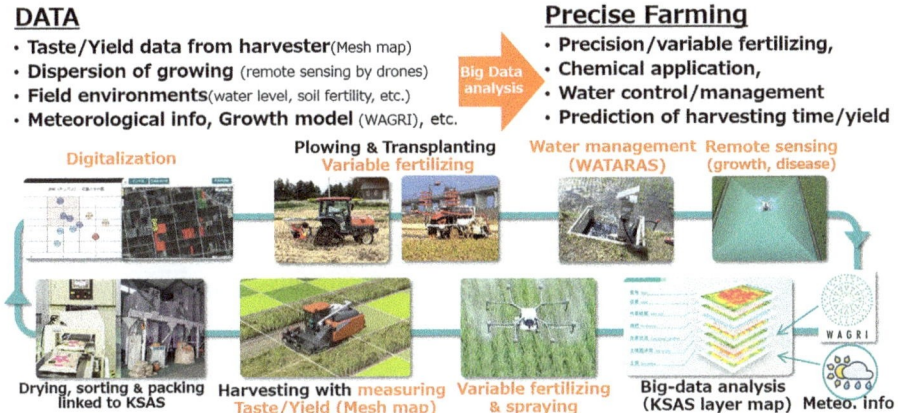

Fig. 5 TKSAS Step 2

weather forecasts, this technology could even be applied to AI-based automation of water management and prediction of optimal harvest times.

For Step 3, with the aim of building an advanced farming support system, in addition to the functions of Step 2, by using AI to analyze and process big data obtained by collaborating with information systems used by farmers (such as accounting and sales systems) and external data (such as market information and information from distribution networks), we plan to promote the evolution of advanced farming simulator technology that can support business planning and cropping using the most suitable crops to maximize the profits of arable farmers. In addition, we hope to be able to support the creation of optimal work plans detailing where, when, and by whom each item of machinery should be operated to maximize efficiency (Fig. 6).

By making KSAS as useful as possible for farmers, the number of farmers that use this system should be increased. To this end, it is essential to collaborate on the use of public and private data such as farmland data and maps, weather information, soil analysis results, and growth models. It is also important to collaborate with other agricultural machinery and information systems. For this reason, Kubota is participating in the Agricultural Data Collaboration Platform Council WAGRI and is working on the implementation of a common infrastructure for agricultural data. Through the work carried out at WAGRI, the findings of research organizations such as the National Agriculture and Food Research Organization are expected to be utilized.

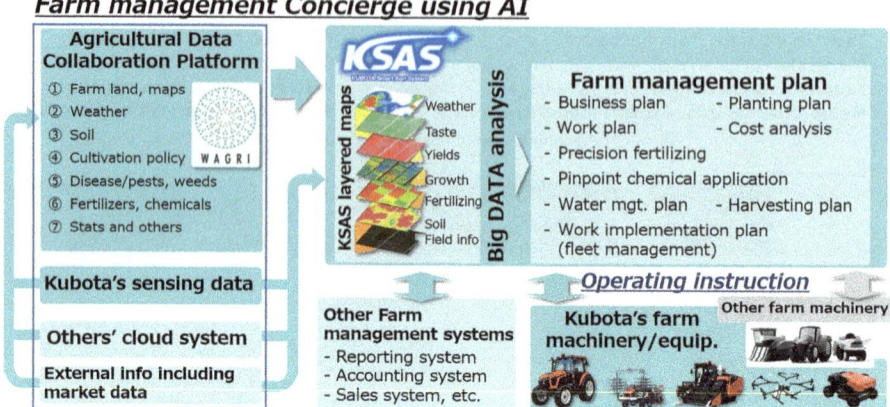

Fig. 6 Future vision of KSAS (Step 3)

4 Automation for Super Labor Savings

4.1 Automatic/Unmanned Agricultural Machinery

In addition to KSAS, which uses data to improve the efficiency of cultivation process and farm management, Kubota is also working on autonomous and unmanned agricultural machinery based on robot technology. The aim is to achieve further increases in the efficiency of work such as tilling and reaping that has already been mechanized, so that precise operations can be performed with super labor savings. The Ministry of Agriculture, Forestry and Fisheries (2015) defines three automation levels for agricultural machinery (Fig. 7), and Kubota is also working on the following themes:

Level I auto-steer technology uses the global navigation satellite system (GNSS) to perform automatic steering. Kubota released the M7 series tractors (130–170 HP) for upland farming with auto-steering functionality (RTK-GNSS), in 2015.

In 2016, the first rice transplanters equipped with go-straight functions were launched in Japan. Existing auto-steering equipment was bulky and expensive, but by developing an own control mechanism combining a cost-efficient sub-meter GNSS (D-GNSS) and an IMU (inertial measurement unit), we were able to implement a compact, low-cost auto-steering system (Noguchi, 2018). This enables even an inexperienced operator to plant rice seedlings with the same precision as a veteran, making the task much less stressful. This system is not only well accepted by farmers but has also received many awards in Japan. We also launched small- and medium-sized tractors equipped with this system. Furthermore, in December 2018, we launched a combine harvester with an automatic driving assist function (WRH1200A; Fig. 8, left).

Level II corresponds to autonomous or unmanned machinery under manned monitoring and includes cooperative work involving unmanned and manned machinery. Kubota has pioneered developments in this field, and in the fall of

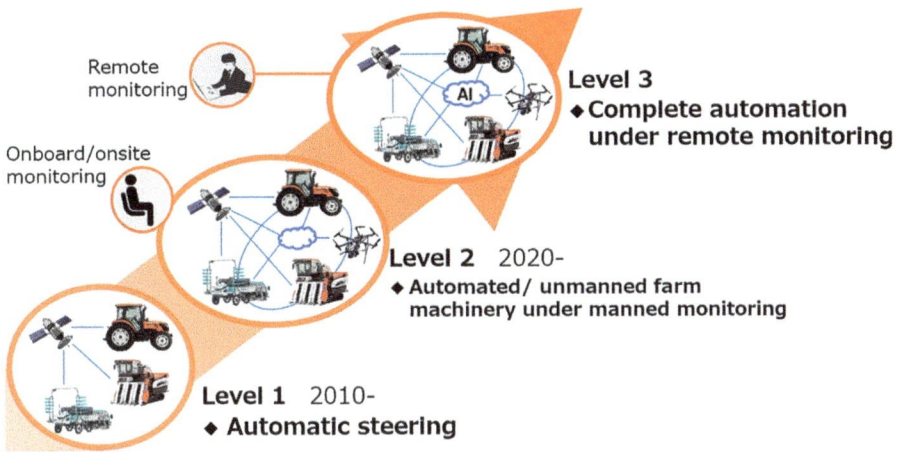

Fig. 7 Automation level of agricultural machinery

Fig. 8 Autonomous agricultural machinery

2017, we started test marketing of a Level II autonomous tractor (SL60A, 60 HP). A high-precision GNSS system called RTK-GNSS, which we manufacture in-house, made it possible to perform automatic driving with a single unmanned machine, cooperative driving with a pair of manned and unmanned machines working together, and automatic steering of a manned machine (Fig. 9).

Demonstrations have shown that this level of autonomous and unmanned operation improves the efficiency of conventional operations in tilling by around 1.5 times.

Concerning safety features, these machines have laser scanners and ultrasonic sonar to ensure reliable stopping when they detect humans and obstacles (Fig. 10) and a system that constantly monitors the surroundings with four cameras. Consequently, it complies with the autonomous agricultural machinery safety guidelines defined by the Ministry of Agriculture, Forestry and Fisheries.

At the end of 2019, an autonomous tractor with 100 HP was launched (MR1000A; Fig. 8, center), then an autonomous rice transplanter was successively launched at the end of 2020 (NW8SA; Fig. 8, right), and our product portfolio has expanded.

Fig. 9 Coordinated operation by manned and unmanned tractors (SL60A)

Fig. 10 Safety devices (laser scanner and ultrasonic sonar)

4.2   Future Plans for the Evolution of Autonomous and Unmanned Agricultural Machinery

With regard to the evolution of autonomous and unmanned agricultural machinery, the first aim is to complete Level II. In addition to the tractors and the rice transplanter that have already been launched, we are also developing autonomous combine harvesters. We are making further advances in farm automation, including the autonomous operation of work at field boundaries, and expansion of applicable implements by enhancing the control systems. When doing so, safety awareness is essential. We believe that the use of AI will be the key to identifying humans and animals in crops, detecting obstacles in the dark, and recognizing boundaries and further more functionalities.

At the same time, we are preparing to provide compatibility with the quasi-zenith satellite system promoted by the Japanese government. We expect it to become more widely used if it is possible to reduce the cost of receivers and achieve 5–6 cm accuracy without requiring a base station.

Next, Level III corresponds to completely unmanned machinery with remote monitoring, where the aim is to perform unmanned work in multiple fields, including driving on farm roads. To achieve this, it will be necessary to make further developments in farmland infrastructure, including farm roads and safety systems, and to install high-speed communication infrastructure (such as 5G) to increase the speed of monitoring and control. If tractors are completely unmanned, they must be capable of driving on roads with their implements attached. This requirement raises issues other than technological developments, such as relaxation of the road traffic law. Thus, achieving Level III requires not only research and development but also the creation of standards and infrastructure in cooperation with government and industry organizations. These approaches are proceeding under the nation's Strategic Innovation Promotion Program (SIP) and some consortia of demonstration projects in cooperation with the government, industry, and academia.

In single operations, autonomous and unmanned agricultural machines have a limited effect. For this reason, Kubota is developing an operating support system for autonomous agricultural machinery. This system is linked to KSAS to enable the optimal operation and management of multiple agricultural machines, including those that are not fully autonomous.

5 Outlook

Smart agriculture will bring in significant value according to the Sustainable Development Goals, such as shown in Fig. 11. That is the path we see for Japanese agriculture to be profitable, easy, environmentally friendly, robust, and sustainable. Kubota will continue to tackle the challenges farmers are facing in Japan and contribute to solving global agricultural challenges by starting to introduce KSAS to rice farming and upland farming in Asia (smart agriculture research group web page).

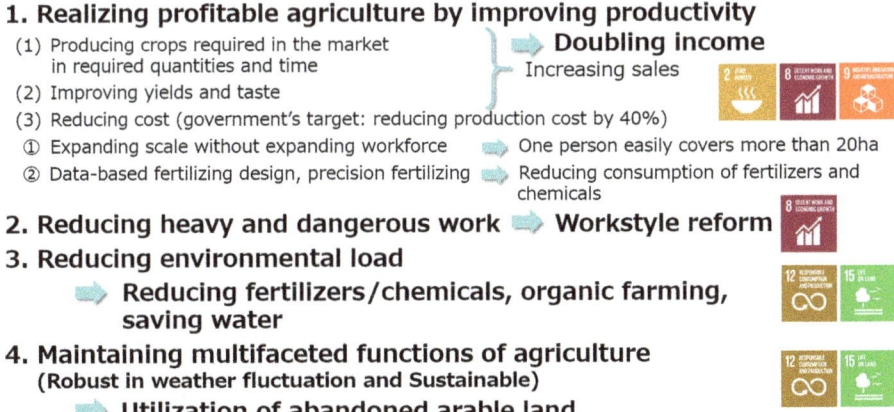

Fig. 11 Value of Smart agriculture

Conflict of Interest The author of this chapter works for the Kubota Corporation, which developed the KSAS digital tool for rice production.

References

2015 Agriculture and Forestry Census (Ministry of Agriculture, Forestry and Fisheries), 2016/2017.

Agricultural Data Collaboration Platform Council (WAGRI) web page. https://wagri.net/

Noguchi, N. (2018). Agricultural vehicle robot. *Journal of Robotics and Mechatronics, 30*(2), 165–172.

Smart agriculture research group web page. http://www.maff.go.jp/j/kanbo/kihyo03/gityo/g_smart_nougyo/

AgSkyNet: Harnessing the Power of Sky and Earth for Precision Agriculture

Suryakant Sawant, Sanat Sarangi, and Srinivasu Pappula

Abstract Climate change is posing a serious threat to sustainable food production. An interdisciplinary approach comprising agriculture, computer science and geographic information science is needed to solve critical problems in the crop–soil–water–weather continuum. The AgSkyNet platform has been developed to address the information needs of growers during crop cultivation and post-harvest. This section provides insights into innovative solutions for integrated pest assessment and crop growth simulation to estimate yield and to assess the severity of the effects of stubble burning and pesticide residue detection. A first case on the prediction and detection of biotic stress (i.e. pest estimation) shows the effective use of data analytics to identify the pests, thereby reducing the use of agro-chemicals. A second case covers insights into the integration of remote sensing and crop growth simulation models to estimate crop yield and the severity of stubble burning. Finally, a case study on the detection of pesticide residue using hyperspectral sensing demonstrates the potential for rapid assessment and is a key enabler to increase awareness among the farming community.

1 Introduction

The world is being buffeted today by a variety of forces ranging from a rapidly growing population, increasing urbanization, decreasing arable land area, reduced water availability and disastrous climate changes which threaten to destroy the planet which we call our home. The imbalance caused due to these forces could manifest itself in many ways with the recent COVID-19 pandemic being a possible example of this. Sustainability has become increasingly important for saving the planet, people and profits.

S. Sawant · S. Sarangi · S. Pappula (✉)
Digital Food Initiatives Group, Tata Consultancy Services Research & Innovation, Hyderabad, India
e-mail: srinivasu.p@tcs.com

© The Author(s), under exclusive license to Springer Nature Switzerland AG 2023
D. Cammarano et al. (eds.), *Precision Agriculture: Modelling*, Progress in Precision Agriculture, https://doi.org/10.1007/978-3-031-15258-0_14

At the TCS Digital Food Initiatives group, sustainability, scalability and suitability (3Ss) have always occupied centre stage, and all our offerings have been designed with these in mind. Through the AgSkyNet platform – the Sky and the Earth are literally brought together by image processing, machine learning/deep learning and blockchain, so that the underlying platform can be easily adopted even by a small and marginal farmer with less than a hectare of land in Asia or Africa or by a farmer owning thousands of hectares in the Americas.

Here, we give examples of innovations that harness the forces at the Earth, Sky, Image Processing and Deep Learning while satisfying the 3Ss.

1.1 Integrated Pest Assessment with Data-Driven Insights

For most crops, timely management of pests is key to maximizing both quality and quantity of harvest. One useful step in this direction is integrated pest management (IPM) which advocates a balance between various techniques – cultural practices to select the most suited varieties, crop sanitation and application of organic nutrition and protection, biological controls with beneficial insects, and responsible application of synthetic fertilizers. Implementing these principles has challenges with operational and cost implications. Moreover, with changing ambient conditions in open cultivation scenarios due to the variation in weather, soil, water and various other factors, stress on the crop is almost unavoidable despite all the measures. Given that pests are an inevitable part of a crop's life cycle, an integrated strategy is needed not just with the most suited agronomic practices but also with technology that can predict, process and react quickly to help manage the crop effectively in its volatile environment.

At TCS Digital Food Initiatives group, we have been working on a technology-led integrated pest assessment (IPA) approach that involves the prediction of pest risk together with pest detection at the early or any of the subsequent cultivation stages in order to provide a precise course of action to be taken to manage the crop. At the heart of this approach is the data that allows models giving specific insights to be configured and customized to the requirements of the local conditions for the crop.

Crops often have very distinct ecosystems within which they are produced, harvested and consumed. The need for insights as part of IPA is closely linked to key stakeholders in the ecosystem and their role in maximizing the effectiveness of their impact on crop management. For example, paddy rice (*Oryza sativa*), a staple crop of India, is cultivated over large open stretches by farmers with landholdings of all sizes – big or small. Given that most farmers are independent small growers, farming here is largely unorganized where farmers take the local rice varieties based on recommendation from the regional agricultural universities. Efforts have started to bring about some organization to this system through the establishment of Farmer Producer Organizations (FPOs) (Pappula, 2013). The largest players in this system are agri-input companies that supply essential pesticides to help contain the biotic stress on the crop. Such organizations, therefore, have a keen interest to digest IPA insights to be able to position their products strategically for farmers. Let us take another major crop, tea (*camellia sinensis*), where India is the second largest producer of the crop after China. A significant fraction of this sector is organized around large tea gardens that are managed with a dedicated staff of workers and managers. Here, maximizing the quality and quantity of crop is essential for the organization (and processor) owning the garden and processing the harvest to produce tea. This drives their interest in IPA insights to monitor closely the potential and current stress incidents.

Our proposed IPA has two broad areas – pest prediction and pest detection. Pest prediction refers to the use of ambient information around a crop such as the weather in open farms to forecast the risk of occurrence of a particular pest in the next few days from a given day. In a closed farming environment such as a greenhouse, the source of data could be a set of sensors that capture the ambient micro-climatic conditions (Elijah et al., 2018). Further, more than awareness of a particular risk, it is important to know the severity level of the risk to plan a course of action whether it is mobilizing supplies close to the region of agri-input sales or taking a stocking decision for a garden. This severity is therefore usually tied to an economic threshold level (ETL) where a pesticide spraying decision is made when the ETL is breached (Carlson, 1970). Otherwise, the cost incurred in pesticides does not offset the market value of the final produce. While the prediction of pest risk gives a probability to help prepare for pest onset with any prophylactic measures, pest detection refers to effectively assessing the severity and spread of an infestation that has already started to facilitate the adoption of curative measures. An essential component of managing farm operations involves regularly scouting the farms by field staff with phones or with other equipment such as drones. This involves capturing all relevant events of interest that includes of course pest-related stress incidents. Each event typically includes a picture of the crop accompanied by supporting information such as textual and audio meta-data. Computer vision models work on the images from such crowdsourced data to generate insights on the condition and severity. This could translate to automated demarcation of specific hotspot regions that need immediate attention. Without technology, collating all this information and interpreting it to recognize a pattern on which some decision can be taken in a timely way is non-trivial – especially with only human effort. Such simplified and timely recommendations associated with pest infestations, therefore, help to significantly bolster the IPA objectives of appropriate curative actions to protect crop.

Development of pest prediction models started by carrying out control studies in a small defined region often with a weather station in proximity that measured various dependent ambient variables at a time when computing capabilities were limited. Common properties that pest growth depends on include temperature, humidity, rainfall and sunshine hours. Models, therefore, work with some of these input properties over a historical period and generate as output a risk of pest appearance. Other properties such as duration of leaf wetness which has been shown to favour pest growth have also been studied and used for modelling. This is also often coupled with inputs from process models associated with the crop or the pest. The approach to modelling itself has evolved over many years. It started with thumb rules and moved towards linear regression as a popular approach. With advances in nonlinear modelling and better computing capabilities, methods such as neural networks for (nonlinear) curve fitting and support vector machines that transform the data to a higher plane, with kernel functions, where the data becomes linearly separable have proved useful. Each modelling approach has some inherent advantages and constraints. One of the key constraints is the amount of data available to work with, which may not be much to begin with but grows over time and crop seasons. In our deployments, we take a data-adaptive approach where the model elements evolve as more data become available and the risk estimates become local and personalized to the region of interest under observation. Figure 1 depicts risk prediction at different severity levels in paddy fields for multiple pests in a district of Telangana, India, on a particular day in the Rabi (winter) season of 2019–2020 (Fig. 2).

To detect pest infestation of a certain severity accurately over a region for decision-making on pesticide spraying is non-trivial and therefore often subjective. Given that periodic monitoring of an indoor or outdoor farm often involves regular capture of events by field supervisors and equipment, where each event consists of one or more images, we have developed computer vision-based models (Szegedy et al., 2013) that identify affected parts of the plant and the degree of infestation in those parts to assess the severity of the condition. These insights collated from a set of geo-tagged images could be fused to create a variable-rate application (VRA) map to guide the application of pesticides precisely. Computer vision development has also evolved rapidly over the years. With advances in computing capacity and convolutional neural networks (CNNs), more commonly understood as deep learning, it has become possible to extract a robust set of features from multilevel neural

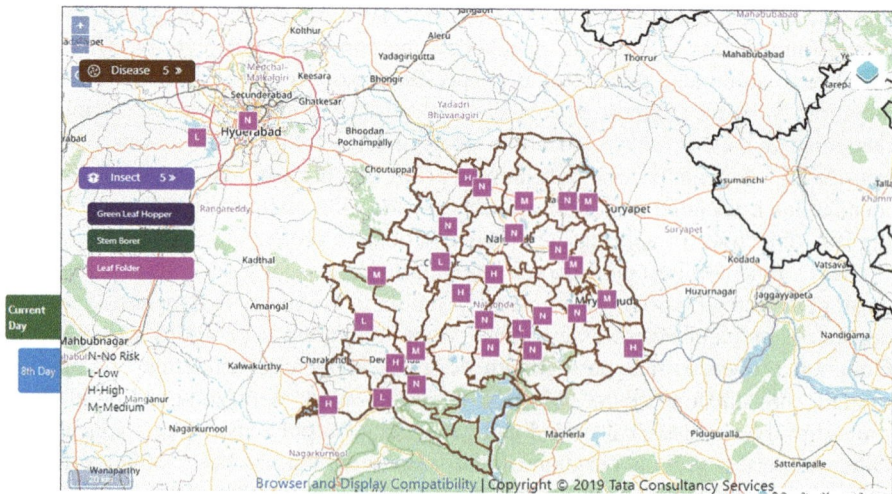

Fig. 1 The overall risk predictions at different severity levels for multiple pests in the Rabi season of 2019–2020

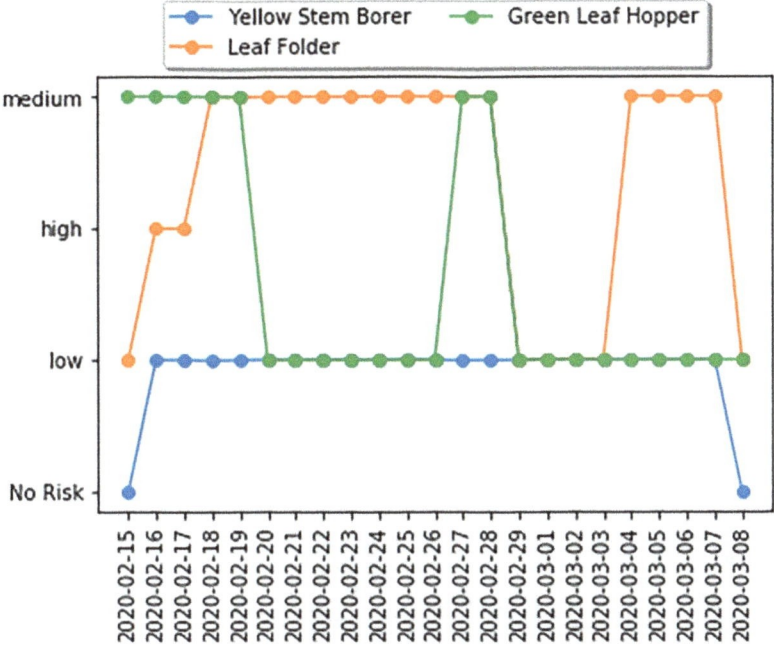

Fig. 2 Risk prediction for multiple pests for a given window across the Rabi 2019–2020 season

network architectures trained on very large image datasets. This has helped considerably to improve accuracy of (say) recognizing a key entity in an image (Krizhevsky et al., 2012). Detecting the severity of a pest infestation involves working to address various aspects of this challenge with models that depend on (data-light) traditional image processing techniques to (data-heavy) deep learning. The volume of available data has a significant bearing on how these models are configured for the best possible results while leaving enough room for the models to evolve as the data accumulates over time (Bhatt et al., 2019). Figure 3 shows the detected pests – *Helopeltis theivora* or tea mosquito bug (TMB) and *Tetranychus urticae* or red spider mites (RSM) – with corresponding confidence scores on tea leaves. Figure 4 shows the probability of each pest in the image.

1.2 Integration of Remote Sensing and Crop Growth Simulation Models to Estimate Crop Yield and Stubble Burning Severity

Cropping systems are constantly under pressure to feed the growing population. Irrigated cereal growing regions are intensively cultivated across the year leaving a very short window for pre-sowing activities such as stubble removal and tillage. As a

Fig. 3 Convolutional neural network-based pest detection on tea leaves with corresponding confidence scores

Fig. 4 Image-based pest identification and confidence score

result, after each cropping season, a large amount of crop residue is burnt in the fields, thereby increasing harmful emissions and damage to soil microbiome. Our earlier study on the monitoring of stubble burning in 2019 showed that a maximum number of fire incidences were reported from Fatehabad, Sirsa, Jind, Kaithal and Karnal in Haryana, India. Reports poured in from Sangrur, Bathinda, Ferozpur and Patiala. The Malwa region of Punjab and Hisar division of Haryana were more prone to burning. Moreover, analysis showed that 26 October–8 November is a peak time of burning in Punjab and is also the maximum period of burning in Haryana. In this study, we integrated the crop growth simulation model with the remote sensing observations to estimate biomass yield. The quantity of grain and biomass yield were estimated in conjunction with the crop phenology parameters such as start, peak and end of the season. The study was conducted in rice–wheat cropping systems of Punjab and Haryana, India. During the cropping season, information was

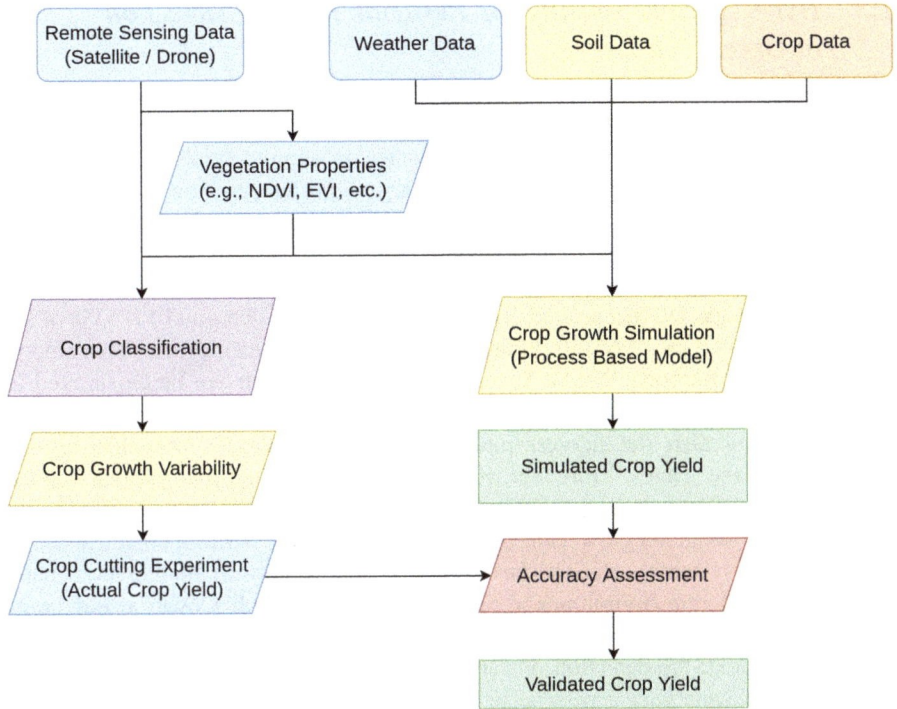

Fig. 5 Framework for crop yield estimation

disseminated to the local agencies and government departments to understand the extent of stubble burning (Fig. 5).

Technique: Enhancing the crop growth simulation models using remote sensing observations (spatialization of crop growth simulation models using remote sensing observations)

Usage: Early estimate of potential harvest. Effective crop residue management

Advantage for the grower: Fair price for the produce

Effect on the environment: Reduction in CO_2 emissions due to reduction in stubble burning

Food Security: Positive. Proactive corrective measures can be taken throughout the season

Transparency: Comply with Crop Certification Agencies, which track crop health, yield and residue management

Adoption obstacles: Unavailability of cost-effective and efficient equipment. Need for public policy decisions to improve adoption by the farming community

1.3   Hyperspectral Sensing for Pesticide Residue Detection in Green Tea Leaves

Tea is consumed worldwide due to its antioxidant properties and other beneficial health effects. More than 36 countries cultivate tea on a commercial scale (Tea Board of India, 2017). China and India are the largest producers; however, changing climate is causing a huge risk of pest/diseases in the tea industry. With changing weather conditions, tea plantations are vulnerable to biotic (insect pest, weeds, etc.) and abiotic (water, nutrient, etc.) stresses.

It has been reported that around 167 species of pests cause around 11–55% of the annual yield loss (Das, 1965). In India, pesticide usage for tea is greater than among all other crops (Barooah, 2011). Several synthetic chemicals are being sprayed to control pest infestation. However, uncontrolled use of such chemicals due to lack of knowledge and also the harvest interval are responsible for pesticide residue remaining on the leaves. Moreover, it has been also noticed that various banned chemicals/pesticides which are not recommended by the Tea Research Board are sprayed by the small tea growers as they are cheaper. These harmful chemicals enter various natural systems such as surface waters, groundwater, soil, human food chain, etc. In addition to the environmental risk, high pesticide residue levels also degrade the quality of manufactured tea.

In this context, the detection of various pesticide residues is crucial. The minimum residue levels vary across pesticides from few μg/L to few mg/L. To detect such small concentrations precisely, researchers have established methods using analytical instruments such as liquid chromatography–mass spectroscopy (LC–MS), gas chromatography–mass spectroscopy (GC–MS), high-performance liquid chromatography (HPLC) and immunoassay (ELISA) (Li et al., 2013). These methods are accurate but involve the destruction of the sample to be tested and take time as well as skilled labour. Considering the time and costs involved in such destructive techniques, researchers have developed various multi-residue analysis methods to test multiple residues in a single analysis (Chen et al., 2014, 2016; Hou et al., 2013, 2014). Chen et al. (2016) developed a gas chromatography–tandem mass spectroscopy-based method for residue analysis of 70 pesticides using a modified QuEChERS (quick, easy, cheap, effective, rugged and safe).

Recently, attempts have been made to use non-destructive techniques such as NIR spectroscopy and hyperspectral sensing. Although technologies such as hyperspectral sensing (HS) are not as accurate as LC–MS and GC–MS, they provide the results in a very short time and in a non-destructive way, so the techniques can be used in near real/real time for the detection of pesticide residues. Hyperspectral sensing (HS) has been used to detect pesticide residue in crops such as mulberry (*Morus alba*), grapes (*Vitis vinifera*), apple (*Malus domestica*) surface and vegetables (Lv et al., 2018; Jun et al., 2016; Mohite et al., 2017; Dhakal et al., 2014). Although HS has been applied with multiple crops and residues, very few attempts have been made to use the method for tea. The objective of this study is to detect three banned pesticides – acetamiprid, cypermethrin and monocrotophos – using a noninvasive hyperspectral sensing-based method and propose an operational approach to monitor the green tea leaves across various processing phases to enable the traceability (Mohite et al., 2019).

To make the system reliable and operational, we have collected the data in a closed environment using light-emitting diodes (LED) as the source of radiation. A deep learning-based approach was used for the detection of individual pesticides. We claim that the device and method can be easily implemented for near real-time operational use to detect multiple pesticide residues (Fig. 6).

Technique: Using cost-effective hyperspectral sensing for pesticide residue detection

Usage: Non-destructive and faster assessment

Advantages for the grower: Awareness regarding effects of uncontrolled agro-chemical usage Residue-free production. Increased price. Greater acceptance and uptake in the market

Effect on the environment: Reduction in pesticide residue in environment

Food safety: Reduction in pesticide residue

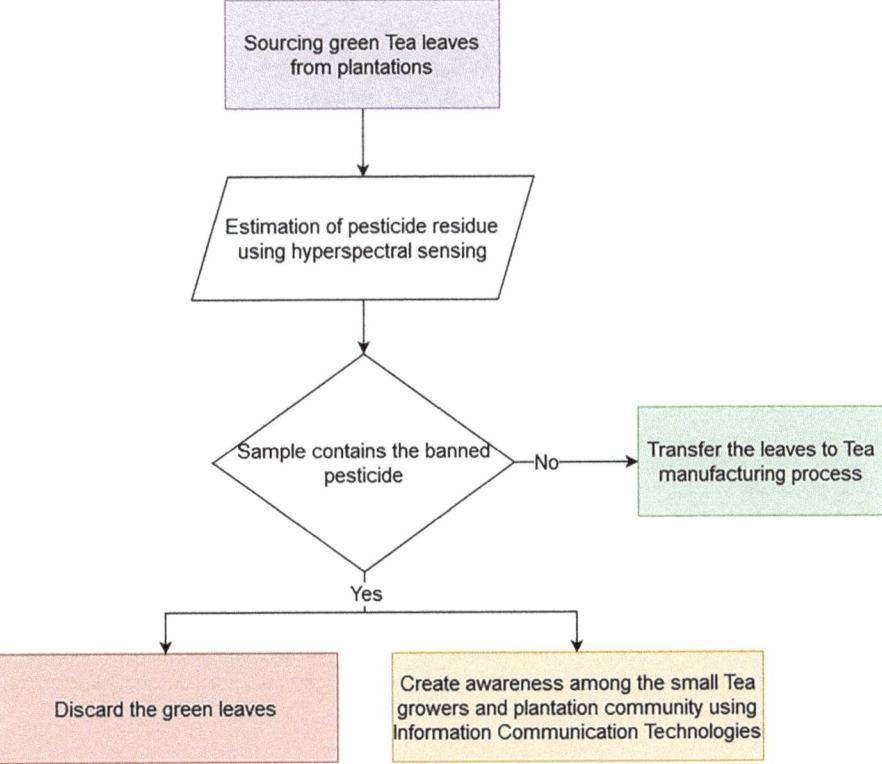

Fig. 6 Use of hyperspectral sensing for pesticide residue detection in tea

Transparency: Comply with the food safety standards and maintain traceability
Adoption obstacles: Certification by implementation agencies

Conflict of Interest The authors declare that they are employed by the company which owns the commercial product that is being discussed.

References

Barooah, A. (2011). Present status of use of agrochemicals in tea industry of eastern India and future directions. *Science and Culture, 77*(9–10), 385390.

Bhatt, P. V., Sarangi, S., & Pappula, S. (2019). Detection of diseases and pests on images captured in uncontrolled conditions from tea plantations. In *Autonomous air and ground sensing systems for agricultural optimization and phenotyping IV* (Vol. 11008). International Society for Optics and Photonics.

Carlson, G. A. (1970). A decision theoretic approach to crop disease prediction and control. *American Journal of Agricultural Economics, 52*(2), 216–223.

Chen, H., Yin, P., Wang, Q., Jiang, Y., & Liu, X. (2014). A modified QuEChERS "sample preparation method for the analysis of 70 pesticide residues in tea using gas chromatography-tandem mass spectrometry". *Food Analytical Methods, 7*(8), 1577–1587.

Chen, H., Hao, Z., Wang, Q., Jiang, Y., Pan, R., Wang, C., Liu, X., & Lu, C. (2016). Occurrence and risk assessment of organophosphorus pesticide residues in Chinese Tea. *Human and Ecological Risk Assessment: An International Journal, 22*(1), 28–38.

Das, G. (1965). *PESTS of tea in north-east India and their control*, 27.

Dhakal, S., Li, Y., Peng, Y., Chao, K., Qin, J., & Guo, L. (2014). Prototype instrument development for non-destructive detection of pesticide residue in apple surface using Raman technology. *Journal of Food Engineering, 123*, 94103.

Elijah, O., Rahman, T. A., Orikumhi, I., Leow, C. Y., & Hindia, M. N. (2018). An overview of internet of things (IoT) and data analytics in agriculture: Benefits and challenges. *IEEE Internet of Things Journal, 5*(5), 3758–3773.

Hou, R. Y., Jiao, W. T., Qian, X. S., Wang, X. H., Xiao, Y., & Wan, X. C. (2013). Effective extraction method for determination of neonicotinoid residues in tea. *Journal of Agricultural and Food Chemistry, 61*(51), 12565–12571.

Hou, X., Lei, S., Qui, S., Guo, L., Yi, S., & Liu, W. (2014). A multi-residue method for the determination of pesticides in tea using multi-walled carbon nano tubes as a dispersive solid phase extraction absorbent. *Food Chemistry, 153*, 121–129.

Jun, S., Shuying, J., Meixia, Z., Hanping, M., Xiaohong, W., & Qinglin, L. (2016). *Detection of pesticide residues in mulberry leaves using vis-nir hyper-spectral imaging technology.* Journal of Residuals Science Technology.

Krizhevsky, A., Sutskever, I., & Hinton, G. E. (2012). Imagenet classification with deep convolutional neural networks. In *Advances in neural information processing systems.*

Li, X., Zhang, Z., Li, P., Zhang, Q., Zhang, W., & Ding, X. (2013). Determination for major chemical contaminants in tea (Camellia sinensis) matrices: A review. *Food Research International, 53*(2), 649–658.

Lv, G., Du, C., Ma, F., Shen, Y., & Zhou, J. (2018). Rapid and non-destructive detection of pesticide residues by depth-profiling fourier transform infrared photoacoustic spectroscopy. *ACS Omega, 3*(3), 3548–3553.

Mohite, J., Karale, Y., Pappula, S., Shabeer, A., Sawant, S., & Hingmire, S. (2017). Detection of pesticide (cyantraniliprole) residue on grapes using hyper-spectral sensing. In *Sensing for agriculture and food quality and safety IX* (vol. 10217, p. 102170P). International Society for Optics and Photonics.

Mohite, J., Sawant, S., Borah, K., & Pappula, S. (2019). Temporal detection of pesticide residues in tea leaves using hyperspectral sensing. In *IEEE International Geoscience and Remote Sensing Symposium (IGARSS 2019)* (pp. 7274–7277). IEEE.

Pappula, S. (2013). The concept of PRIDE™ empowering farmers to live with pride. *CSI Communications, 37*(8).

Szegedy, C., Toshev, A., & Erhan, D.. (2013). Deep neural networks for object detection. In *Advances in neural information processing systems.*

Tea Board of India 64th annual report 2017–18, URL: http://www.teaboard.gov.in/pdf/64th_Annual_Report_2017_18_pdf4214.pdf. Accessed on 20 May 2020.

Dacom Precision Agriculture System

Frenk-Jan Baron and Lud Uitdewilligen

Abstract The Dacom Farm Management Information System (FMIS) provides farmers with detailed advice on crop protection, crop growth, irrigation management and precision farming. The data are collected at one central platform which is connected to many partners. The user has all data required in one place, and using the different advice modules, the data are transformed into management information. This information helps the user to apply chemicals, water and fertilizer at the right time, in the right place and in the right amount. The user can save on agri-inputs, and with the best possible crop care, the yields will increase. By connecting the Dacom platform to the farm machinery, the data can be exchanged via the cloud, and working with variable-rate application maps becomes easier.

1 Introduction

Dacom provides an integrated precision agriculture software system for their customers. Among these customers are farmers, contractors, food processing companies, the agri-supply industry and crop advice companies. The Dacom crop recording platform is used to collect the necessary data to run the different models. Based on the location of the fields, the crops, the varieties and the operations (e.g. seeding, fertilizer), the models can generate advice for the customers. At this moment, the two following models are operational and used by customers.

F.-J. Baron · L. Uitdewilligen (✉)
Dacom Farm Intelligence, Groningen, The Netherlands
e-mail: info@dacom.nl

1.1 Fungal Disease Prediction Model

This model is used to generate timely advice to prevent an infection of the crop. This model is available for more than 40 fungal diseases in more than 20 crops. The history of this model goes back to 1980 when it was developed together with the Wageningen University and Research. Since then, the model has been constantly under development and has been expanded with new modules and technologies. The core of the model is calculation of the life cycle of the fungus. Based on the location of the field, the nearest weather station is selected. In Western Europe, Dacom has more than 100 weather stations placed in regions where potatoes are cultivated. In other parts of the world, Dacom works with the weather data of private weather station. The measured and predicted weather are combined with the amount of unprotected leaf area and the disease pressure to calculate the infection risk (Fig. 1). Where a large infection risk is predicted in the coming days, the farmer gets an advice to spray a preventive chemical. If the advice is ignored and the infection has taken hold, a curative chemical is advised.

The model for Late Blight in potato is the one most used. The crop observations (crop stage and crop growth) needed for reliable advice are generated automatically by the crop growth model. So, it is no longer necessary to fill in these observations manually. In addition, the application of chemicals can be recorded automatically using ISOBUS machinery. Recently, the model was extended to be able to give an advice on zones within the field when conditions differ between them. Examples are

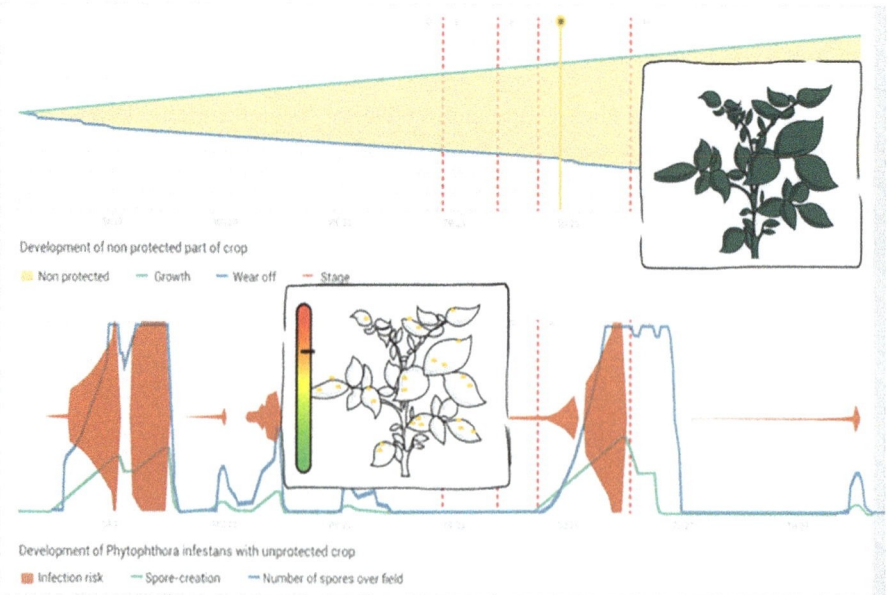

Fig. 1 Visual representation of the different parts of the Dacom Fungal disease model

less growth in shaded areas under trees, microclimate under trees and parts of the field under irrigation (this is automatically recorded using, for example, Raindancer irrigation technology). This information is combined with crop growth observed by satellite to give advice on variable-rate application (VRA) of chemicals.

1.2   Crop Growth Model

The Dacom crop growth model is based on the well-known LINTUL and WOFOST models (Kooman and Haverkort, 1995, Van Diepen et al., 1989). The necessary inputs come from the crop recording, weather stations and weather prediction. The model is used to generate weekly crop observations for the disease model. It is also used to predict the final yield and to analyse the effect of different weather scenarios on the final yield. This service was developed together with food processing companies that wanted insight into the expected yield. Those companies already have the Dacom Farmlook software in which they collect all the crop recording data of their growers for food safety programmes and data analysis. This platform has been expanded with the yield prediction model and possibilities to study the effects of different weather scenarios. Satellite images are used to validate the model predictions for the above-ground biomass. When, for example, the calculated emergence date does not match the leaf area index (LAI) measured by satellite, the model is adjusted. The same is done at the end of the season when the calculated senescence does not match the LAI from satellite images. This model adjustment is done manually by looking at the modelled and measured LAI graphs. This is a time-consuming process which will be automated in 2022. The model also has an elaborate water balance component that can also be used to give irrigation advice. Where there are differences in growth within the field, variable-rate irrigation based on the predicted water uptake is possible.

1.3   Development of Dacom Software System

The outcomes of the aforementioned models are combined with task maps in the Cloudfarm software. In this way, the VRA of agri-input is no longer based only on the experience of the farmers and advisors but on scientific models. In this way, precision agriculture can be combined with decision support systems (DSS) to result in decision agriculture. The strong point of the Dacom software is the full integration between the different parts, so a farmer no longer needs stand-alone apps for each application (Fig. 2). The complete farm can be managed in one software package which combines state-of-the-art technology both on the model and task map generation sides.

Dacom Farm Intelligence was established in 1980. During the past 40 years, Dacom developed into a Software as a Service (SaaS) company. Dacom software is

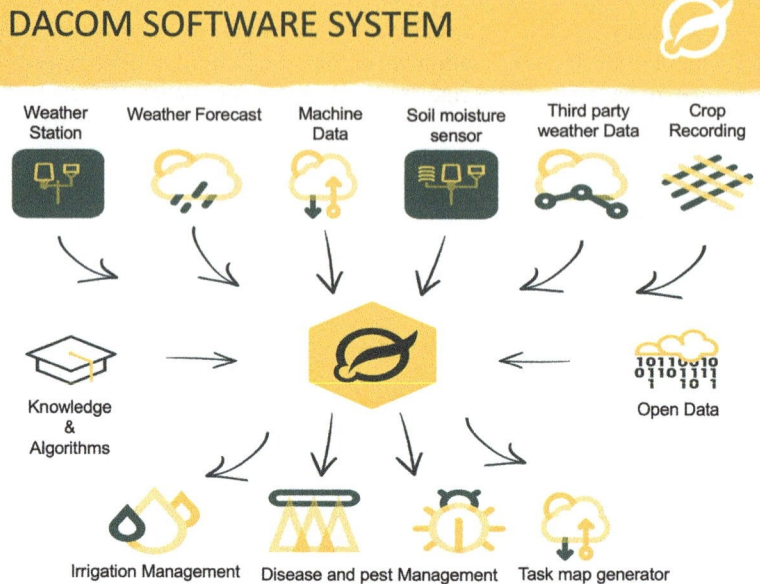

Fig. 2 Overview of the Dacom software system

currently used on approximately 26,000 farms and 800,000 crop fields and in 40+ countries around the world. In 2021, the company was bought by CropX. The SaaS solutions from Dacom are applicable worldwide as it is designed with internationalization in mind. The platform is currently translated into eight languages. However, it also offers dedicated options for localization to provide the best possible crop recording experience for the farmer. An example is the use of open datasets provided by government institutions to make it easy to enter data of crop field boundaries.

The company focus in Western Europe has shifted to arable farms aiming to grow their business based on data-driven farming. To boost the adoption of digital farming, customers can start with a relatively simple product 'Boer&Bunder'. This product shows ready-to-use data for all crop fields in the Netherlands, Belgium, France and Denmark based on open data, satellite data and weather and disease models. Farmers that want to log and apply farm data on their own farm can upgrade to Dacom crop recording. Dacom has a feature-based subscription model which allows customers to scale up with product as they expand and grow to the next level of data-driven farming. This includes more options for automatic crop recording and the use of advice models in combination with task maps.

Dacom also believes that to be able to communicate, process and analyse farm data, the data should be stored in a strongly structured way. Dacom maximizes the use of standardized crop data. The crop recording data are stored in the EDI-crop file format[1]; this standard provides coding lists for chemicals, fertilizers and methods like spraying, ploughing, etc. As-applied machine data are mainly stored in the ISO-xml file format. This also makes connections with other farm management systems (FMS) more viable. This global approach is unique and especially interesting for international farm enterprises and other agribusiness partners of Dacom.

Digital crop recording is mainly used by customers to comply legislation and food safety programmes from food processing companies. Often the same crop data are shared with multiple supply chain partners saving on time by sharing data from one single source. A key objective of Dacom is to maximize automatic data collection at the farm level. The feature-based subscription model makes it possible for farmers to scale in features that contribute to time savings and more accurate data collection. Examples are the following:

- Open data makes it easier and faster to enter data of crop field boundaries and soil types.
- Third-party integrations allow farmers to import automatically soil analysis, chemical or fertilizer stocks or contracts from suppliers.
- Satellite data are used to monitor crop growth and detect structural or in-season variation.
- Weather data and forecast are used throughout the Dacom platform. One example is that crop recordings are enriched with local weather conditions as a requirement for GLOBALG.A.P. certification.
- For planting and irrigation purposes, the soil moisture content and temperature are predicted by models or measured more accurately by connecting soil moisture sensors.
- Automatic time tracking (with mobile devices) is possible for employees that carry out scheduled tasks for larger farms and contractors.
- A recent development is the connection of machinery in order to send task data wirelessly to tractor devices and logging as-applied data to automate crop recordings.

In addition to fast and accurate data collection, the use of advice models in combination with task maps for tractors results in several advantages:

- Optimal usage of agri-inputs like fertilizers, chemicals, water and fuel.
- Yield and quality improvement.

Advantages of variable-rate applications in arable farming may not be the same in all years and on all farms due to changing conditions: crop rotation, crop varieties, treatments and weather and soil conditions. Some examples of variable-rate applications in the Netherlands (like variable planting of potatoes) can be found on www.proeftuinprecisielandbouw.nl.

The use of advice models in combination with task maps is important to contribute to circular agriculture, which is the vision of the Dutch Ministry of Agriculture,

---

[1] https://www.agroconnect.nl/en-us/productgroups/cropcultivation.aspx

Nature and Food Quality.[2] Variable-rate application and soil improvement focus on good yields and the sparing use of agri-inputs. Furthermore, it puts less pressure on the environment, nature and climate. Although precision agriculture is in the farm management software of Dacom, it is still not used by all farmers who use the Dacom system. Many farmers are still not convinced that the benefits outweigh the cost of the software. Dacom also discovered a gap between this new highly technical precision agriculture system and a large part of the older and less digitally minded user group. Despite all the efforts on data standardization and exchange of machine data, many farmers still struggle with getting the variable-rate application maps working on all different tractor terminals. Support through machinery manufactures and their dealers is key in the adaptation of precision technology by this large group of farmers.

Conflict of Interest The authors declare that they are employed by Dacom Farm Intelligence.

References

Kooman, P., & Haverkort, A. (1995). *Modelling development and growth of the potato crop influenced by temperature and daylength: LINTUL-POTATO. Potato ecology and modelling of crops under conditions limiting growth* (pp. 41–59). Springer.
Van Diepen, C. V., Wolf, J. V., Van Keulen, H., & Rappoldt, C. (1989). WOFOST: A simulation model of crop production. *Soil Use and Management, 5*, 16–24.

---

[2] https://www.rijksoverheid.nl/actueel/nieuws/2018/09/08/minister-schouten-wil-omslag-naar-kringlooplandbouw-nu-inzetten

Akkerweb and farmmaps: Development of Open Service Platforms for Precision Agriculture

Thomas H. Been, Corné Kempenaar, Frits K. van Evert, Idse E. Hoving, Geert J. T. Kessel, Willem Dantuma, Johan A. Booij, Leendert P. G. Molendijk, Fedde D. Sijbrandij, and Koen van Boheemen

Abstract The development of the Akkerweb service platform (https://akkerweb.eu) was started around 2010. It is an open platform in precision farming, providing the maps, services, data, and connections required, in principle, for any smart farming application envisioned. This includes background maps, services for weather data, satellite images, soil maps, crop polygons, etc., but also visualization tools, an app store, a task map generator, and crop growth models. Akkerweb provides the infrastructure needed to develop an application easily using the available services and to publish it on the Akkerweb platform. Moreover, Akkerweb applications can also run on other websites, seemingly as stand-alone applications with the look and feel of the customer's website.

A unique point of the Akkerweb service platform is the availability of several science-based agronomic models which are currently made available as APIs for use in smart farming applications. Examples of these models are those to calculate water availability (Watbal model), potato crop growth (Tipstar model), late blight infection (Blight module), and nematode management (Nemadecide) at individual field and within-field levels. Other models are available for variable-rate application of soil

T. H. Been · C. Kempenaar (✉) · F. K. van Evert · F. D. Sijbrandij · K. van Boheemen
Agrosystems Research, Wageningen University & Research, Wageningen, The Netherlands
e-mail: corne.kempenaar@wur.nl

I. E. Hoving
Livestock & Environment, Wageningen University & Research, Wageningen, The Netherlands

G. J. T. Kessel · J. A. Booij · L. P. G. Molendijk
Open Field Crops, Wageningen University & Research, Wageningen, The Netherlands

W. Dantuma
Dobs Automatisering b.v, Wageningen, The Netherlands

© The Author(s), under exclusive license to Springer Nature Switzerland AG 2023   269
D. Cammarano et al. (eds.), *Precision Agriculture: Modelling*, Progress in Precision Agriculture, https://doi.org/10.1007/978-3-031-15258-0_16

herbicides and fungicides against blight, nitrogen top-dress application in potato and potato haulm killing. In total, Akkerweb has more than 30 apps.

farmmaps (https://farmmaps.eu) is the next version of Akkerweb. It has a new data repository and management system as well as a new, more intuitive dashboard design running on all devices. It became available first in 2021.

Abbreviations

AWS	Amazon web services
CI	chlorophyll index
FMIS	farm management information system
GIS	geospatial information system
GPU	graphical processing unit
ICT	information and communications technology
NSS	nitrogen side dress system
UAV	unmanned aerial vehicle
WDVI	weighted difference vegetation index

1   Introduction

Precision agriculture (PA) is a farming management concept based on observing, measuring, and responding to inter- and intra-field variability in crops (Wikipedia, 2020). It is expected to lead to an increase in productivity and quality, as well as increased sustainability (Van Evert et al., 2017a, b). Adoption of PA is slower than expected because it is complex to apply, among other issues (EIP, 2015). Kempenaar et al. (2018) listed at least 13 different PA applications in potato production. All these PA applications have data, a decision support, and an actuation component and cannot be carried out without information and communications technology (ICT). Many hard- and software tools have been developed and marketed in the last 25 years to capture data on variation in crops. The technologies enable measuring and viewing of site- and time-specific soil, crop, climate, and input and output variables and using these data in operational, tactical, and strategic decisions. To start with, a farmer must have a suitable farm management information system (FMIS) (Fountas et al., 2015) to use the data. The FMISs have developed from simple crop-recoding software tools on stand-alone computers to web-based database platforms with geo-spatial information system (GIS) features and connections to data suppliers. In the Netherlands, we identified at least 25 of such platforms in arable farming. They are marketed by different types of institutions such as advisory, sensor, crop protection and fertilizer companies, machine companies, farmers' cooperatives, and knowledge institutes. The Wageningen University & Research also developed such a platform in cooperation with the Agrifirm Plant, a farmers' cooperative, and DOBS automatisering, a software company, with the aim of developing decision support for precision faming as close as possible to Dutch farming practices and to work as much as possible with the state-of-the-art of data that are available on farms.

In this chapter, we describe the development of the Akkerweb and **farm**maps precision agriculture service platforms. In Sect. 2, we give a brief history and technical implementation and features of the platforms and in Sect. 3, a selection of models and apps that can be used on the platforms. We end in Sect. 3 with an outlook on how data platforms will contribute to mature data-driven precision agriculture.

1.1 *Development of Akkerweb and farmmaps Platforms*

The development of Akkerweb was set in motion by two projects more than 10 years ago. The first was a pilot developed for *Boerenbond Helden*, a farmers' cooperative that needed a GIS to match tenants with landlords. The prototype system depicted one's own fields as well as fields available for rent. Requests could be made by interested farmers to open these fields for review, presenting all necessary information, such as soil type, fertilization history, cropping scheme, diseases, etc. The second initiative came from the NemaDecide project (Been & Schomaker 2004; Been et al., 2005) and the development of a decision support system (DSS) for potato cyst nematodes and root knot nematodes, both quarantine organisms. This required a geographical component identifying the field and the exact location and dimension of the sampled areas, regardless of being disease free or infested, to produce the correct advice needed to manage these pests.

The Wageningen University & Research joined forces with Agrifirm Plant, a farmers' cooperative, in 2010 to develop a service platform for precision agriculture, named Akkerweb. Software programming was done by DOBS. By 2016, the partners concluded that Akkerweb was mature enough to be launched on the Dutch market as an open service platform for precision agriculture. Users can create an account for no charge in which they can manage public and private data, do crop monitoring, and apply precision agriculture applications, for example, to make variable-rate prescription maps. Today Akkerweb has over 5.000 users, mainly Dutch arable farmers, farm advisors, and researchers. It has undergone several technical modifications, and currently, Akkerweb 3.0 named **farm**maps with a new data repository and management system with a new, more intuitive dashboard design, has become available in 2021.

1.2   Features of the Platforms

Akkerweb is a cloud-based service platform on which apps provided by different developers can run. It provides basic apps and services (e.g., weather data) together with many specific applications. Table 1 provides a list describing the services and features that are available to every user, while Table 2 lists the applications Akkerweb provides free of costs. These are considered the basic instrumentation every user and developer needs. The landing page and dashboard of Akkerweb are shown in Fig. 1.

Table 1 List of services and features freely available to use

Feature	Description
Account grower	Akkerweb offers an individual login for accounts; this means that only the user (grower) has access to the files he owns. Via "single sign-on," the user is directly logged into other services and applications that are relevant for him/her; this avoids repeated entering of user account credentials
Background maps	Displaying geographical information against a freely chosen background: Google maps, OpenStreetMap, Bing, Top10, or any other background
Crop polygons	Can be downloaded using various web services, directly from the Dutch government (RVO) or by public reference map layers of crop fields (most EU countries)
Contact details	Once entered, data are available in all applications the grower wishes to use
Downloading or drawing fields	Fields can be uploaded from Dutch FMIS CROP-R and CropVision (upload xml message) or added using drag and drop. (formats supported: Shape files, Google polygons, etc.). If necessary, crop polygons can be drawn by hand
Crop cultivation data	If not available by upload from an FMIS, Akkerweb provides a simple crop rotation app to provide advisory systems with cropping information (crops, varieties, planting date, growing period)
Sensor data	Tractor logs, data via .csv, .xlsx, .dbf, .dat, but also satellite data (NVDI, WDVI, yield maps (geotiff)) can be uploaded. Addition of a new format will take half a day
Editing	Standard functionality for editing geographical information; edit, merge, or split fields, add objects, and convert grid files and other formats into visual information
Links	Web services to agro partners are available, a.o. ISAcert, Eurofins, WEnR, NEO, Dacom and AgroVision. VanderSat, Agrirouter, etc
Storage	All data remain property of the grower/owner and can be used or stored for later application. Data are stored on Amazon AWS in Ireland
Output	Generic output is possible. As .csv file, as .pdf for pictures, etc. this functionality is available to each application
Applications	The grower can use the free and the commercial applications with the plot as his starting point

(continued)

Table 1 (continued)

Feature	Description
ISOBUS, shape, and surfer	Task maps can be generated for ISOBUS, shape, and surfer format
Sharing information	Growers can share their geographical information, time-limited if necessary, completely or partially, with third parties (neighbor, cooperation, or payroll employee
Send/retrieve data	Edi-crop v 4.0 and the eLab messages are supported. Using services to business management systems prevents repetitive data input
Uptime and scalability	The Akkerweb platform is hosted on Amazon servers under European law. This ensures an uptime of almost 100% and instant scalability (number of users) and provides the need for fast processing of geographic data
Weather	Available are 40 years of historical data (30 by 30 km grid) and observed and forecasted weather (500 by 500 m) worldwide. Rain radar information is included for the Netherlands
Available information and models	Availability of soil maps, crop growth models, etc. crop polygons from 2009–now, agricultural acreage 2013

Table 2 Freely available Akkerweb applications (apps)

App	Description
Cropping plan	Entering grower's fields plus crops, cultivars, soil type, planting, and harvesting date either manually or via web services connecting to official bodies and farm management systems.
Map	Overview of all of a grower's fields on a map. Add layers of information from your applications, for example, all fields that have been sampled, all fields with sensor data, etc. click on a field polygon and all information available will be listed.
Sensor data	Upload all types of sensor data and visualize the information. Use these data in various applications on Akkerweb.
Satellite	Download satellite imagery – WDVI and NVDI biomass maps for further use in Akkerweb, for example, haulm killing (currently WEnR and NEO WCS services).
Contractor	Order sensor information, EM38 (v/d Borne), Veris scan (Agrometius) and NVDE, and WDVI drone images (dronewerker.nl) directly from Akkerweb, and receive the results in your account.
My advices	Advice generated in the advisory apps used by third parties and not available to the farmer can be stored and retrieved for later use.
Charing information	Connect with another Akkerweb user and share information in different ways (time bound, editable or not, etc.).
Store	Here all available applications can be found and downloaded to the account holders desktop, application can be for free, licensed, sponsored, etc.

Akkerweb apps can be integrated in several ways within the framework. In its most simple form, a third-party app runs in an i-frame and can be found in the Akkerweb Store and added to the dashboard if needed. The next step is an external

(a)

(b)

Fig. 1 Impression of Akkerweb 2.0. (**a**) Landing page and (**b**) a user dashboard

app that uses some of the services (Akkerweb APIs), for instance, the farmer's cropping scheme, but has no connection to the other apps. When fully "integrated," Akkerweb apps can share their information with other apps. For example, new layers generated by a fully integrated app are available for use by other apps.

The data model of Akkerweb is the reference model Agro (rmAgro) developed by Goense (2017). This standard facilitates the connections with other platforms, portals, and agribusiness companies that comply with rmAgro for laboratory results, cropping schemes, advice, and cultivation messages (e.g., fertilizers, irrigation, crop protections, and yields).

Several functions and services are encapsulated in Angular software components and can easily be added to a new app. Developers can now concentrate on the desired interface and implementation of their own knowledge. In **farm**maps, using Angular by default, a component library is available with interface objects that can be reused.

1.3 Technical Implementation

Akkerweb is a web-based platform based on a client-server architecture. There are currently two generations of Akkerweb applications, using a slightly different technology stack.

1.3.1 Authentication/Authorization

Akkerweb uses a claim-based authentication and authorization scheme; user identities are provided in the API requests as JSON Web Token (JWT), provided by an OpenID Connect compliant identity server.

1.3.2 Server Side

The business logic of the first generation was implemented in .Net C# using the NetTopology suite for the spatial operations. Some computation-intensive parts were offloaded to parts implemented in C / C++ on a CUDA server. The server side of second-generation Akkerweb applications is also implemented in C# (.Net Core), but NetTopology suite has mostly been replaced with GDAL (GDAL/OGR contributors, 2021).

1.3.3 Server Side, Data Persistence Layer

Akkerweb runs on Amazon AWS servers under European law. It also allows for robust availability and scalability. Amazon was chosen specifically because only its servers were equipped with GPUs for graphical calculations. Most of the Wageningen University & Research apps run on Amazon, but app developers can also run their apps on their own servers.

Apps can run on Akkerweb and can be stand-alone. A digital request for soil sampling for instance runs on Akkerweb, where it is connected with the crop polygons of the farmer, but is also on the website of the soil sampling agency offering this service. Data persistence is implemented using a Postgres/Postgis database server for the meta-data.

1.3.4   Client Side

The client side of Akkerweb applications of the first generation is implemented in a multiple page pattern using JavaScript and HTML5/CSS supported by toolkits as jQuery, Knockout js, and OpenLayers 2. Applications of the second generation are implemented as a SPA (single page application) using Angular 9+, typescript, HTML5 /CSS, bootstrap 4, and OpenLayers 3+/Cesium. Cesium is used for the 3D visualizations.

1.3.5   Client Server Communication

In Akkerweb applications, all communication (between the client and the server and between applications) is based on RESTful services with the payload formatted as Json/GeoJson. Most geographical data are transported following different OGC standards such as WMS, WFS, WMTS, and WCS, but in the APIs, a GeoJson-based representation is used. SignalR on a WebSocket transport is used for push communications (events).

2   Models and Apps on Akkerweb/farmmaps

There are currently more than 30 apps available on Akkerweb. See Table 3 for an overview of the most used apps. Some of these are Akkerweb apps which are free for use and provide basic functionality. The other apps are provided by WUR and external partners who use the platform to enable their applications. Some agro-ecological models in webservices are linked to Akkerweb and **farm**maps. Hereafter, we describe two models and five apps that run on Akkerweb and **farm**maps. The models deliver information on soil and crop status to the apps.

2.1   WatBal Soil Moisture Balance Model

WATBALsig is a simple water balance model for an unsaturated or saturated soil profile. It is derived from the WATBAL model (Berghuis-van Dijk, 1985) which was primarily developed to produce dynamic hydrological data, such as moisture content, over a long time scale for a separate agricultural nitrate leaching model. Vinten (1999) used WATBAL to simulate nitrate leaching from soils with different textures. These qualities make WATBAL suitable for other modeling purposes such as irrigation advice and grass or crop growth prediction (see Sect. 3.4.2, Chap. 3).

Complex agro-hydrological models for predicting moisture content like SWAP (Kroes et al., 2008) have the disadvantage that rather detailed input is needed. The

Table 3 List of WUR and third-party applications available in the Store

Application name	Short description
Digital sampling request	Nematodes, phosphate, and white rot
Stripbuilder	Program to subdivide fields into sampling units for nematodes, phosphate, white rot, tracks, and buffers, in order to supply sampling maps and display results
NemaDecide Geo	Free version of a decision support system for the potato cyst nematode
NemaDecide Geo PLUS	Version of NemaDecide including root knot nematodes (*Meloidogyne chitwoodi*) and the root lesion nematode (*Pratylenchus penetrans*) with extra features
Agrifirm mineral	Calculate fertilizer needs, based on crop, soil type, and acreage for your whole farm
AgrifirmGBM	Calculate crop protection product needs, based on crop, soil type, and acreage for your whole farm
Task map nematostats	Based on soil sampling results, a task map will be calculated including official delimitation
Potato information	More than 400 potato cultivars and all their properties, including partial resistance against potato cyst nematodes, blight, etc., data querying and links to the breeders
Task map haulm killing	Biomass-dependent haulm killing of potatoes based on NDVI and WDVI originating from satellite and E-bee
Task map side dress nitrogen	Biomass-dependent task map for nitrogen side dress for potatoes based on NDVI and WDVI originating from satellite, E-bee, or Yara images
Task map herbicides	An application map based on lutum (clay particles <2 µm) and organic matter content of the soil
Get hold of your grass production	Measure your grass height georeferenced using your mobile and calculate the Feed Wedge on Akkerweb
Bioscope	A service that provides a farmer with different products, among which is an WDVI-green biomass map, every 10 days. Imagery originates from satellites or from drones, when satellite images are unavailable
Late Blight	A completely new state-of-the-art decision support system to avoid yield losses by *Phytophthora infestans.*
Task map Blight treatment	The late blight app can be extended with the variable-rate app
Task map lime application	Calculate task map for site-specific need for potassium based on pH map (Veris sensor)
Grip on Data	Drop word, excel, and pdf files on your plot and make them georeferenced. Click on your plot and have a look at all your plot data
Mycokey App	Comprehensive monitoring of mycotoxins at critical control points in the food/feed chains for wheat and maize (DON, Aflatoxin B1, and Fumonisin)
Tipstar	Crop growth model for potato (growth is limited by the availability of water and nitrogen)
WatBalSig	Dynamic prediction of moisture content depending on precipitation, evapotranspiration, irrigation, and drainage
GAOS	Calculate the optimal driving path of your tractor over the field, avoiding obstacles, and download it to your tractor
Zoning	Read in sensor data for your field, calculate treatment zones, and make a task map yourself

precision and very small time-step ($\gg$ 0.1 day) with which they describe the water balance are superfluous compared to the precision that is needed to simulate crop transpiration on a daily basis depending on soil moisture content and practical precision farming purposes. Therefore, a model is required that describes the water balance of a soil profile more globally. Such a model should meet the following demands (adapted from Berghuis-van Dijk (1985)):

- Input of only the most important soil characteristics
- Good simulation of water balance of the root zone, from which crop transpiration is subtracted at a rate which is affected by soil moisture content
- Good simulation of movement of the groundwater table, which is a condition for simulating boundary water fluxes and a proxy for judging the quality of water balance simulations
- Quantification of real evapotranspiration, capillary rise, drainage (subsurface), irrigation, and upward and downward seepage, which are important for correct simulation of the soil water balance and management
- Short computing time for facilitating real-time interactive exchange with the user and the possibility of developing a future self-learning system for enhancing simulation quality

The WATBAL model simulates the water balance of a (cropped) soil in a simple and fast way, dividing the soil profile into two layers. The first layer represents the root zone and the second layer the subsoil to at least the depth of the deepest groundwater table. The model calculates analytically on a daily basis per sub-time step ($\leq$ 1 day) the changes in moisture content of the two layers and the movement of the groundwater table. Calculations depend on precipitation, evapotranspiration, capillary rise from layer two to layer one, water transport from layer one to layer two, and possibly drainage to differently defined drainage systems (including a bottom boundary).

WATBALsig is developed from WATBAL by implementing some new features and adaptations to meet the specific requirements for the intended use of the model. The most important of these are the following:

The dependence of crop transpiration on soil moisture is described with the well-known Feddes model (Feddes et al., 1978)

The first layer is divided into a root zone with dynamic depth and a zone without roots

The model is made more robust for special conditions such as perched groundwater (layer one saturated and layer two unsaturated)

For precision farming purposes, WATBALsig has been made operational by a user interface that runs on Akkerweb/**farm**maps (Fig. 2). To arrange field-specific input, the web application provides weather data and the crop concerned from Akkerweb/**farm**maps and soil data and hydrologic features from digital geographic maps of the Wageningen University & Research, department Environmental Research. The next step is to build into WATBALsig all kinds of precision farming Internet devices.

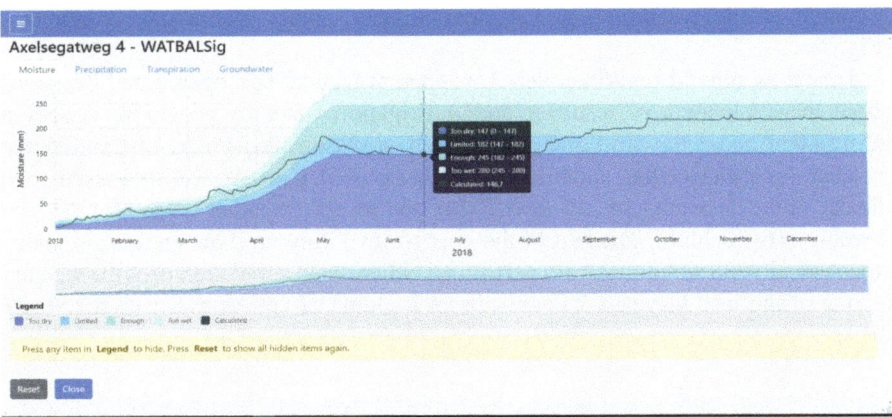

Fig. 2 WatBal provides the water requirement related to the crop grown, here potato. The dark line represents the calculated water requirement of the crop that should be maintained at the "Enough" level. Currently, water requirement can be calculated for more than 10 major crops and, when using weather prediction, irrigation advice, and both the amount and time to apply, can be provided

2.2 Tipstar Water- and Nitrogen-Limited Potato Crop Growth Model

Tipstar is a potato-specific model that simulates crop growth, soil water dynamics, and nitrogen dynamics for crops and soil. It was originally developed to support decision-making in starch potato production (Jansen, 2008) and has recently been updated (Van Oort et al., 2020). The model is unique in that growth of the canopy takes precedence over growth of other organs. Thus, tuber initiation and growth of potato tubers take place when more assimilates are created than are needed for canopy growth. In adverse growing conditions, assimilates can be remobilized and translocated from the tuber to the canopy.

Tipstar-based recommendations for the application of nitrogen and irrigation water lead to yields similar to those obtained by the best farmers (Jansen et al., 2003). In spite of this, the model has not found practical application, possibly because of the effort required to gather input data, perform simulations, and present results in a meaningful way.

In 2020, Tipstar was integrated into Akkerweb 3.0 where soil, weather, and crop management information can be provided. Now, the user can use a mouse to indicate on a map on the computer screen the field(s) for which Tipstar should simulate production. The location of the field is used by Akkerweb to retrieve weather data and biophysical properties of the soil. Planting date, maturity class of the cultivar, irrigation dates and amounts, and the dates and amounts of fertilizer application can be provided by the user or retrieved from the user's farm management information system (FMIS). With this information, Tipstar produces time series of several important variables that describe the status of the soil (e.g., water content profile,

N content profile) and crop (e.g., LAI, leaf biomass, canopy biomass, tuber biomass) (Fig. 3).

Tipstar as offered on Akkerweb 3.0 supports tactical and operational decisions about the application of fertilizer and/or irrigation water by providing in-season information about the crop and soil. For an in-season decision, a mix of current weather, forecast weather, and historic weather is used. Current weather is used up to the day of the decision (e.g., 15 June 2020), and forecast weather is used for the next 2 weeks (16–30 June). The rest of the year (from 1 July until harvest) is simulated stochastically: 30 simulations are performed, where each simulation uses the weather of one of the last 30 years. This results in a plume of curves rather than a single

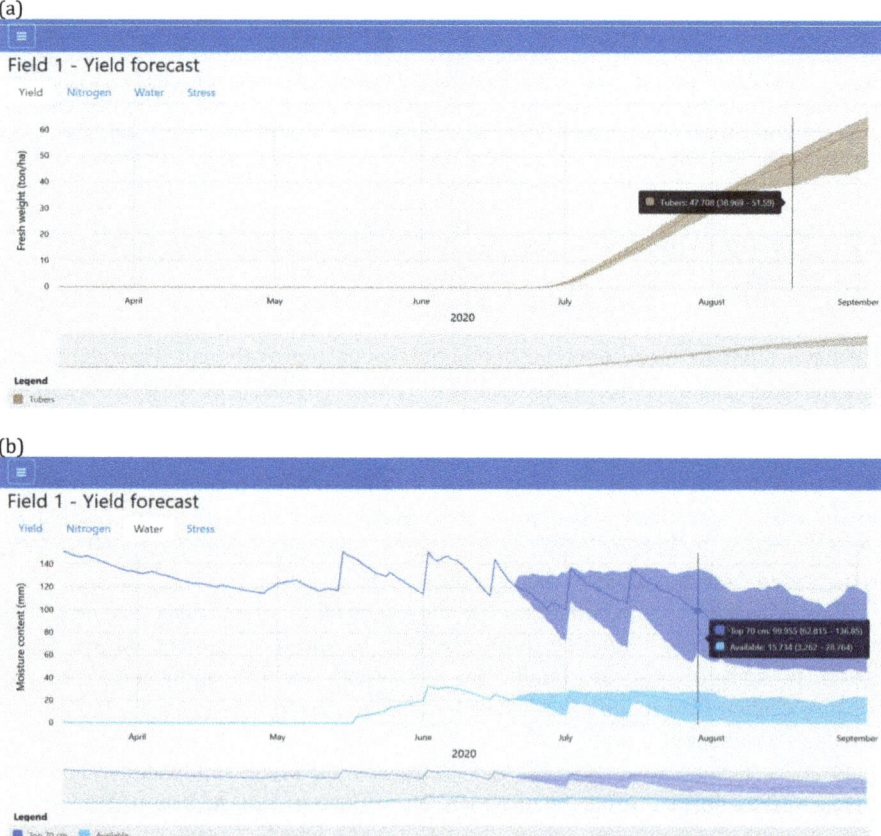

Fig. 3 Tipstar forecast for a potato crop near Wageningen. (**a**) Fresh tuber yield. The crop was planted on 15 April 2020 and the forecast was made on 4 June. From planting to 4 June, observed weather was used for the simulation. For the 2 weeks following 4 June, forecast weather was used. From 20 June, a stochastic simulation using 30 years of historic weather was performed; thus, from 20 June, the line becomes a plume. (**b**) Total amount of water in the upper 70 cm of the soil and plant available water. The figure shows that the crop will run out of water with the irrigation schedule that is proposed

curve; the median of the curve is taken as the expected value, and the spread around the median indicated the uncertainty of the simulation. Once a field has been registered, Tipstar is run every day, and the application creates an alert when the simulation indicates that N or water will become limiting to growth within the next few days.

2.3   NemaDecide

NemaDecide is a DSS for the management of plant parasitic nematodes that started as a PC version (Been et al., 2004, 2005) but has now been converted to an Akkerweb application. Models for population dynamics, yield reduction, partial resistance, soil sampling, chemical control, etc. are included in the system. It combines several data sources and models to enable strategic and operational decisions at the farm level (see Fig. 4a, b). The quantitative information system provides growers with the possibility to estimate risks of yield loss, population development, detection probabilities of infestation foci by soil sampling, and calculation of cost/benefit of control measures and provides adequate advice for farmers to optimize financial returns. Farmers can compare cropping scenarios and ask "what if" questions. The DSS provides answers to the top 10 questions an extension officer is exposed to. A database with more than 500 potato cultivars is available with information on susceptibility for fungal and viral diseases, cultivar properties, and partial resistance, expressed as relative susceptibility, of the cultivar against both species of the potato cyst nematode.

NemaDecide is targeted to control potato cyst nematodes (*Globodera* spp.), *Pratylenchus penetrans*, and *Meloidogyne chitwoodi*. The latter two species have a broad range of hosts which necessitated the inclusion of a suitable number of crops to enable simulations of crop rotations over successive years. Almost 50 arable, green manure and vegetable crops, grasses, and some flower bulbs have been added. The system was extended with the population dynamics of *P. penetrans* and *M. chitwoodi*, both in the absence and presence of hosts. In addition, new population dynamic models and yield loss models had to be developed to include the competition between species when attacking the same host. NemaDecide has been developed in close cooperation with extension officers and was tested "on-farm" with more than 60 farmers. The system is connected to soil sampling companies to retrieve nematode sampling results.

2.4   Soil Herbicide Variable-Rate Application App

Christensen et al. (2009) describe site-specific weed control and variable-rate application (VRA) of herbicides as a way to reduce herbicide use in agriculture. The WUR developed and validated a regression model for VRA of soil herbicides

Fig. 4 (**a**) The NemaDecide digital chain. In the cropping scheme app, a farmer can download field boundaries from the government database. Selecting a field, the farmer can order a nematode sampling. The sampling agency uses the Stripbuilder app to generate the sampling units, and after collection and processing of the soil samples, the results will be returned to the farmer. He/she can now obtain advice from NemaDecide, choose the most suitable potato cultivar, or generate a task map for a granular nematostatic. (**b**) NemaDecide GEO application. On the right, a map of the parcel with the sampled strips and the detected infestation. To the left, the app dashboard

(SH) (Kempenaar et al., 2014a). The amount of soil herbicides used in a crop can be reduced by adjusting the dosage to site-specific soil conditions. Soil herbicides are more effective on parts of the field where soil organic matter and clay content are small. The application rate of soil herbicides can be decreased in those areas without

Fig. 5 Three steps in the soil herbicide apps. (**a**) Upload of soil organic matter map, (**b**) herbicide prescription map, and (**c**) machine-ready task map

affecting herbicide efficacy. The SH VRA model is integrated into a herbicide app on Akkerweb. The app requires a soil map that shows the variation of soil organic matter and/or clay of the field. The user of the app selects the dosing algorithm of the herbicide of interest to make a prescription map. The last step is to make the VRA task map set to the conditions of the user's sprayer. In Fig. 5, we show the steps from soil map to prescription map to task map. These steps also occur in the other VRA apps on Akkerweb. On-farm testing of this VRA soil herbicide system with sprayers that can vary the dose at a scale of ca 50 m^2 showed reductions in herbicide use of between 20 and 40%, with an average of 27% (Kempenaar et al., 2018). An additional benefit of site-specific optimization of soil herbicide use is less reduction in the growth of the crop, leading to increases in crop yield of up to 5%.

2.5 Variable-Rate Application App to Kill Potato Haulm

Modern potato farmers have to kill the aboveground green haulm of their crops before they can mechanically harvest the belowground potato tubers. Most commonly, they do this by spraying a nonsystemic defoliant or leaf desiccant herbicide some weeks before harvest (Kempenaar & Struik, 2008). Variable-rate application (VRA) of the herbicide is a way to minimize herbicide use in a rational way. The WUR developed and validated a regression model for VRA of potato haulm killing (PHK) herbicides (Kempenaar et al., 2014b). The more biomass present on a part of the field, as expressed in a crop reflection biomass index, the larger the dose of the defoliant should be. The PHK model is integrated in a potato haulm killing app on Akkerweb. The steps to make a PHK VRA map are similar to those for the SH VRA app. In this case, the app requires a biomass map that shows the variation in aboveground crop biomass (NDVI, WDVI, or other crop reflection index). The other steps to make the prescription and task maps are similar to those for the SH VRA app (see Sect. 3.4.3, Chap. 3). Figure 6 shows the data flow and output of the app. On-farm experiments with the PHK dosing model and app showed reductions in herbicide use of between 20 and 47%, with an average of 38% (Kempenaar et al., 2014b, 2018). The PHK model can also be used on variable-rate spraying on-the-go with sensors mounted on the sprayer and direct control of the dosing by the computer on the sprayer.

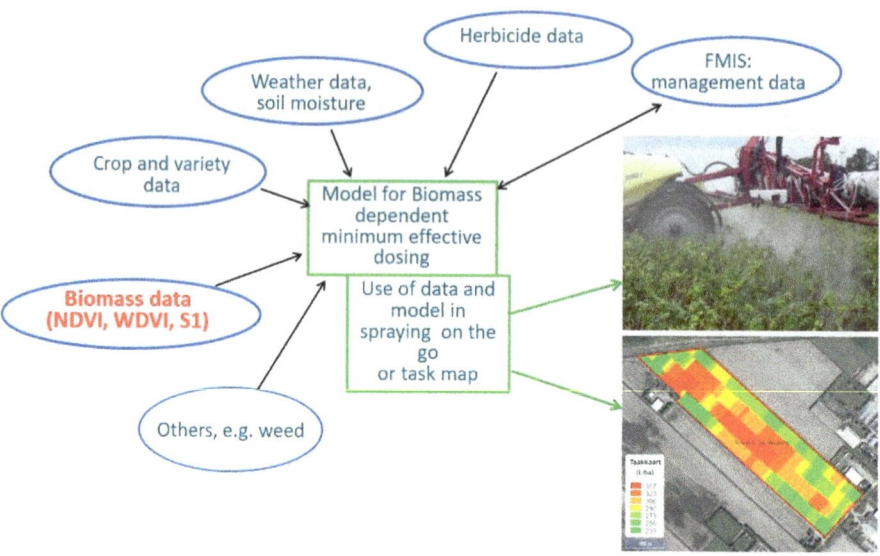

Fig. 6 Data input (blue text balloons) into PHK model (green text boxes) for biomass-dependent dosing and output of the model for direct control of the on-the-go sprayer or to a variable-rate prescription map that can be used to make a task map for a sprayer. For potato haulm killing, biomass data (in red) is the main driver in the model

2.6 Late Blight App

Potato late blight, caused by the oomycete *Phytophthora infestans*, is one of the most economically damaging diseases in potato. Annual losses are conservatively estimated at one billion euros for Europe and three billion euros globally (Haverkort et al., 2008). Current control methods rely primarily on frequent (calendar-based) fungicide applications (Cooke et al., 2011).

The Late Blight app helps farmers to optimize the timing of fungicide applications to prevent potato late blight. For this purpose, local measured and local forecasted weather data, obtained from the Akkerweb weather service, are analyzed to identify infection events in the near future and recent past (see Fig. 7a). Preventive fungicide applications are advised just prior to predicted infection events in the near future. Curative fungicide applications are advised to treat young, latent infections identified in the recent past. Eradicant fungicide applications aim to remedy older latent and active infections. The principles of the infection risk model in the app are described by Kessel et al. (2018).

The Late Blight app uses fungicide-specific ratings for rain intensity and protection of the foliage and tubers, for example, as published by Euroblight (www. euroblight.net), to calculate the time- and weather-dependent protection level of foliage and tubers. These ratings have been established through multiyear and

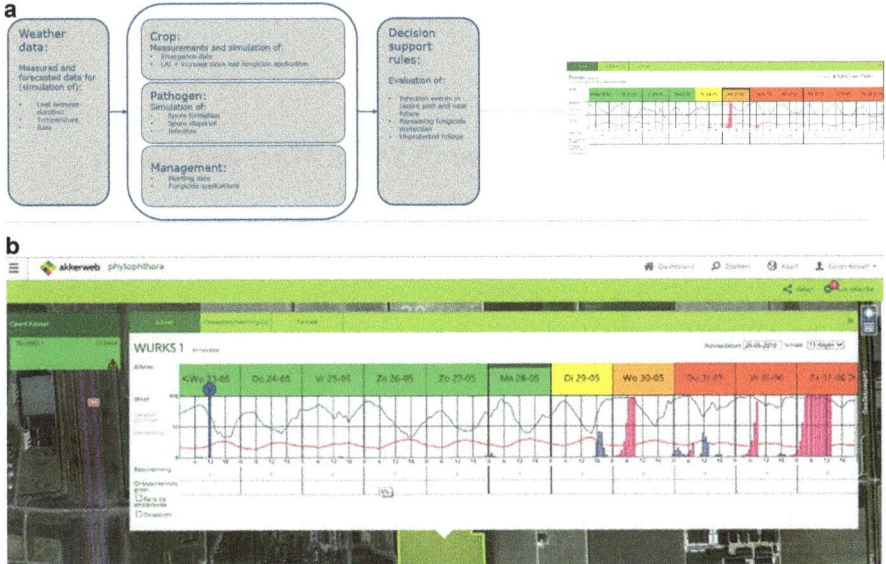

Fig. 7 (**a**) Data use and flow in the potato blight App: Forecasted and measured local weather data are used to calculate crop emergence, crop growth (LAI), pathogen infection events and (remaining) fungicide protection. Preventive fungicide applications are advised when an infection event is predicted in the near future AND the remaining fungicide protection is insufficient. (**b**) Screenshot of a Late Blight app advice on 28 May. On 30 May, an infection event is predicted by the infection risk model. That is why on 29 May, a preventive fungicide application is recommended (yellow). If this advice cannot be carried out, a curative advice (orange) or an eradicant advice (red) is issued for 30 and 31 May. The marker on 23 May indicates crop emergence. Spray advice is not issued before crop emergence

multicountry field trials and are updated when necessary. A further fungicide application is advised only when protection from the previous application is insufficient and an infection event is identified in the near future. The results are displayed in a simple interface which summarizes visually the key parameters and the advice (Fig. 7b). The Late Blight App operates at field level. It allows farmers to enter multiple fields. Advisors can view the results of multiple clients (farmers), each with multiple fields, once permission is granted by the client. The interface is built on a digital map of the farm or region with the potato fields highlighted in different colors representing the spray advice:

Green: No action is necessary
Yellow: A preventive fungicide application is advised
Orange: A curative fungicide application is advised
Red: An eradicant fungicide application is advised

In the next version, of the Late Blight app, the model is extended in a way that the resistance of the potato variety can be taken into account and variable-rate prescription maps can be made using crop biomass maps. This version has been tested in on-farm research since 2019.

2.7 Nitrogen Side Dress System (NSS) in Ware and Starch Potatoes

This application supports farmers to apply N side dress with variable-rate technology to maximize growth of the potatoes while minimizing the use of N (Booij et al., 2017, Booij and Uenk, 2004; Van Evert et al., 2012). In a side dress system, N is applied twice. Before planting, two-thirds of the nitrogen recommendation is applied as a flat rate by chemical and/or organic fertilizers (manure, compost, etc.). In late June or early July, a site-specific amount of side dress N is applied according to the crop's needs. To achieve this, canopy reflectance is measured with a satellite, an unmanned aerial vehicle (UAV), or a tractor-mounted sensor, and a vegetation index is derived. The application has the possibility to import a map of the vegetation index into the farmer's account on Akkerweb, or it can make use of already available data from other services on Akkerweb, like the Bioscope application, satellite application, or sensor data application.

The NSS app consists of three parts. First, the vegetation index map is converted to a nitrogen uptake map. Using relations derived from field research, a vegetation index like weighted difference vegetation index (WDVI) (Clevers, 1988) or chlorophyll index (CI) (Gitelson and Merzlyak, 1994; Gitelson et al., 2005), which give relative values, are converted to nitrogen uptake of the aboveground biomass of potatoes, which are absolute values in kg N per hectare. Second, an empirical growth curve is used to determine the amount of nitrogen that would have been taken up by a crop that is not limited by the availability of N. This curve takes into account the temperature sum between planting date and scouting date, expected yield, and the aim of the cultivation (ware or starch potatoes). Last, from the optimal and the measured nitrogen uptake, a site-specific recommendation is calculated (van Evert et al., 2012; Booij et al., 2017).

The nitrogen side dress system leads to an average reduction in the use of N of about 40 kg nitrogen per hectare with no reduction in yield (Van Evert et al., 2012) (Fig. 8 and Table 4). Additionally, in wet seasons, leaching of N is reduced (van Evert et al., 2017a, b).

Figure 9 shows the architecture of the NSS application. The user of the application must select a field within the crop rotation plan. These data are retrieved from the crop field app in Akkerweb. The variety and planting date are also included. Then, the user imports a biomass map, either manually from his/her own PC or from other applications on Akkerweb, such as the sensor data app, Bioscope app, or satellite app. The user must also fill in the expected yield. The application, in return, shows a nitrogen uptake map, the target nitrogen uptake derived from the optimum growth curve, and the recommended amount of side dress N. In the background, the

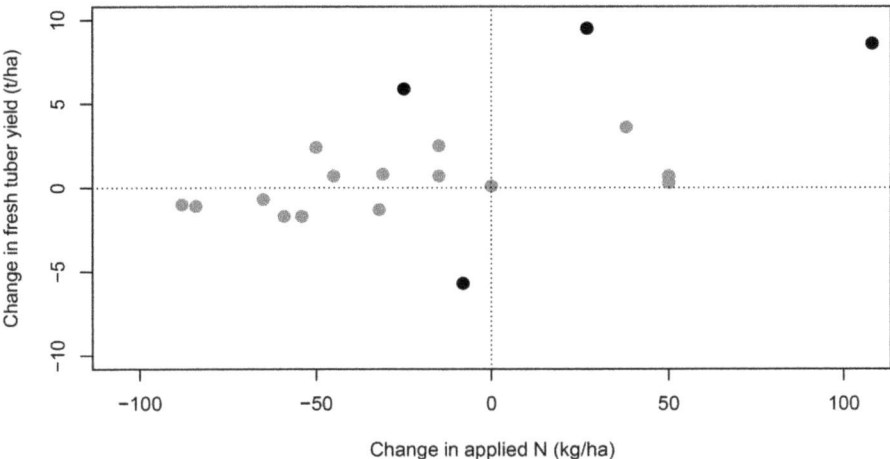

Fig. 8 Comparison of applied N and fresh tuber yield obtained with the recommended N rate and with the N side dress system. On the horizontal axis, the change in applied N with NSS relative to the recommended N rate is given. On the vertical axis, the change in fresh tuber yield is given. Each symbol represents an experiment. For black symbols, the yield difference was statistically different; for gray symbols, the yield difference was not statistically different. The overall trend is that if NSS recommends a large N rate, then yield is high, too, whereas if NSS recommends a small N rate, then yield is not affected

application connects to a webservice which contains the growth model and a connection to weather data. Last, the user can tune the recommendation by filling in the nitrogen content of the specific fertilizer that will be used, the minimum and maximum rate of N to be applied, and the grid size. A task map is then generated, which can be used in an up-to-date tractor terminal.

2.8 farmmaps Dashboard

The models and apps presented in Sects. 2.1 and 2.7 also run on the **farm**maps platform, the third version of Akkerweb, currently in development and available in spring 2021. **farm**maps features several improvements which resulted from observation on the interaction between farmers and Akkerweb and their needs observed during the last few years: data files are getting bigger and bigger, farmers like to be "taken by the hand," and everything should also be available on the mobile phone.

First of all, **farm**maps has an updated data repository which can process and store data independent of their size. Data can be dragged and dropped onto the map and analyzed to recognize the type of data and provide an image for visualization in one go. The properties of each file will be stored in a metafile for the rapid retrieval of files, appropriate for use in a certain application. Non-GEO data can be dropped on a

Table 4 Summary of tests with NSS

Year	Location	Soil	Cultivar	Use	N rate (kg ha⁻¹)		Fresh yield (t ha⁻¹)		Starch (t ha⁻¹)		Sign.	Ref.
					Recomm.	Side dress	Recomm.	Side dress	Recomm.	Side dress		
2002	Colijnsplaat	Clay	Agria	Ware	80+54	80+0	62.5	60.8				1,2
2002	Colijnsplaat	Clay	Felsina	Ware	131+88	131+0	49.5	48.5				1,2
2003	Colijnsplaat	Clay	Agria	Ware	102+68	102+60	57.5	51.8			y	1,3
2003	Colijnsplaat	Clay	Felsina	Ware	141+94	141+35	44.9	43.2				1,3
2002	Rolde	Sand	Seresta	Starch	225	150+30	54.6	55.3	12.2	12.4		1,4
2002	Rolde	Sand	Mercator	Starch	155	120+20	53.2	53.9	11.3	11.5		1,4
2003	Rolde	Sand	Seresta	Starch	250	115+70	52.1	51.4	11.1	10.9		1,4
2003	Rolde	Sand	Mercator	Starch	185	90+70	49.3	55.2	9.9	11.4	y	1,4
2010	Valthermond	Sand	Seresta	Starch	225	175+100	47.8	48.5	10.0	10.2		5
2010	Colijnsplaat	Clay	Victoria	Ware	250	150+50	44.4	46.8				5
2011	Valthermond	Sand	Seresta	Starch	120+60	120+45	46.6	49.1	10.3	10.5		6
2011	Reusel	Sand	Fontane	Ware	165	219+54	78.6	87.2			y	6
2011	Reusel	Sand	Fontane	Ware	229	229+27	91.4	100.9			y	6
2011	Biddinghuizen	Clay	Milva	Ware	150+54	150+54	70.9	71.0				6
2012	Vredepeel	Sand	Fontane	Ware	266	145+90	78.0	78.8				7
2012	Hulsberg	Löss	Fontane	Ware	182	150+0	63.5	62.2				7
2013	Vredepeel	Sand	Fontane	Ware	287	150+53	70.6	69.5				7
2013	Hulsberg	Löss	Fontane	Ware	171	150+59	58.3	61.9				7
2014	Valthermond	Sand	Seresta	Starch	205	205+50	49.6	49.9	10.9	10.7		8

The difference between yields in an experiment is statistically significant if a "y" is noted in column "Sign." Reports cited are [1]Van Evert et al. (2012), [2]Slabbekoorn (2002), [3]Slabbekoorn (2003), [4]Van Geel et al. (2004), [5]Van Evert et al. (2011), [6]Van der Schans (2012), [7]Van Geel et al. (2014), [8]and Van Geel and Van der Schans (2015)

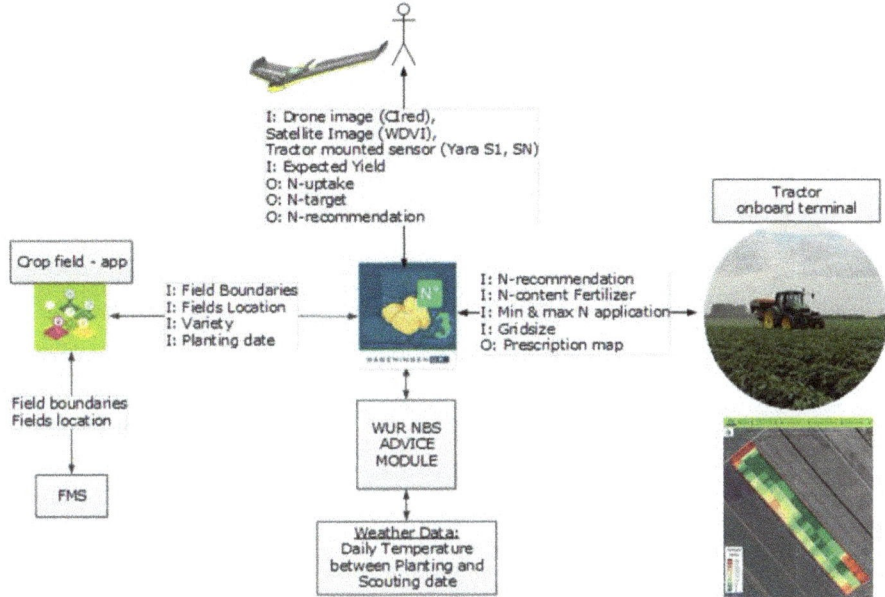

Fig. 9 Schematic flow of inputs (I) and (O) in Akkerweb NSS app. A task map based on CI map and NSS model and expected yield of 55 tons per ha is shown

parcel and will be georeferenced and are viewable (Excel, Word, PDF, images). The farmer can now store and retrieve all information concerning his parcel.

The concept of "workers" is introduced, small pieces of code that carry out automatic and, if necessary, scheduled tasks that can also be triggered by events and which fuel the dashboard by constantly updating the information presented.

The interface is created in Angular which means that **farm**maps can run on all devices, including mobile phones. The most important interface change is the use of information widgets in a dashboard (Fig. 10a) that is eventually intended to provide the farmer with an instant overview concerning the current state of his farm, including the necessary alerts to take action if required. The widgets present condensed information and alerts. More specific information can be obtained when opening the widget (Fig. 10b). Clicking on the icon will take the user to the app that will solve the problem. This three-tier approach should provide a more intuitive approach for a farmer than a selection of 30+ different single app tiles on a dashboard.

The models running on Akkerweb are also available on **farm**maps. However, the new engines are implemented as web services that will be available to third parties who like to use these engines but prefer to use and present them on their own GEO-platform with their own interfaces.

(a)

(b)

Fig. 10 farmmaps look and feel. (**a**) Dashboard with widgets giving soil and crop information. Detailed information can be provided by opening widgets. (**b**) Soil moisture data as measured with radar remote sensing. Widget information on water availability and deficiency is shown in Fig. 2 on the crop growth of potato in Fig. 3

3   Outlook

It is likely that precision agriculture will become more mainstream in the coming years contributing to more sustainable and circular crop production and that farming and value chain optimization will be based on data to a much greater extent. This transition will be much influenced by the availability of service platforms like Akkerweb that bring together all relevant data in a safe farm data space where the farmer has control over his data. He wants to use the data to monitor and benchmark his crops, to report to chain partners, and to make better management decisions. In addition, digital agri-food chains can be optimized, and tracking and tracing become possible when we have mature value chains. The question is when will this all become mainstream, for example, more than 50% of the farms in countries like the Netherlands use data to do all aforementioned farming activities. A few issues still have to be solved to reach this stage (today, we estimate that less than 1% of the farms in the Netherlands are mature data-driven farms). Farmers must become convinced that data-driven farming is not too complex, that the investments bring significant added value, and that data will not be misused. Ease of use, equal sharing of benefits across the value chain, and trust by good governance will be crucial in making data-driven farming mainstream.

Conflict of Interest The authors declare that they are employed by the Wageningen University & Research, which is the sole member of the not-for-profit foundation that owns and exploits Akkerweb and **farm**maps.

References

Been, T. H., & Schomaker, C. H. (2004). A geo-referenced decision support system for nematodes in potatoes. In D. K. L. Mac Kerron & A. J. Haverkort (Eds.), *Decision support systems in potato production* (pp. 154–167). Bringing Models to Pratice Wageningen.

Been, T. H., Schomaker, C. H., & Molendijk, L. P. G. (2005). Nema decide: A decision support system for the management of potato cyst nematodes. In A. J. Haverkort & P. C. Struik (Eds.), *Potato in progress* (pp. 154–167). Wageningen Academic Publishers.

Berghuis-van Dijk, J. T. (1985). *WATBAL: A simple water balance model for an unsaturated-saturated soil profile*. Institute for Land and Water Management Research. Note Nr. 1670.

Booij, R., & Uenk, D. (2004). Crop-reflection-based DSS for supplemental nitrogen dressings in potato production. In D. Mackerron & A. J. Haverkort (Eds.), *Decision support systems in potato production: Bringing models to practice* (pp. 46–53). Wageningen Academic Publishers.

Booij, J. A., van Evert, F. K., & van Geel, W. C. A. et al.. (2017, June). Roll-out of online application for N sidedress recommendations in potato. In *Proceedings of EFITA. Montpellier.* http://library.wur.nl/Web Query/wurpubs/fulltext/445495

Christensen, S., Sogaard, H. T., Kudsk, P., et al. (2009). Site-specific weed control technologies. *Weed Research, 49*(3), 233–241.

Clevers, J. G. P. W. (1988). The derivation of a simplified reflectance model for the estimation of leaf-area index. *Remote Sensing of Environment, 25*, 53–69.

Cooke, L. R., Schepers, H. T. A. M., Hermansen, A., et al. (2011). Epidemiology and integrated control of potato late blight in Europe. *Potato Research, 54*, 183–222.

EIP. (2015). *EIP-AGRI focus group on precision farming: Final report.* https://ec.europa.eu/eip/agriculture/en/publications/eip-agri-focus-group-precision-farming-final

Euroblight. (2020, May 13). http://www.euroblight.net

Feddes, R. A., Kowalik, P. J., & Zaradny, H. (1978). *Simulation of field water use and crop yield.* (189 p). Simulation Monographs Pudoc.

Fountas, S., Carli, G., Sorensen, C. G., et al. (2015). Farm management information systems: Current situation and future perspectives. *Computers and Electronics in Agriculture, 115*, 40–50.

GDAL/OGR contributors. (2021). *GDAL/OGR geospatial data abstraction software library*. Open Source Geospatial Foundation. https://gdal.org

Gitelson, A., & Merzlyak, M. N. (1994). Quantitative estimation of chlorophyll-A using reflectance spectra: Experiments with autumn chestnut and maple leaves. *Journal of Photochemistry and Photobiology B: Biology, 22*, 247–252.

Gitelson, A. A., Vina, A., Ciganda, V., Rundquist, D. C., & Arkebauer, T. J. (2005). Remote estimation of canopy chlorophyll content in crops. *Geophysical Research Letters, 32*. https://doi.org/10.1029/2005gl022688

Goense, D. (2017). *rmAgro/drmAgro/drmCrop*. https://edepot.wur.nl/408327

Haverkort, A. J., Boonenkamp, P. M., Hutten, R. C. B., & Jacobsen, E. (2008). Societal costs of late blight in potato and prospects of durable resistance through Cisgenic modification. *Potato Research, 51*(1), 47–57.

Jansen, D. M. (2008). *Beschrijving van TIPSTAR: hét simulatiemodel voor groei en productie van zetmeelaardappelen* (Nota 547). Plant Research International.

Jansen, D. M., Davies, J., & Steenhuizen, J. (2003). *Gevoeligheid van TIPSTAR voor de waarden van situatie-specifieke invoergegevens* (Nota 258). Plant Research International.

Kempenaar, C., & Struik, P. C. (2008). The canon of potato science: Haulm killing. *Potato Research, 50*, 341–345.

Kempenaar, C., Heiting, S., & Michielsen, J. M. (2014a). Perspectives for site specific application of soil herbicides in arable farming. In *Proceedings of ICPA conference*, Sacramento, USA, July 2014. Paper 1414, https://www.ispag.org/icpa

Kempenaar, C., van Evert, F.K., & Been, Th. (2014b). Use of vegetation indices in variable rate application of potato haulm killing herbicides. In *Proceedings of ICPA conference*, Sacramento, USA, July 2014. Paper 1413, https://www.ispag.org/icpa

Kempenaar, C., Been, T., Booij, J. A., van Evert, F. K., Michielsen, J. M., & Kocks, C. G. (2018). Advances in variable rate technology application in potato in The Netherlands. *Potato Research, 60*(3–4), 295–305.

Kessel, G. J. T., Mullins, E., Evenhuis, A., et al. (2018). Development and validation of IPM strategies for the cultivation of cisgenically modified late blight resistant potato. *European Journal of Agronomy, 96*, 146–155. https://doi.org/10.1016/j.eja.2018.01.012

Kroes, J. G., van Dam, J. C., Groenendijk, P., Hendriks En, R. F. A., & Jacobs, C. M. J. (2008). *SWAP version 3.2: Theory description and user manual*. Wageningen University & Research.

Slabbekoorn, H. (2002). *Stikstofbijmestsystemen in consumptieaardappelen, 2002 = N sidedress systems in ware potatoes, 2002*.WUR-PPO, Westmaas.

Slabbekoorn, H. (2003). *Stikstofbijmestsystemen in consumptieaardappelen, 2003 = N sidedress systems in ware potatoes, 2003*.WUR-PPO, Westmaas.

Van der Schans, D.A. (2012). Sensorgestuurde advisering van stikstof bijbemesting in aardappel: implementatie en integratie Praktijkonderzoek Plant & Omgeving, Business Unit Akkerbouw, Groene Ruimte en Vollegrond[s]groenten, .

Van Evert, F. K., Van der Schans, D. A., Malda, J. T., Van den Berg, W., Van Geel, W. C. A., & Jukema, J. N. (2011). *Geleide N-bemesting voor aardappelen op basis van gewasreflectie-metingen: Integratie van sensormetingen in een N-bijmestsysteem* (PPO Rapport 423). Praktijkonderzoek Plant & Omgeving (PPO), Lelystad.

Van Evert, F. K., Booij, R., Jukema, J. N., Ten Berge, H. F. M., Uenk, D., Meurs, E. J. J., et al. (2012). Using crop reflectance to determine sidedress N rate in potato saves N and maintains yield. *European Journal of Agronomy, 43*, 58–67. https://doi.org/10.1016/j.eja.2012.05.005

Van Evert, F. K., Gaitán-Cremaschi, D., Fountas, S., & Kempenaar, C. (2017a). Can precision agriculture increase the profitability and sustainability of the production of potatoes and olives? *Sustainability, 9*(10), 1863–1886. https://doi.org/10.3390/su9101863

Van Evert, F. K., Fountas, S., Jakovetic, D., et al. (2017b). Big data for weed control and crop protection. *Weed Research, 57,* 218–233. https://doi.org/10.1111/wre.12255

Van Geel, W. C. A., & Van der Schans, D. A. (2015). *Toepassing van NBS-aardappelsensing in de teelt van zetmeelaardappelen: IJkakker, veldproef 2014 't Kompas Praktijkonderzoek Plant & Omgeving, onderdeel van Wageningen UR.* Business Unit Akkerbouw, Groene ruimte en Vollegrondgroenten.

Van Geel, W. C. A., Wijnholds, K. H., Grashoff, C.. (2004). *Ontwikkeling van geleide bemestingssystemen bij de teelt van zetmeelaardappelen 2002–2003* (Development of guided fertilization for starch potatoes in 2002–2003). WUR-PPO.

Van Geel, W. C. A., Kroonen-Backbier, B. M. A., Van der Schans, D. A., & Malda, J. T. (2014). *Nieuwe bijmestsystemen en -strategieën voor aardappel op zand- en lössgrond. Deel 2: resultaten veldproeven 2012 en 2013* (p. 66). PPO-AGV.

Van Oort, P.A.J., Van Evert, F. K., & Kempenaar, C. (2020). *Calibration of the Tipstar potato model using remote sensing data.* Second International Crop Modelling Symposium (iCROPM2020). Montpellier.

Vinten, A. J. A. (1999). Predicting nitrate leaching from drained arable soils derived from glacial till. *Journal of Environmental Quality, 28,* 988–996.

Wikipedia. (2020, May 13). https://en.wikipedia.org/wiki/precision_agriculture

Index

Milton Keynes UK
Ingram Content Group UK Ltd.
UKHW022234090124
435726UK00001B/2